Sequencing of proteins and peptides

LABORATORY TECHNIQUES IN BIOCHEMISTRY AND MOLECULAR BIOLOGY

Edited by

T.S. WORK – *Cowes, Isle of Wight (formerly N.I.M.R., Mill Hill, London)*
R.H. BURDON – *Department of Biochemistry, University of Glasgow*

ELSEVIER/NORTH-HOLLAND BIOMEDICAL PRESS
AMSTERDAM · NEW YORK · OXFORD

SEQUENCING OF PROTEINS AND PEPTIDES

G. Allen

The Wellcome Research Laboratories
Langley Court
Beckenham
Kent BR3 3BS
England

1981

NORTH-HOLLAND PUBLISHING COMPANY
AMSTERDAM · NEW YORK · OXFORD

ISBN – series: 0 7204 4200 1
Volume 9 paperback: 0 444 80254 1
hardbound: 0 444 80275 4

Published by:

ELSEVIER/NORTH-HOLLAND BIOMEDICAL PRESS
335 JAN VAN GALENSTRAAT, P.O. BOX 211
AMSTERDAM, THE NETHERLANDS

Sole distributors for the U.S.A. and Canada:

ELSEVIER/NORTH-HOLLAND INC.
52 VANDERBILT AVENUE
NEW YORK, N.Y. 10017

Library of Congress Cataloging in Publication Data

Allen, Geoffrey, 1928-
Sequencing of proteins and peptides.

(Laboratory techniques in biochemistry and molecular biology ; v. 9)
"Pocket edition."
Bibliography: p.
Includes index.
1. Amino acid sequence. 2. Proteins--Analysis. 3. Peptides--Analysis. I. Title. II. Series: [DNLM: 1. Amino acid sequence. 2. Peptides--Analysis. 3. Proteins--Analysis. W1 LA232K v. 9 / QU 55 A425s]
QP551.A42 574.19'245 80-26924

Printed in The Netherlands

Preface

For investigations on the structure and function of a protein, knowledge of the primary structure is of fundamental importance, especially when combined with crystallographic analysis, which is, by itself, insufficient for the elucidation of the complete covalent structure. Steady progress has been made towards the theoretical prediction of protein secondary structure from the primary structure, but the accuracy is at present still too low to be of great value. With the accumulation of more structural information further advances in this field are likely.

When chemical modification is performed as part of an investigation into the function of a protein, knowledge of the position in the amino acid sequence of the sites of modification is essential, and this will usually require partial sequence determination of modified peptides. The structures of 'active-site peptides' may be of particular interest.

The knowledge of the primary structures of proteins has been indispensible for an understanding in molecular terms of many physiological processes. In variant haemoglobin, for example, changes in oxygen binding have been shown to result from single amino acid substitutions. Likewise, knowledge of the primary structures of immunoglobulins has been of importance for the understanding of antibody formation and specificity. Investigation of the primary structures of other classes of variant proteins, such as the histocompatibility antigens, may be expected to provide further insight into immunological problems.

The study of precursors of excreted proteins has led to the identification of 'signal peptide' sequences which cause those ribosomes synthesizing these proteins to attach to the endoplasmic reticulum. Information on primary structures has also helped in the understanding of the conversion of precursor molecules to active enzymes (for example, in blood-clotting) and of prohormones to hormones.

With the advent of rapid methods for the specific isolation and sequence analysis of DNA it might be thought that the direct determination of protein sequences would decline in importance. However, in eukaryotic organisms the structure of the DNA is not reflected directly in the protein structure: untranslated regions (leader sequences and intervening sequences) are present. Direct determination of at least the partial primary structure of proteins is still required. In addition, little can be learned about post-translational modifications from the analysis of DNA sequences. In fact, the ease with which DNA sequences may be determined could well increase the demand for analysis of protein primary structure.

The purpose of this monograph is to provide in one volume detailed descriptions of the techniques used in the determination of amino acid sequences of proteins by manual methods, together with a discussion of the theoretical background. The size of the field has necessitated a high degree of selectivity, but references are given to alternative methods which may occasionally prove useful. The selection of material for inclusion has been based mainly on the widespread use of the techniques, but partly on the personal experience of the author. Preference is given to the more sensitive methods. The choice between equivalent techniques by an investigator will depend to some extent upon the equipment available and the techniques already established in his own laboratory.

Most of the apparatus and materials required are relatively inexpensive; the amino-acid analyzer is the exception. Automated sequencing instruments and advanced chromatographic techniques are discussed. The use of these techniques adds considerably to the cost of setting up a laboratory for the sequence analysis of proteins. Detailed descriptions of the construction and use of the automated spinning-cup sequencer, originally described by Edman and Begg

(1967), or the automated solid-phase sequencer, described by Laursen (1971), are not included, since it is felt that purchasers of such instruments will have access to the information provided by the manufacturers and available in the specialist literature. Excellent detailed descriptions of these automated techniques have recently been given (Waterfield and Bridgen, 1975; Niall, 1977; Laursen et al., 1975; Niall, 1973).

Techniques of chemical modification of proteins and amino acid analysis have been described in a recent volume in this series (Glazer et al., 1975), and only a limited description of these techniques is included here. A large number of techniques used in protein sequence determination have been described in *Methods in Enzymology*, Vols. *11, 25* and *47*, and in *Protein Sequence Determination* (Needleman, 1975), and these are valuable sources of information. The subject of sequence analysis has recently been reviewed by Konigsberg and Steinman (1977).

List of abbreviations

Atz-amino acid	anilinothiazolinone derivative of an amino acid
BNPS-skatole	2-(2-nitrophenylsulfenyl)-3-methyl-3-bromo-indolenine
butyl-PBD	2-(4′-*t*-butylphenyl)-5-(4″-biphenylyl)-1,3,4-oxadiazole
CM-	carboxymethyl-
Dabitc	4,4-dimethylaminoazobenzene-4′-isothiocyanate
Dabtc-	4,4-dimethylaminoazobenzene-4′-thiocarbamyl-
Dabth-amino acid	4,4-dimethylaminoazobenzene-4′-thiohydantoin derivative of an amino acid
DEAE-	diethylaminoethyl-
Dnp-	2,4-dinitrophenyl-
Dns-	1-dimethylaminonaphthalene-5-sulphonyl-
Dns-Cl	1-dimethylaminonaphthalene-5-sulphonyl-chloride (dansyl chloride)
DTNB	5,5′-dithio*bis*(2-nitrobenzoic acid)
E. coli	*Escherichia coli*
EDTA	ethylenediamine tetraacetic acid
glc	gas-liquid chromatography
hplc	high-performance liquid chromatography
Pitc	phenylisothiocyanate
Ptc-	phenylthiocarbamyl-

Pth-amino acid	phenylthiohydantoin derivative of an amino acid
SDS	sodium dodecyl sulphate
tlc	thin-layer chromatography
TLCK	3-tosylamido-7-amino-1-chloroheptan-2-one
TPCK	1-chloro-3-tosylamido-4-phenylbutan-2-one
ϵ	molar extinction coefficient in litre mol^{-1} cm^{-1}

Contents

Preface . . . . *v*

List of abbreviations . . . . *ix*

Chapter 1. Introduction *1*

1.1. Protein structure 1
1.2. Historical background 7
1.3. Outline of the strategy for sequence determination 9

Chapter 2. Preliminary characterization of the protein 17

2.1. Purification of the protein 17
2.2. Separation of the constituent polypeptide chains of heterooligomeric proteins 18
2.2.1. Separation of light and heavy chains of immunoglobulin G 19
2.2.2. Separation of α- and β-chains of haemoglobin 20
2.2.3. Separation of light and heavy chains of myosin 20
2.3. Determination of the molecular weight of the polypeptide chains 21
2.4. Removal of salts and non-covalently bound molecules 23
2.4.1. Removal of lipids 23
2.4.2. Removal of haem from cytochrome b_5 24
2.5. Amino acid analysis 24
2.5.1. Determination of phosphate 26
2.5.2. Determination of the extinction coefficient 26
2.5.3. Spectrophotometric determination of tryptophan 27
2.6. Chemical modification of the protein for sequence analysis 28
2.6.1. Conversion of cysteine residues to stable derivatives 28
2.6.2. Deletion or introduction of points of enzymic cleavage 30

2.6.3. Introduction of labels to facilitate the detection and isolation of peptides containing specific residues 30
2.6.4. Conversion of the protein to a soluble form, facilitating digestion and isolation of peptides 30
2.6.5. Reduction and carboxymethylation 30
2.7. Determination of the amino-terminal residue 31
2.7.1. Determination of the N-terminal residue using dansyl chloride 33
2.7.2. Determination of the N-terminal residue using 4-dimethylamino-benzene-4′-isothiocyanate (Dabitc) 37
2.8. Determination of the carboxy-terminal residue 38
2.8.1. Carboxypeptidase A 39
2.8.2. Carboxypeptidase B 40
2.8.3. Carboxypeptidase C 40
2.8.4. Carboxypeptidase Y 41

Chapter 3. Specific cleavage of the protein 43

3.1. General approach to cleavage 43
3.2. Selective cleavage of the polypeptide chain to produce large fragments . . 44
3.2.1. Limited proteolysis 44
3.2.1.1. Ribonuclease A 45
3.2.1.2. Glycogen phosphorylase 46
3.2.1.3. Immunoglobulin G 47
3.2.1.4. Sarcoplasmic reticulum ATPase 47
3.2.1.5. Myosin 48
3.2.2. Selective chemical cleavage for the production of large fragments . . 49
3.2.2.1. Cleavage at Asn-Gly bonds with hydroxylamine 49
3.2.2.2. Cleavage at Asp-Pro bonds in dilute acid 50
3.3. Enzymic cleavage of denatured polypeptides 51
3.3.1. Preparation of the protein for digestion 52
3.3.2. Trypsin 53
3.3.2.1. Use of a pH-stat to follow tryptic digestion 54
3.3.2.2. Tryptic digestion in volatile buffers 55
3.3.3. Cleavage by trypsin at arginine residues after modification of lysine residues 55
3.3.4. Cleavage by trypsin at lysine residues after modification of arginine residues 57
3.3.5. Staphylococcal protease 58
3.3.6. α-Chymotrypsin 59
3.3.7. Pepsin 59
3.3.8. Thermolysin 60

3.3.9. Subtilisin . . . 61
3.3.10. Papain . . . 61
3.4. Specific cleavage of proteins by chemical methods . . . 62
3.4.1. Cleavage with cyanogen bromide . . . 62
3.4.2. Cleavage at cysteine residues after cyanylation . . . 65
3.4.3. Cleavage at tryptophan residues by BNPS-skatole . . . 68
3.4.4. Cleavage at tryptophan residues with *o*-iodosobenzoic acid . . . 69
3.4.5. Cleavage with dimethylsulphoxide/HCl/HBr . . . 70
3.4.6. Partial acid hydrolysis . . . 70

Chapter 4. Separation and purification of peptides . . . 73

4.1. Outline of methods used . . . 73
4.2. Amounts of digested protein required . . . 76
4.3. Preliminary study of the peptides in a digest . . . 77
4.3.1. Analytical SDS-polyacrylamide gel electrophoresis . . . 78
4.3.2. Analytical polyacrylamide gel electrophoresis in the presence of urea and SDS . . . 80
4.3.3. Analytical isoelectric focusing of peptides in polyacrylamide gels . . . 81
4.3.4. Analytical thin-layer peptide mapping . . . 81
4.4. Separation of peptides on the basis of solubility . . . 83
4.5. Gel filtration . . . 84
4.5.1. Chromatographic columns . . . 86
4.5.2. Eluents for gel filtration of peptides . . . 86
4.5.3. Packing of gel filtration columns . . . 89
4.5.4. Loading the sample, and elution . . . 90
4.5.5. Recovery of peptides from fractions . . . 91
4.5.6. Examples of the separation of peptides by gel filtration . . . 92
4.6. Ion-exchange chromatography . . . 93
4.6.1. Cellulosic ion-exchangers . . . 95
4.6.1.1. DEAE-cellulose . . . 97
4.6.1.2. CM-cellulose . . . 101
4.6.1.3. Phosphocellulose . . . 101
4.6.2. Ion-exchange derivatives of cross-linked dextrans . . . 103
4.6.3. Ion-exchange derivatives of cross-linked polystyrene . . . 104
4.6.3.1. A standard procedure for chromatography on Dowex-50-type resins 107
4.6.3.2. Chromatography on strongly cationic anion-exchange resins . . . 109
4.7. Paper electrophoresis . . . 109
4.8. Paper chromatography . . . 116
4.9. Peptide mapping on paper . . . 119
4.10. Thin-layer methods for purification of peptides . . . 121

4.10.1. Thin-layer electrophoresis 122
4.10.2. Thin-layer chromatography 126
4.10.3. Preparative peptide mapping on thin layers 127
4.11. Preparative polyacrylamide gel electrophoresis 127
4.12. High-performance liquid chromatography 128
4.13. Affinity chromatography 130
4.14. Selective isolation of peptides by covalent binding to an insoluble matrix 131
4.15. 'Diagonal' methode 131
4.16. Selective purification of blocked N-terminal peptides 132

Chapter 5. Methods for the detection of peptides 135

5.1. General methods for the detection of peptides in column eluates 135
5.1.1. Low-wavelength UV-absorbance 135
5.1.2. Reaction with ninhydrin in solution 139
5.1.2.1. Alkaline hydrolysis and reaction with ninhydrin 140
5.1.3. Reaction with fluorescamine 141
5.1.4. Reaction with *o*-phthalaldehyde and 2-mercaptoethanol 142
5.1.5. Hydrolysis followed by amino-acid analysis 143
5.1.6. Reaction with dansyl chloride 144
5.1.7. Thin-layer chromatography or electrophoresis 144
5.1.8. Polyacrylamide gel electrophoresis 145
5.2. General methods for the detection of peptides adsorbed to paper or thin-layer supports 146
5.2.1. Reaction with ninhydrin 146
5.2.1.1. Ninhydrin/acetic acid reagent 147
5.2.1.2. Ninhydrin/collidine 147
5.2.1.3. Ninhydrin/cadmium reagent 148
5.2.2. Fluorescamine 148
5.2.2.1. Fluorescamine for analytical peptide maps 149
5.2.2.2. Fluorescamine for preparative peptide maps 149
5.2.3. Hydrolysis of peptides followed by reaction with ninhydrin 150
5.2.4. Chlorination 150
5.2.4.1. Reaction with chlorine 150
5.2.4.2. Reaction with *t*-butyl hypochlorite 151
5.2.4.3. Detection with *o*-tolidine 151
5.2.4.4. Detection with starch/KI 151
5.3. Detection methods specific for certain amino acid residues, in column fractions 152
5.3.1. UV-absorbance 152
5.3.2. Fluorescence 152

5.3.3. Liquid-scintillation counting or gamma-counting 153
5.3.4. Detection of cystine residues 153
5.3.5. Detection of glycopeptides 154
5.3.6. Other specific detection methods 154
5.4. Detection methods specific for certain amino-acid residues, on paper or thin-layer supports 155
5.4.1. Fluorescence emission 155
5.4.2. Radioautography 155
5.4.3. Phenanthrenequinone reagent for arginine 157
5.4.4. Ehrlich test for tryptophan residues 157
5.4.5. Pauly test for histidine residues 158
5.4.6. Nitrosonaphthol test for tyrosine residues 158
5.4.7. Iodoplatinate test for methionine residues 159
5.4.8. Isatin test for N-terminal proline residues 159
5.4.9. The technique of dipping chromatograms 159

Chapter 6. Determination of peptide sequences 161

6.1. Methods for the estimation of the purity of peptides 161
6.1.1. Elution profiles from chromatographic columns 162
6.1.2. Analytical chromatography or electrophoresis 162
6.1.2.1. The use of electrophoretic mobilities of peptides 164
6.1.3. Determination of the N-terminal residue 165
6.1.4. Amino-acid analysis 165
6.2. Determination of the N-terminal residues of peptides 168
6.2.1. Dansyl chloride method 168
6.2.2. Dabitc method 173
6.3. Determination of peptide sequences by the Edman degradation 174
6.3.1. Chemistry of the Edman degradation 174
6.3.1.1. Coupling 175
6.3.1.2. Cleavage 178
6.3.1.3. Conversion 180
6.3.2. Some problems in the use of the Edman degradation 181
6.3.2.1. Incomplete coupling and cleavage reactions 181
6.3.2.2. Histidine residues 182
6.3.2.3. Asparaginyl-glycine peptide bonds 183
6.3.2.4. Glutamine and *S*-carboxymethylcysteine residues 184
6.3.2.5. Tryptophan residues 185
6.3.2.6. Glycosylated residues 185
6.3.2.7. Other blocking reactions 186
6.4. Experimental procedures for the Edman degradation 186

6.4.1. Manual methods and identification of phenylthiohydantoins 187
6.4.1.1. High-performance liquid chromatography 188
6.4.1.2. Gas-liquid chromatography 191
6.4.1.3. Back hydrolysis and amino-acid analysis 191
6.4.1.4. Thin-layer chromatography 192
6.4.2. Automated liquid-phase sequence determination 199
6.4.2.1. Application to peptides 201
6.4.2.2. High-sensitivity applications 202
6.4.3. Solid-phase sequencing 205
6.4.3.1. Types of solid support 205
6.4.3.2. Attachment of peptides to the solid support 208
6.4.3.3. Edman degradation procedures 214
6.4.4. The 'dansyl-Edman' method 216
6.4.5. Subtractive Edman degradation 221
6.5. Alternative isothiocyanate reagents 221
6.5.1. Liquid-phase sequencing with Dabitc 223
6.5.2. Manual solid-phase sequencing with Dabitc 226
6.6. Other chemical degradation methods for peptide sequence determinations 227
6.7. Enzymic methods for peptide sequence determination 228
6.7.1. Carboxypeptidase A 229
6.7.2. Leucine aminopeptidase 230
6.7.3. Aminopeptidase M 230
6.8. Mass spectrometry 230
6.9. Subfragmentation of peptides 231
6.10. Summary of possible difficulties in peptide sequence determination . . . 234

Chapter 7. Structures introduced by post-translational modification 237

7.1. γ-Carboxyglutamic acid 238
7.2. Phosphorylated amino-acid residues 238
7.3. Methylated amino acids 240
7.3.1. Trimethylalanine and ϵ-*N*-trimethyllysine 240
7.3.2. *N*-Methylalanine 241
7.3.3. N^{τ}-Methylhistidine 241
7.4. Identification of points of attachment of carbohydrate in glycoproteins . 242
7.4.1. Erythrocyte glycophorin 244
7.4.2. Antifreeze glycoproteins 244
7.4.3. Haemoglobin A_{1c} 245
7.4.4. Immunoglobulin G 245
7.4.5. Human chorionic gonadotropin 245

7.4.6. Other exemples 245
7.5. Amino-terminal acyl groups 245
7.5.1. Acetylated N-terminal residues 246
7.5.2. Pyrrolidone carboxylic acid residues 249
7.5.3. Formylated N-terminal residues 250
7.6. Examples of other modified residues 250
7.7. Cross-linking structures 251
7.7.1. Cystine residues 251
7.7.2. Lysine-derived cross-links in collagen and elastin 256

Chapter 8. Deduction of the total primary structure of a protein from the sequences of constituent peptides 257

Chapter 9. Integral membrane proteins 263

9.1. Purification of integral membrane proteins 264
9.2. Specific cleavage methods 266
9.3. Purification of peptides from integral membrane proteins 268
9.4. Determination of the sequences of membrane peptides 271

Chapter 10. Applications of partial sequence analysis 273

10.1. Identification of positions within the sequence of selectively modified residues 273
10.2. Identification of residues involved in chemical cross-linking 273
10.3. Comparison of related proteins by peptide mapping 274
10.3.1. Use of methyl-[^{14}C]acetimidate 276
10.3.2. Use of radioiodination 277
10.4. Comparison of partial sequences of apparently unrelated proteins . . . 278
10.5. Amino-terminal sequences 279
10.6. Correlation with nucleotide sequences in DNA 280

Appendix I. Manufacturers and suppliers of equipment and reagents for protein sequence analysis 283

(1) Equipment 283
(2) Chromatographic materials 286
(3) Enzymes 288
(4) Other reagents 289

Appendix II. Purification of solvents and reagents *291*

Appendix III. Safety in the protein chemistry laboratory *294*

Appendix IV. Notes on work with very small amounts of proteins and peptides *295*

(1) Cleanliness 295
(2) Adsorption to surfaces 296
(3) Transfer of samples 296
(4) Reactions 297
(5) Chromatography 298
(6) Radioactive proteins 299

References *299*

Subject index *319*

CHAPTER 1

Introduction

1.1. Protein structure

The structures of proteins may be discussed in terms of a hierarchy of the following concepts. The *primary structure* is the arrangement of covalent bonds, which is equivalent to the sequential arrangement of the constituent amino acid residues (the amino acid sequence, or simply 'sequence'), together with the covalent cross-links between different residues in the polypeptide chain. The *secondary structure* consists of the arrangement of the polypeptide chain into such hydrogen-bonded features as the α-helix and the β-sheet, the β-turn, or a more random conformation. The *tertiary structure* is the arrangement in space of these secondary structural features, to form, for example, the typical globular shape of soluble proteins. The *quaternary structure* is the arrangement in space of the constituent folded polypeptide chains of a multimeric protein.

This monograph is concerned with the first of these levels of structure, although aspects of the secondary and tertiary structure will be discussed where such structure affects chemical reactivity, susceptibility to enzymic digestion, or isolation of peptides.

Twenty amino acids, all of the L-configuration (except for glycine, which is not disymmetric), are incorporated into nascent protein molecules during biosynthesis on the ribosome. However, many of these amino-acid residues may be post-translationally modified, for example by methylation, phosphorylation, glycosylation, hydroxylation, oxidation, or linkage to prosthetic groups. The structures of the twenty protein amino-acid residues, their residue weights and their accepted abbreviations are given in Table 1.1. Structures of some of the more common modified residues, and examples of some proteins in which they are found, are given in Table 1.2. Far more extensive lists of biosynthetically modified residues have recently been given by Uy and Wold (1977) and by Horáková and Deyl (1978).

TABLE 1.1
The twenty common protein amino acids

Amino acid	Three-letter abbreviation of amino acid residue	Single-letter abbreviation	Structure of amino acid residue in proteins	Residue weight
Alanine	Ala	A	—NH—CH—CO— | CH_3	71.1
Arginine	Arg	R	—NH—CH—CO— | $(CH_2)_3$ | NH | C=NH | NH_2	156.2
Asparagine	Asn	N	—NH—CH—CO— | CH_2 | $CONH_2$	114.1
Aspartic acid	Asp	D	—NH—CH—CO— | CH_2 | COOH	115.1
(Asparagine or aspartic acid:	Asx	B)		
Cysteine	Cys	C	—NH—CH—CO— | CH_2SH	103.1
Glutamic acid	Glu	E	—NH—CH—CO— | CH_2 | CH_2COOH	129.1

TABLE 1.1 *(continued)*

Amino acid	Three-letter abbreviation of amino acid residue	Single-letter abbreviation	Structure of amino acid residue in proteins	Residue weight
Glutamine (Glutamic acid or glutamine:	Gln Glx	Q Z)	—NH—CH—CO— │ CH_2 │ CH_2CONH_2	128.1
Glycine	Gly	G	—NH—CH_2—CO—	57.1
Histidine	His	H	—NH—CH—CO— │ CH_2 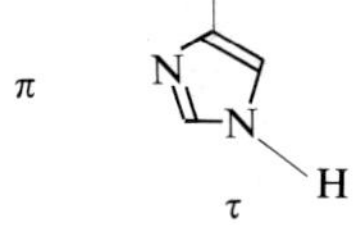	137.2
Isoleucine	Ile	I	—NH—CH—CO— │ CH—CH_3 │ CH_2CH_3	113.2
Leucine	Leu	L	—NH—CH—CO— │ CH_2 │ CH—CH_3 │ CH_3	113.2
Lysine	Lys	K	—NH—CH—CO— │ $(CH_2)_4$ │ NH_2	128.2
Methionine	Met	M	—NH—CH—CO— │ CH_2 │ CH_2 │ S—CH_3	131.2

TABLE 1.1 *(continued)*

Amino acid	Three-letter abbreviation of amino acid residue	Single-letter abbreviation	Structure of amino acid residue in proteins	Residue weight
Phenylalanine	Phe	F	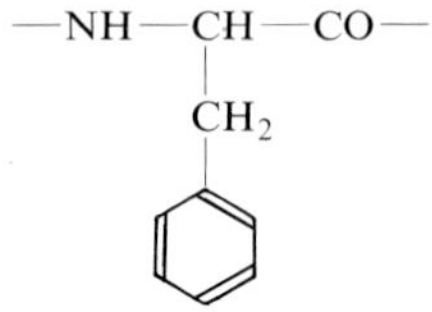	147.2
Proline	Pro	P	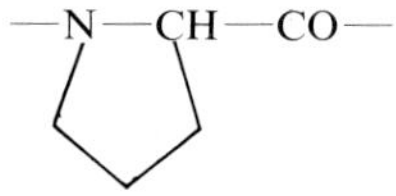	97.1
Serine	Ser	S	$-NH-CH-CO-$ CH_2OH	87.1
Threonine	Thr	T	$-NH-CH-CO-$ CH_2OH CH_3	101.1
Tryptophan	Trp	W	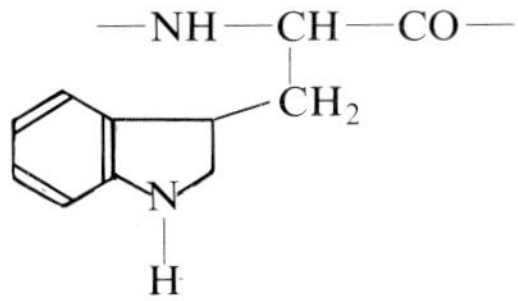	186.2
Tyrosine	Tyr	Y	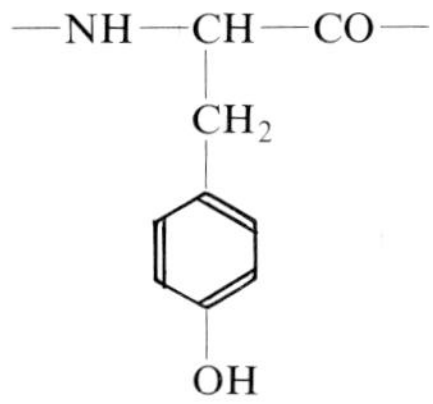	163.2
Valine	Val	V	$-NH-CH-CO-$ $CH-CH_3$ CH_3	99.1

TABLE 1.2
Some examples of naturally occurring modified residues in proteins

Amino acid	Example of protein
N-Trimethylalanine	*E. coli* ribosomal protein L11 (Dognin and Wittmann-Liebold, 1977; Lederer et al., 1977)
N^{ω}-Methylarginine	Histones (Paik and Kim, 1971)
N^{ω}-(ADP-ribosyl)arginine	*E. coli* DNA-dependent RNA polymerase, modified by a bacteriophage T4 enzyme (Rohver et al., 1975)
N^4-(*N*-acetylglucosaminyl)-asparagine	Many glycoproteins (Spiro, 1970), including ovalbumin (Marshall and Neuberger, 1964)
Cystine	Many proteins
S-Cysteinyl-haeme	Cytochrome *c* (Margoliash and Schejter, 1967)
S-(Diacylglyceryl)cysteine	*E. coli* outer membrane murein lipoprotein (Hantke and Braun, 1973)
γ-Carboxyglutamic acid	Blood-clotting proteins (Stenflo and Suttie, 1977)
O^{γ}-(ADP-ribosyl)glutamic acid	Rat liver lysine-rich histone (Riquelme et al., 1979)
τ-Methylhistidine	Actin (Elzinga et al., 1973)
N^{ϵ}-Methyl-lysine, N^{ϵ}-dimethyl-lysine, N^{ϵ}-trimethyl-lysine	Several ribosomal proteins and histones (Paik and Kim, 1971)
N^{ϵ}-(γ-glutamyl)lysine	Fibrin, wool proteins (Folk and Finlayson, 1977)
N^{ϵ}-Biotinyllysine	Several carboxylases (Wood and Barden, 1977)
Lysinonorleucine, desmosine, pyridinoline	Collagen and elastin (Gallop et al., 1972; Fujimoto et al., 1978)
5-Hydroxylysine, 4-hydroxyproline	Collagen (Bornstein, 1974)
O^3-Galactosaminyl serine	Many glycoproteins (Spiro, 1970)
O^3-Phosphonoserine, O^3-phosphonothreonine	Many phosphoproteins, including casein, glycogen phosphorylase and histones (Taborsky, 1974)

Amino acids are linked in peptides and proteins through amide bonds ('peptide bonds') between the α-carboxyl group of one residue and the α-amino group of the adjacent residue. The accepted abbreviation for a peptide consists of the three-letter codes of the amino-acid residues in order, starting with the amino-terminal residue, separated by hyphens; thus the tripeptide L-alanyl-glycyl-L-lysine is abbreviated Ala-Gly-Lys and has the structure:

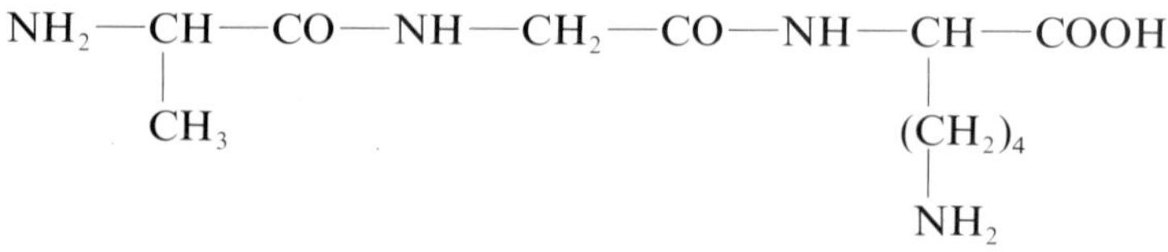

In neutral aqueous solution, amino and carboxyl functions are protonated and dissociated, respectively, so that, like amino acids, peptides are usually zwitterionic. The average pK_a of the α-amino group in peptides is, however, significantly lower than that in amino acids, and is close to pK_a 7.8 (Steinhardt and Beychok, 1964).

While some workers use the single-letter code for amino acid residues (Table 1.1), this is not recommended: it is difficult to include abbreviations for modified residues, confusion may arise on occasion with the code for nucleosides, most workers outside the field are not conversant with it, and the three-letter code is more easily comparable with the nucleotide sequence in nucleic acids. Internationally accepted abbreviations and nomenclature are described in *Biochemical Nomenclature and Related Documents* published for the International Union of Biochemistry by The Biochemical Society.

Most proteins of which the structure has been determined have a single homogeneous sequence, but occasionally two different residues are identified at a single position in the sequence. Such *microheterogeneity* may be due to differences in the genome in different individuals from which the protein was isolated (or heterozygosity in a single organism), or may be due to different extents of post-translational chemical changes, especially the partial conversion of asparagine to aspartic acid residues, or biosynthetic modifications as mentioned above.

1.2. Historical background

The modern approach to protein sequence determination had its origins in careful work by many investigators over several decades. By 1940 it was agreed that protein molecules were built up from amino acid residues bound together by peptide bonds to form polypeptide chains. However, it was by no means agreed that all molecules of a given protein had an identical amino acid sequence. The studies of Sanger and his collaborators showed that at least one protein, insulin, had a unique molecular structure (Ryle et al., 1955; Sanger, 1959).

During these studies several methods for sequence analysis were developed, including amino-terminal labelling with fluorodinitrobenzene (Sanger, 1945). 1-Fluoro-2,4-dinitrobenzene reacts with amino groups in peptides to give dinitrophenyl derivatives; acid-catalyzed hydrolysis cleaves peptide bonds, releasing the dinitrophenyl amino acid from the N-terminus of the peptide:

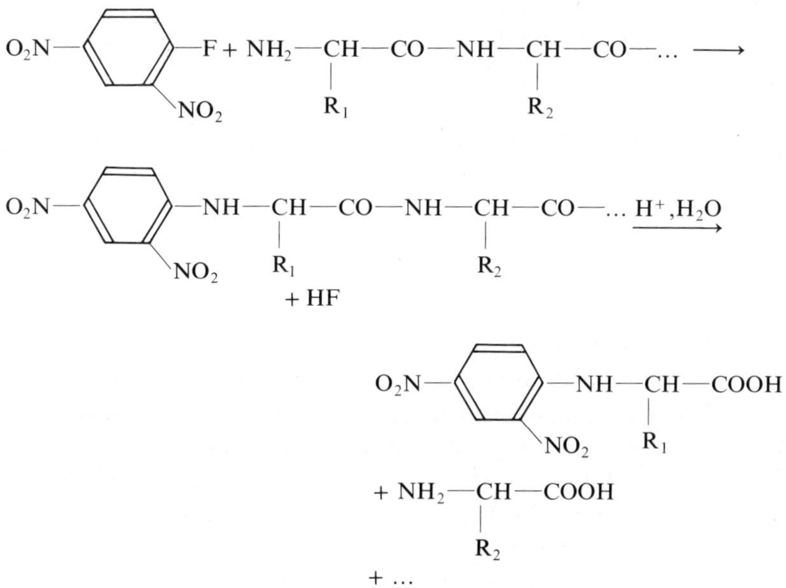

The dinitrophenyl amino acids are yellow compounds which could be identified by partition chromatography.

Partial acid hydrolysis of the dinitrophenyl B chain of insulin yielded a mixture of dinitrophenyl peptides [Dnp(Phe,Val), Dnp(Phe,Val,Asp) and Dnp(Phe,Val,Asp,Glu)] in addition to Dnp-Phe. Separation of these derivatives and determination of their amino acid compositions enabled the sequence of the amino-terminal four residues to be deduced. Similar analyses were performed on peptides isolated from partial acid and enzymic digests of the insulin chains. The total structure could be deduced by alignment of the sequences of many small peptides. A large number of separations of peptides and dinitrophenyl derivatives were performed, and the technique of paper chromatography (Consden et al., 1944) was particularly important.

A more efficient approach to the determination of peptide sequences was the use of sequential degradation methods, from the amino-terminus using phenylisothiocyanate (Edman, 1950; 1953; 1956; § 6.3) and from the carboxy-terminus using carboxypeptidases (§ 6.7). The early methods for peptide sequence determination were described thoroughly by Fraenkel-Conrat et al. (1955).

The same underlying principle, of chemical or enzymic cleavage of the protein, followed by separation of the resulting peptides, determination of their structures and deduction of the total sequence by alignment of the peptides, still forms the basis of protein sequencing strategy. Techniques for all the stages in sequence determination have, however, improved greatly in speed and sensitivity.

Dansyl chloride (Gray and Hartley, 1963) replaced fluorodinitrobenzene as an amino-terminal reagent, since the fluorescence of the dansyl amino acids allows a hundred-fold increase in sensitivity.

Methods for the specific enzymic and chemical cleavage of polypeptide chains were gradually improved. A major advance was the introduction of cleavage at methionine residues with cyanogen bromide (Gross and Witkop, 1961, 1962). The proteases trypsin, chymotrypsin and pepsin were joined by other proteases, including thermolysin and staphylococcal protease, and methods for chemical modification of proteins to limit tryptic cleavage to either lysine or arginine residues were developed.

Improvements in methods for the separation of amino acids and peptides and their derivatives were particularly important. The amino acid analyzer (Spackman et al., 1958) has been dramatically improved since its introduction. High-voltage paper electrophoresis (Michl, 1951, 1952), ion-exchange chromatography, gel filtration and thin-layer methods were introduced and developed.

The phenylisothiocyanate (Edman) degradation has proved especially valuable. Improvements in the technique in the three decades since its introduction have made possible the determination of amino acid sequences of peptides the size of the insulin chains in only a day or so.

With the steady improvement in techniques, the size of the polypeptides studied has increased up to the recent determination of the complete sequence of β-galactosidase of *Escherichia coli*, consisting of 1021 amino acid residues (Fowler and Zabin, 1977).

Additionally, highly sensitive radiochemical methods have been developed for the study of proteins available in only very small amounts.

1.3. Outline of the strategy for sequence determination

Most determinations of the amino acid sequences in proteins at present utilize the phenylisothiocyanate degradation method of Edman, or methods closely related to this (§ 6.3), in which amino acid residues are cleaved one at a time from the amino-terminus of the protein or from peptide fragments. The amino acids released are identified as their phenylthiohydantoins or related derivatives; alternatively, the newly released amino-terminal residues may be identified as dansyl derivatives using small aliquots of the peptide removed after each cycle of degradation.

The entire sequence of a small polypeptide may be determined by sequential degradation from the amino-terminus through to the carboxy-terminus (e.g. eel calcitonin, Table 8.1). The procedure may be automated (Edman and Begg, 1967; Laursen, 1971), and commercially produced automated sequencers are widely used. The stepwise yield of the automated Edman degradation is high, up to 98%, but the maximum number of residues which may be placed in sequence in a single run is about 70, and often only half that number can be identified. The lower efficiency of manual techniques imposes an upper limit usually between 15 and 30 residues, depending on the peptide or protein.

For typical protein chains (200–500 residues) direct sequential degradation provides only a small proportion of the total information required. Cleavage of the protein at specific points, separation of the resulting peptides, and determination of the sequences of each peptide are required.

The classical approach, still generally the most efficient for relatively small proteins of up to 300 residues, uses cleavage of the protein with trypsin to generate small peptides which are usually easy to isolate and sequence. The order of the tryptic peptides is then found using overlapping peptides isolated from other digests of the protein, often using chymotrypsin, staphylococcal protease, thermolysin and pepsin. An example of this approach was the determination of the sequence of ribosomal protein S 8 of *E. coli* (Allen and Wittmann-Liebold, 1978). The sequence of this small protein (129 residues) was determined from less than 8 mg protein.

With larger polypeptides, these cleavage methods produce complex mixtures of peptides and the isolation of all of these may present considerable difficulty. In addition, determination of the alignment of a large number of small peptides becomes rapidly more difficult with increasing size of the parent polypeptide chain. Initial cleavage into a few relatively large fragments is preferred, either with cyanogen bromide or, after modification of lysine residues, with trypsin. The resultant large fragments are separated from one another and their sequences are determined, after subfragmentation into smaller peptides. The large fragments are then aligned with the help of overlapping peptides from a different digest of the protein.

In favourable cases, automated determination of the sequences of a few large peptides may provide most of the information required for the determination of the total sequence of the protein. It is still necessary in most cases to isolate and determine the structures of small peptides for the completion of the sequence, and manual methods are generally more efficient than automated methods for this purpose. For example, the almost exclusive use of automated methods for determining the sequence of rabbit β_2-microglobulin (99 residues) required almost 30 mg protein, even though homology with the human protein was used to indicate which cleavage methods would probably be most suitable (Gates et al., 1979).

The combined use of automated and manual methods may be illustrated by the determination of the sequence of coelacanth triose phosphate isomerase (Fig. 1.1) (Kolb et al., 1974). Comparison with the homologous rabbit muscle enzyme was useful in the selection of cleavage methods. Amino acid analysis revealed the presence of two methionine and seven arginine residues. Preliminary inspection of cyanogen bromide cleavage products by polyacrylamide gel electrophoresis in SDS showed that the methionine residues were suitably placed for the generation of large fragments. Other large fragments were derived from tryptic cleavage at arginine residues after reaction of the protein with citraconic anhydride. About 70% of the protein sequence was determined by automated methods; the remaining part being determined by manual degradation of smaller peptides. Not all peptides were aligned unambiguously; homology with the rabbit muscle enzyme was used for aligning a few peptides.

Experience reveals that to avoid errors it is preferable to determine each part of an amino acid sequence more than once, using independent methods. Thus, even if almost the whole protein sequence may be determined by automated degradation of large peptides it is good strategy to isolate small peptides, such as tryptic peptides, as well, and ensure that the total sequence deduced is consistent with the structures, or at least the amino acid compositions, of these peptides.

Ala-Pro-Arg-Lys-Phe-Phe-Val-Gly-Gly-Gly-Asn-Trp-Lys-Met-Asn-Gly-Asp-Lys-Lys-Lys-Ser-Leu-Gly-Glu-
10
20
CNBr I
Leu-Ile-Gln-Thr-Leu-Asn-Ala-Ala-Lys-Val-Pro-Phe-Thr-Gly-Gln-Ile-Val-Cys-Ala-Pro-Pro-Glu-
30
40
II T_1
Ala-Tyr-Leu-Asp-Phe-Ala-Arg-Leu-Lys-Val-Asp-Pro-Lys-Phe-Gly-Val-Ala-Ala-Gln-Asn-Cys-Tyr-
50
60
CNBr II
Lys-Val-Ser-Lys-Gly-Ala-Phe-Thr-Gly-Glu-Ile-Ser-Pro-Ala-Met-Ile-Lys-Asp-Cys-Gly-Val-Thr-
70
80
II T_2
III T_1
Trp-Val-Ile-Leu-Gly-His-Ser-Glu-Arg-Arg-His-Val-Phe-Gly-Glu-Ser-Asp-Glu-Leu-Ile-Gly-Gln-
90
100
110
Lys-Val-Ala-His-Ala-Leu-Ala-Glu-Gly-Leu-Gly-Val-Val-Ala-Cys-Ile-Gly-Glu-Lys-Leu-Asp-Glu-
120
130
III T_2
Arg-Glu-Ala-Gly-Ile-Thr-Glu-Gly-Val-Val-Phe-Glu-Val-Thr-Glu-Val-Ile-Ala-Asp-Asp-Val-Lys-
140
150

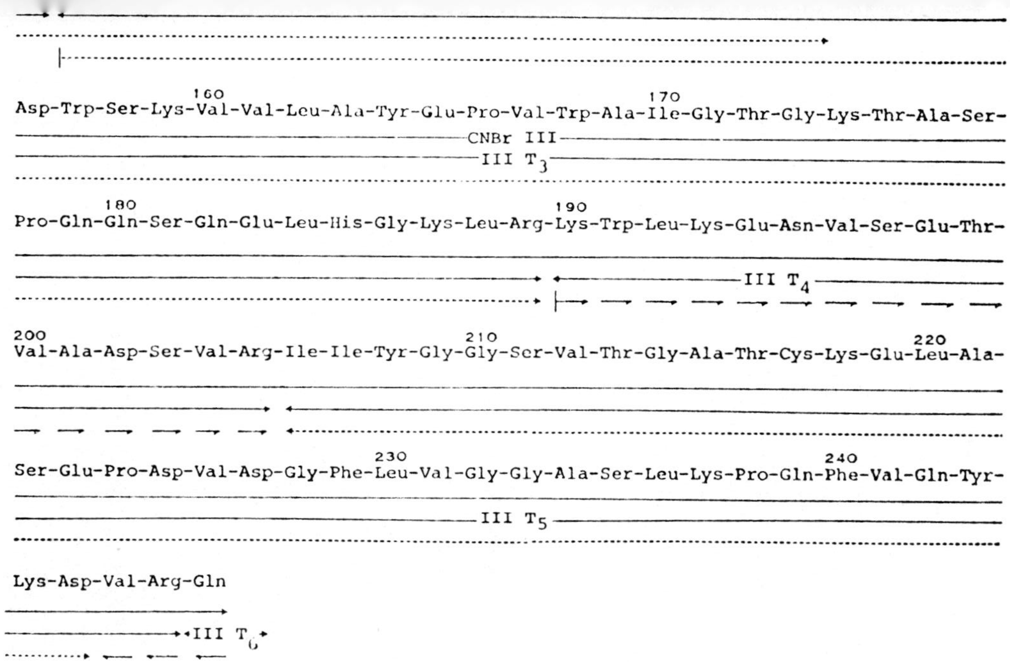

Fig. 1.1. Amino acid sequence of triose phosphate isomerase from coelacanth muscle. The CNBr and tryptic arginine fragments are shown as solid lines. Residues identified by sequencer analysis (– – – –), by dansyl-Edman degradation (———→), by carboxypeptidase digestion (←———). Serine at position 20 and glutamine at position 26 could not be identified during sequencer analysis of fragment CNBr II. Overlaps were not obtained between residues 58 and 59, residues 68 and 69 and residues 71 and 72. Residues are numbered as for the rabbit muscle enzyme (Corran and Waley, 1973) and include a deletion at position 3 in the coelacanth enzyme.

Reproduced with permission from Kolb, E., Harris, J.I. and Bridgen, J. (1974) Biochem. J. *137*, 185–197.

In addition, the study of small peptides will generally be required for the determination of disulphide bond cross-links and the identification of other modified residues.

The efficient operation of an automated liquid-phase sequencer, with ancillary equipment, such as a high-performance liquid chromatograph, and possibly an additional amino acid analyzer, requires a large capital outlay; running costs are also large. Such equipment is thus restricted to a minority of laboratories. Automated solid-phase sequencers are considerably cheaper, but the solid-phase technique is not routinely as effective as the liquid-phase technique. The manual methods described in this volume are far more cost-effective, and are more sensitive than the automated liquid-phase methods currently used routinely. Technical improvements in the automated instruments are likely to turn the balance in their favour in this respect, in the future. For some specialized applications, such as the determination of the amino-terminal sequences of radio-labelled proteins available in only minute quantities from cell-free protein synthesis systems, the automated techniques have particular advantages (Chapter 10).

To summarize, the steps required for the determination of the complete sequence of a protein are: characterization of the protein, including amino acid analysis (Chapter 2), specific cleavage of the polypeptide chain, by at least two different methods (Chapter 3), isolation of peptides from the mixtures resulting from the cleavage (Chapter 4), using suitable detection methods (Chapter 5), determination of the structures of the peptides (Chapter 6), including structures introduced into the protein by post-translational modification (Chapter 7), and deduction of the total sequence from the sequences of the constituent peptides (Chapter 8). The determination of the sequences of integral membrane proteins presents unique difficulties, and special approaches to this problem are discussed in Chapter 9. Some techniques used for limited sequencing objectives are discussed in Chapter 10, together with an alternative approach to protein structure determination, based on the determination of the nucleotide sequence of DNA coding for the protein. Appendices list manufacturers and suppliers of apparatus and reagents for protein sequence determination, purification of solvents and reagents, aspects of safety in the protein sequencing laboratory and some notes on the handling of small quantities of peptides.

CHAPTER 2

Preliminary characterization of the protein

2.1. *Purification of the protein*

The extraction and purification of proteins are outside the scope of this monograph. Methods routinely employed include salt fractionation, gel filtration, ion-exchange chromatography, affinity chromatography, ultracentrifugation, isoelectric focusing and crystallization. For sequence analysis the retention of secondary and tertiary structure is not required (apart from the use of limited proteolysis, § 3.2), so purification in denaturing media, such as 8 M urea or sodium dodecyl sulphate solution, is possible. Polyacrylamide gel electrophoresis in the presence of dodecyl sulphate is a method of high resolving power, but has only rarely been used, except for the determination of amino-terminal sequences by radiochemical methods.

The protein should be at least 95% pure on a weight basis, and no single polypeptide impurity should be present in an amount greater than 3% on a molar basis. Impurities present in amounts smaller than this will not generally interfere with sequence analysis. The determination of the purity of proteins may be made by several methods, of which the most useful are probably polyacrylamide gel electrophoresis and isoelectric focusing (cf. Gordon, this series, 1975, and Righetti and Drysdale, this series, 1976).

The amount of protein required for the complete sequence determination depends on many factors. The size of the molecule is a major factor, as is the availability of a high-sensitivity amino acid analyzer. An unfavourable distribution of residues such as arginine and methionine, which provide suitable points for cleavage of the peptide chain, and the presence of unusual residues may increase significantly the amount required. In favourable cases, the complete sequence of a small protein, of molecular weight 10,000, may be determined from 1 mg of protein. For a protein of molecular weight 20,000, about 20 mg will be required, and for a large polypeptide of molecular weight 100,000 more than 1 g. The amount of protein necessary is discussed further in Chapter 4.

Conditions deleterious to the covalent structure of proteins should be avoided during the purification and subsequent analysis. Exposure to sunlight causes photooxidation of tryptophan residues (Pirie, 1971), and extremes of pH lead to loss of amide groups and hydrolysis of some peptide bonds. Strongly alkaline conditions ($pH > 10$) lead to destruction of cystine residues (Asquith and Carthew, 1972). Microbial contamination must be prevented. All reagents, including water, should be the purest available. Nitrogen gas, frequently used to prevent autoxidation, should be oxygen free.

2.2. Separation of the constituent polypeptide chains of heterooligomeric proteins

Many proteins consist of oligomers of non-identical polypeptide chains, and separation of these is the first step in the determination of their primary structures. The chains may be held together by non-covalent interactions (e.g. haemoglobin) or by covalent bonds, especially disulphide bonds linking half-cystine residues (e.g. immunoglobulins). The separation of the polypeptide chains may be achieved by gel filtration, if their molecular weights differ by a factor of 1.5 or more, or by ion-exchange chromatography, affinity chromatography or electrophoresis under conditions where the interaction between the chains is abolished. Such conditions may include denaturing detergents, 8 M urea, 6 M guanidinium chloride, or extremes of pH. Urea, if used, should be free from cyanate, which carbamylates proteins (Stark et al., 1960); a primary amine may be included in the buffers to compete with the protein for any cyanate present.

Several examples of methods for the separation of polypeptide chains are listed in Table 2.1, and a few examples are described in detail below.

TABLE 2.1
Some examples of methods for the separation of constituent polypeptide chains of proteins

Ribulose-1,5-bis-phosphate carboxylase (Holder, 1976)
The protein was incubated in 0.2 M 2-mercaptoethanol, 1% sodium dodecyl sulphate solution at 55 °C for 1 h, then the chains were separated on a column of Sephadex G-100 in 50 mM Tris–HCl, pH 8.5, containing 0.5% sodium dodecyl sulphate. The detergent was removed by dissolving the freeze-dried subunits in 10 mM HCl and precipitating the protein with four volumes of acetone at −18 °C.

Acetyl CoA carboxylase from a thermophilic Bacillus
(Buckley et al., 1979). The biotin-containing subunit of the enzyme was specifically isolated by affinity chromatography on Sepharose-avidin, using elution with denaturing reagents.

Nitrogenase from Azetobacter vinelandii (Lundell and Howard, 1978)
The reduced and carboxymethylated protein was incubated in 7 M urea at pH 5.4 at 45 °C for 45 min, and two subunits were separated by ion-exchange chromatography on a column of sulphopropyl-Sephadex in 7 M urea at pH 5.4.

Fibrinogen (Henschen and Edman, 1972)
Human fibrinogen (or fibrin) was reduced and carboxymethylated in 9 M urea solution. Three major, and two minor, components were separated by chromatography on a column of carboxymethyl cellulose in 8 M urea solution, with a pH gradient from pH 4.8 to pH 6.0 prepared from Tris and acetic acid.

2.2.1. *Separation of light and heavy chains of immunoglobulin G* (Fleischman et al., 1962)

A 2% (w/v) solution of γ-globulin in 0.55 M Tris–HCl buffer, pH 8.2, was reduced with 2-mercaptoethanol (0.75 M) for 1 h at room temperature. The solution was cooled to 0 °C and an equal volume of 0.75 M iodoacetamide was added. Trimethylamine was added cautiously during the alkylation to maintain pH 8. After 1 h, the protein was dialyzed overnight against 100 vol. of cold saline. The reduced protein (100–200 mg) was dialyzed against cold 1 M propionic acid, and chromatographed on a column (3 cm × 65 cm) of Sephadex G-75 in 1 M propionic acid at 4 °C. Two peaks, consisting of heavy and light chains, respectively, were obtained. Separation of the chains in a buffer solution at pH 3.5 was less satisfactory.

The modified conditions of Chen and Poljak (1974) included reduction by 20 mM dithiothreitol at pH 8.5 in the presence of 5 mM EDTA for 2 h at 20 °C, followed by alkylation with 60 mM iodoacetic acid for 20 min, in the dark. Separation of the chains was performed in 1 M propionic acid on a column of Sephadex G-100.

2.2.2. *Separation of α- and β-chains of haemoglobin*

The chains of human haemoglobin were separated by counter-current distribution in 2-butanol/0.5 M acetic acid/10% dichloroacetic acid (9 : 10 : 1, by vol.) (Hill et al., 1962). A more widely available technique is chromatography on carboxymethylcellulose at pH 6.7 in a gradient of sodium phosphate buffer containing 50 mM 2-mercaptoethanol and 8 M urea (Clegg et al., 1966).

Braunitzer and Sorger (1969) were able to separate the chains of carp haemoglobin by gel filtration after oxidation in air for 4 h at pH 8.5 in 8 M urea. The β-chains, which contained sulphhydryl groups, were polymerised and could be separated from the α-chains on a column of Sephadex G-100 in 50% aqueous acetic acid.

2.2.3. *Separation of light and heavy chains of myosin* (Weeds and Lowey, 1971)

Rabbit skeletal muscle myosin contains in addition to the heavy chains three different types of light chains, one of which may be dissociated by treatment with 5,5′-dithio*bis*(2-nitrobenzoate) (DTNB) without loss of ATPase activity, and two of which are removed by treatment with alkali.

DTNB-light chains: Myosin, dissolved at 15 mg/ml in 0.5 M KCl, 10 mM EDTA, at pH 8.5, was treated with DTNB (10 mM) for

10 min at 0 °C. Ten volumes of water were added to precipitate the myosin, which was collected by centrifugation at 8000 rev./min for 20 min. The supernatant was applied to a column of DEAE cellulose at pH 8.0 in 50 mM Tris–HCl, and the DTNB light chains were cluted ahead of DTNB with 1.0 M KCl. The light chains were dialyzed against a solution of 2-mercaptoethanol.

Alkali light chains: Myosin was dissolved at 5 mg/ml in 2 M LiCl, 1 mM dithiothreitol, 0.1 M glycine buffer, pH 11.1, at 0 °C for 30 min. The myosin was precipitated by the addition of neutralized 2.5 M potassium citrate to a final concentration of 0.8 M, and stirring was continued for 10 min. The alkali light chains were recovered from the supernatant by dialysis and freeze-drying after centrifugation at 12,000 rev./min.

2.3. Determination of the molecular weight of the polypeptide chains

An estimate of the molecular weight of a polypeptide is required before sequencing work can be planned. Several methods for the determination of molecular weights are available. The ultracentrifuge may be used; the molecular weight of a polypeptide may be determined by sedimentation velocity or sedimentation equilibrium techniques under disaggregating conditions. Reliable results are only obtained when great care is taken with the alignment of the optical system, control of temperature, etc., and when factors such as hydration of the protein, effective density, solvent density distribution in the centrifugal field, non-ideality of the solution and inhomogeneity of the protein may be allowed for. The techniques are described in detail by Teller (1973), Aune (1978) and Van Holde (1975).

The ultracentrifuge has been largely replaced for the determination of the molecular weights of polypeptides by the technically less demanding methods of gel filtration in denaturing media and electrophoresis in polyacrylamide gels in the presence of sodium dodecyl sulphate, although the latter methods are not given absolute support by hydrodynamic theory. Details of the use of gel filtration for the estimation of the molecular weights of polypeptides are given by Mann and Fish (1972), and of SDS gel electrophoresis by Nielsen and Reynolds (1978) and Weber and Osborn (1975).

These methods, which rely upon calibration with polypeptides of known molecular weight, generally give values within 10% of the actual molecular weight, but anomalous results are often obtained, particularly using SDS gel electrophoresis with glycoproteins or highly charged proteins. Use of two or more sets of conditions, such as different acrylamide gel concentrations, may indicate that anomalies are occurring. It is best to regard molecular weights derived by these methods as apparent until confirmed by other evidence.

Quantitative determination of the amino-terminal group in the polypeptide may be used to estimate the molecular weight, but the difficulties of quantitation, particularly on the micro scale, and the possibility of partial blockage of the amino-terminal residue lessen its usefulness. Similarly, the simplest whole-number ratio of amino acid residues obtained by analysis after acid hydrolysis gives an estimate of the minimum molecular weight, but in practice the errors in the method, especially if there is even slight contamination by other proteins, make this useful in general only for small polypeptides of up to 200 residues.

Peptide mapping methods may be used for the estimation of molecular weights, in conjunction with the amino acid analysis. For example, counting the number of tryptic peptides will give an estimate of the number of lysine and arginine residues, which, coupled with the amino acid analysis, yields the molecular weight. This method is subject to uncertainties resulting from the possible presence of uncleaved bonds (e.g. Lys-Pro), alternative points of cleavage (e.g. Lys-Lys-X), or non-specific cleavage. In addition, simple peptide maps do not always resolve peptides completely, and some peptides may be insoluble. Care is required to prevent partial autoxidation of methionine or alkyl-cysteine residues, which would lead to the formation of multiple spots from a single peptide.

Cleavage at methionine residues with cyanogen bromide may be used, but this method is subject to similar uncertainties. Radioactive labelling of the cysteine residues, followed by peptide mapping, with quantitation by liquid-scintillation counting, is a useful method for determining the number of unique cysteine residues in the polypeptide, and hence the molecular weight. Details of methods for peptide mapping are given in Chapters 4 and 10.

2.4. *Removal of salts and non-covalently bound molecules*

Most salts and low molecular weight substances may be removed from the protein by dialysis against 0.1 M NaCl, followed by exhaustive dialysis against water. However, some substances, such as lipids or haem, adhere strongly to some proteins, and their removal may require additional methods (§§ 2.4.1, 2.4.2). Methods for the removal of lipids and detergents from integral membrane proteins are discussed in Chapter 9. Metal ions or coenzymes which are tightly bound to a native protein may usually be removed by dialysis after denaturation.

2.4.1. *Removal of lipids*

The following method was described by Scanu and Edelstein (1971) for the removal of lipids from serum lipoproteins. A solution of the lipoprotein (2–5 mg) in 1 ml of 0.15 M NaCl, 1 mM EDTA, was added dropwise to 50 ml of stirred absolute ethanol/diethyl ether (3 : 1, v/v) at −10 °C. The precipitated apoprotein was collected by centrifugation at −10 °C and extracted with the cold ethanol/ether solvent, followed by dry ether. Part of the protein was soluble in the initial wet ethanol/ether, and could be precipitated by the addition of dry ether to the supernatant. Alternatively, chloroform/methanol (2 : 1, v/v) could be used, but subsequent dissolution of the protein was less easy. Extraction of the dry lipoprotein with the mixed organic solvents did not remove the phospholipids completely. The organic solvents used must be pure, and free from peroxides.

2.4.2. *Removal of haem from cytochrome* b_5 (Nóbrega and Ozols, 1971)

To a solution of the salt-free cytochrome b_5 (0.1–0.5 μmol) in 0.5 ml H_2O at 0 °C was added 10 ml of cold acetone containing 0.2% (v/v) HCl, with stirring. After 10 min at 5 °C the white precipitate of apo-cytochrome was collected by centrifugation, dried in a stream of N_2, and dissolved in 0.5 ml of 0.1 M NH_4HCO_3 for digestion with trypsin.

For the removal of haem from human haemoglobin (Rossi-Fanelli et al., 1958), one volume of cold 1–3% haemoglobin solution in water was added dropwise with vigorous stirring to 20–30 volumes of acetone, containing 3 ml of 2 N HCl/l, at −20 °C. After 10–20 min at −20 °C, the apoprotein was collected by centrifugation for 15 min at 1060 × *g* at −20 °C. The precipitate was dissolved in cold water, and dialyzed against water for 5–6 h at 2 °C.

2.5. *Amino acid analysis*

This subject is discussed in detail in a recent volume in this series (Glazer et al., 1975), and little will be added here. Analysis for covalently bound phosphate (§ 2.5.1) and carbohydrate (Glazer et al., 1975) should be performed on samples of the salt-free protein. Spectroscopic analysis may reveal the presence of covalently bound chromophoric groups. The determination of an accurate extinction coefficient for the protein (§ 2.5.2) allows absorbance to be used as a measure of protein concentration in later work, and this is far more convenient than weighing. Even freeze-dried proteins may contain significant amounts of water, and drying to constant weight at 105 °C in vacuo is required for accurate quantitation. Unfortunately many proteins are not available in sufficient quantity for thorough characterization by dry weight analysis.

Recent advances in analytical techniques (Hare, 1975, 1977; Benson, 1975), including the use of fluorescamine (Udenfriend et al., 1972) or *o*-phthalaldehyde and 2-mercaptoethanol (Roth and Hampai, 1973) as detection reagents, have made analysis at the level of 1 nmol of each amino acid routine in many laboratories.

The *o*-phthalaldehyde reagent is particularly suitable, and its use is reviewed by Lee and Drescher (1978). Stringent precautions must be taken to avoid contamination of samples, and highly purified reagents and water are required. Alternative methods of amino acid analysis, by chromatography of fluorescent, radioactive, coloured or volatile derivatives, are not widely used, as all suffer from difficulties in the reproducible formation of the derivatives, or separation from excess reagents.

An accurate amino acid analysis of the whole polypeptide chain is extremely important, as it is the yardstick against which the accumulated data from the sequence analysis of peptides are to be compared. Since a single analytical determination may be subject to an inaccuracy of up to 3%, or more for some amino acids, it is essential to perform several analyses, after hydrolysis in vacuo at 110 °C in 6 M HCl containing 0.1% (w/v) phenol or 0.05% (v/v) 2-mercaptoethanol for 24 h, 48 h and 72 h. Serine and threonine residues are estimated by extrapolation to zero time of hydrolysis, and isoleucine and valine are estimated from the values after 72 h hydrolysis. Cysteine and methionine may be determined as cysteic acid and methionine sulphone, respectively, after hydrolysis of the performic acid oxidized protein, or as *S*-carboxymethylcysteine (or other alkyl derivative) and methionine if mercaptoethanol is included in the acid for hydrolysis. Cystine and cysteine may also be determined spectrophotometrically using 5,5′-dithio*bis*(2-nitrobenzoate) (Glazer et al., 1975). Tryptophan is destroyed under these hydrolysis conditions. Mercaptoethane sulphonic acid is more satisfactory (Penke et al., 1974) than methane sulphonic acid for hydrolysis for the determination of tryptophan. Spectrophotometric determination of tryptophan is accurate if no unusual chromophoric groups are present in the protein (§ 2.5.3).

The amino acid analysis charts should be carefully inspected for the presence of unusual peaks which may indicate the presence of modified amino acid residues. The identification and determination of several modified residues are described by Glazer et al. (1975) and by Horáková and Deyl (1978).

2.5.1. Determination of phosphate (McClare, 1971)

The salt-free protein, containing 0.1–1 μmol bound phosphate, is dried in a 1.7 cm × 13 cm glass-stoppered borosilicate glass tube, previously tested by this method to be free from phosphate. $HClO_4$ (sp. gr. 1.70; 0.5 ml) is added. The tube is placed in a heating block at 210 °C for 30 min, with a loosely fitting stopper, then allowed to cool to room temperature. The colour reagent is prepared from $HClO_4$ (40 ml), ammonium molybdate (5 g) and ascorbic acid (1 g), made up to 500 ml with water. The ascorbic acid is dissolved separately in water and added last. 9.5 ml of the reagent is added to the tube, which is then heated at 50 °C for 1 h. The tube is shaken and the absorbance at 825 nm is measured in a 1 cm path-length cell. The blue colour is stable for several hours. A standard curve is constructed using samples of Na_2HPO_4 (0.05–2 μmol) taken through the same procedure. The sensitivity may be increased by reducing the volumes of the reagents.

2.5.2. Determination of the extinction coefficient

The extinction coefficient may be determined most directly by dry weight analysis. A solution of the protein is dialyzed exhaustively against water. The absorbance at 280 nm of an approximately 1 mg/ml solution of the protein is determined, and an aliquot of the solution is placed in a vial dried to constant weight at 105 °C. The solution of the protein is carefully dried in a vacuum dessicator, followed by heating to 105 °C to constant weight. About 10 mg of protein is required. Experimental details leading to high precision are described by Kupke and Dorrier (1978).

A more sensitive method is the use of the amino acid analyzer. Norleucine is added to a final concentration of 0.5 mM to a solution of the protein (about 1 mg/ml) in water. The absorbance at 280 nm is determined. Aliquots (0.01–0.1 ml, depending on the sensitivity of the analyzer) are transferred to small tubes for hydrolysis and analysis, and the concentrations of residues, corrected for losses during transfers using the value for norleucine, in the original solution are calculated; the sum of the weights of each amino acid residue is the weight of the protein.

2.5.3. Spectrophotometric determination of tryptophan (Edelhoch, 1967)

The ratio of the numbers of tryptophan and tyrosine residues in a protein may be determined from the absorbances at 280 nm and 288 nm in neutral 6 M guanidinium chloride solution, using the formula:

$$\frac{\text{Trp}}{\text{Tyr}} = \frac{A_{288} \times 1280 - A_{280} \times 385}{A_{280} \times 4815 - A_{288} \times 5690}$$

The method is not applicable when other chromophoric groups are present. High concentrations of cystine residues interfere, so observations on the reduced and carboxymethylated protein are to be preferred. The method is accurate within the range of ratios of tryptophan to tyrosine of 0.2–1. A more accurate determination at high ratios of tryptophan to tyrosine residues is based on the difference in the absorption spectra in 6 M guanidinium chloride in 0.1 M NaOH and at neutral pH. The tyrosine concentration is determined using $\Delta\epsilon_{295} = 2480$. The absorbance at 280 nm due to tryptophan residues is then determined by subtracting the absorbance due to tyrosine residues (calculated using $\epsilon_{280} = 1280$) from the total absorbance, and the concentration of tryptophan residues calculated using $\epsilon_{280} = 5690$. The absorbance at high pH should be determined immediately upon addition of the NaOH, since gradual absorbance changes were observed with some proteins. The content of tyrosine residues in the protein is determined from the amino acid analyzer; hence the tryptophan content is determined from this value and the ratio of the numbers of the residues.

2.6. *Chemical modification of the protein for sequence analysis*

This subject is discussed in detail by Glazer et al. (1975) and the reader is referred to that volume.

Chemical modification is performed for several purposes.

2.6.1. *Conversion of cysteine residues to stable derivatives*

Cysteine residues are subject to autoxidation, leading to a variety of products, including the random formation of disulphide bridges, during procedures for the isolation of peptides. These residues must therefore be converted to more stable derivatives. Alkylation with iodoacetic acid has often been used, since the reaction is rapid and specific, the reagent is reasonably stable (in the dark) and is available with radiolabels, and the introduction of a charged group tends to enhance the solubility of peptides. Cystine residues may also be modified by alkylation after reduction. Experimental details for use with iodo[^{14}C]acetate are given below (§ 2.6.5).

The cyclization of amino-terminal *S*-carboxymethylcysteine residues under acidic conditions (Bradbury and Smyth, 1973) may lead to difficulties in peptide isolation and sequence determination, particularly if oxidation to the sulphone occurs (Zervos and Adams, 1977). Several alternative means of modification of cysteine residues are also used (Table 2.2); each method has its advantages and disadvantages.

Alkylation with 3-bromopropionate, giving *S*-carboxyethylcysteine, which is more stable, is much slower, and higher concentrations of the reagent are required. Performic acid oxidation leads to the destruction of tryptophan residues; methionine is converted to its sulphone, and cleavage with CNBr is not possible after oxidation. Ethyleneimine is highly reactive and suspected of carcinogenicity, but the reagent may be used for the introduction of additional points of tryptic cleavage. An equivalent reagent, *N*-(iodoethyl)trifluoroacetamide, which is much less reactive, has recently been introduced (Schwartz et al., 1979). The products of

TABLE 2.2
Chemical modification reagents for cysteine residues

Reagent	Product	Reference
Performic acid	Cysteic acid	Hirs, 1967a
Iodoacetic acid	*S*-Carboxymethylcysteine	§ 2.6.5
Iodoacetamide	*S*-Carboxamidomethyl-cysteine	
3-Bromopropionic acid	*S*-(2-Carboxyethyl)-cysteine	Bradbury and Smyth, 1973
Ethyleneimine	*S*-(2-Aminoethyl)cysteine	Lindley, 1956; Raftery and Cole, 1966
(2-Bromoethyl)trimethyl-ammonium bromide	4-Thialaminine	Itano and Robinson, 1972
Methyl-*p*-nitrobenzene sulphonate	*S*-Methylcysteine	Heinrikson, 1971
N-(4-Dimethylamino-3,5-dinitrophenyl)maleimide		Witter and Tuppy, 1960
4-Vinylpyridine	*S*-(2-(4′-Pyridyl)ethyl)-cysteine	Friedman et al., 1970
[^{35}S]cystine	Cystine (mixed disulphide with [^{35}S]cysteine)	Weeds and Hartley, 1968
2-Bromoethane sulphonate	*S*-Sulphoethylcysteine	Niketic et al., 1974
1,3-Propane sultone	*S*-3-Sulphopropylcysteine	Ruegg and Rudinger, 1977

the reactions of *N*-ethylmaleimide or its derivatives with cysteine residues are subject to a variety of further reactions, giving rise to problems in peptide purification (Smyth et al., 1960). The phenylthiohydantoin derivative of the product formed from cysteine and 4-vinylpyridine is readily identified (Hermodson et al., 1972), unlike that of *S*-carboxymethylcysteine, although the latter residue may be identified readily if radio-labelled. The polarity of *S*-sulphoethylcysteine residues helps retention of peptides in the cup during automated liquid-phase sequence determination (Niketic et al., 1974).

2.6.2. *Deletion or introduction of points of enzymic cleavage*

The action of trypsin may be limited to arginine residues by modification of lysine residues (§ 3.3.3), or to lysine residues after modification of arginine residues (§ 3.3.4). Additional points of tryptic cleavage may be introduced by aminoethylation of cysteine residues (Raftery and Cole, 1966), although cleavage of *S*-(2-aminoethyl)cysteine peptide bonds is slow.

2.6.3. *Introduction of labels to facilitate the detection and isolation of peptides containing specific residues*

Active-site residues in enzyme proteins may often be specifically modified with radioactive or chromophoric reagents. Methionine-containing peptides, required for the identification of the order of peptides derived by cleavage with CNBr, may be specifically labelled by alkylation under acidic conditions with radioactive iodoacetate after modification of cysteine residues with non-radioactive reagent. The introduction of radioactive substituents is used in some sensitive peptide mapping techniques (Chapter 10).

2.6.4. *Conversion of the protein to a soluble form, facilitating digestion and isolation of peptides*

Some proteins are highly aggregated or insoluble after denaturation, and especially after freeze-drying, and are resistant to proteolytic attack. This problem is particularly serious with integral membrane proteins (Chapter 9). Conversion of the protein to a polyanionic derivative by reaction of the amino groups with a dicarboxylic anhydride is usually successful in producing a derivative soluble in slightly alkaline conditions and readily digested by proteases. The peptides produced are also usually more soluble than the unmodified peptides.

2.6.5. *Reduction and carboxymethylation*

The following procedure, which is economical in the use of radioactive reagent, has given reliable results with a variety of proteins.

The protein is dissolved at 1–20 mg/ml in a solution containing 6 M guanidinium chloride, 0.1 M Tris and 1 mM EDTA, adjusted to pH 8.3 with HCl. Dithiothreitol is added to a concentration of 2 mM, or in a 2 mM excess over the disulphide bonds in the protein. Nitrogen (oxygen-free) is passed through the solution, which is then incubated at 37 °C for 1 h. A solution of 50 mM iodo[^{14}C]acetic acid, neutralized with NaOH, is added, to give a 1.1-fold molar ratio over total thiol groups in the solution (typically 4.5–5 mM). The iodoacetate solution is stable for several months at −20 °C in the dark. For most purposes, a specific activity of 1500 cpm/nmol is suitable. Nitrogen is blown over the surface of the reaction mixture, the vessel is sealed, and the alkylation is allowed to proceed in the dark at 37 °C for 1 h. A more complete reaction is obtained by repeated addition of dithiothreitol (1 mM) followed by incubation at 37 °C for 1 h, then addition of iodo[^{14}C]acetate (2 mM) and incubation at 37 °C for 30 min.

Mercaptoethanol is added to a concentration of 1% (v/v), and the solution is dialyzed exhaustively against 50 mM NH_4HCO_3/0.01% (v/v) thiodiglycol, and freeze-dried.

The iodoacetic acid used must be colourless; any iodine present, revealed by a yellow colour, causes rapid oxidation of thiol groups, preventing alkylation, and may modify tyrosine residues. Non-radioactive iodoacetate may be used, but the incorporation of a radio-label greatly facilitates the identification of carboxymethylcysteine residues.

2.7. Determination of the amino-terminal residue

A large number of methods have been described for the determination of N-terminal residues in proteins. The dansyl chloride method of Gray and Hartley, as described by Gray (1972) (§ 2.7.1) is sensitive and generally successful. For amino-terminal histidine, arginine, tryptophan or cysteine (either as *S*-carboxymethylcysteine or as cysteic acid) it is not completely satisfactory, and the dicarboxylic acids, Asp and Glu, are not distinguished from their amides, Asn and Gln. Phenylisothiocyanate may be used to determine the N-terminal residue, using the techniques described for sequential degradation in Chapter 6. This reagent is less sensitive, and unsatisfactory for serine and threonine. Derivatives of phenylisothiocyanate which are coloured or fluorescent lend greater sensitivity. The use of 4-dimethylaminoazobenzene-4′-isothiocyanate is described below (§ 2.7.2). Some other methods which have been used for the determination of N-terminal residues in proteins are listed in Table 2.3. In addition, any of the reagents developed for sequential degradation from the amino-terminus (Chapter 6) may be used.

TABLE 2.3

Some reagents used for the determination of amino-terminal residues in proteins

Dansyl chloride (Dns-Cl)	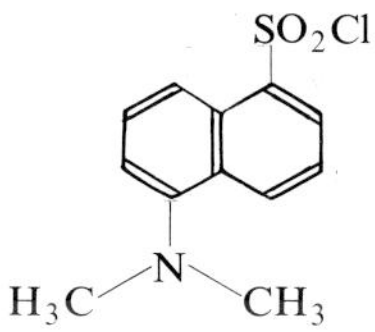	(Gray, 1972; § 2.7.1)
Phenylisothiocyanate (Pitc)	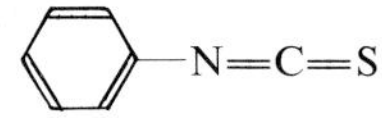	(Chapter 6)
4-Dimethylaminoazobenzene-4′-isothiocyanate (Dabitc)	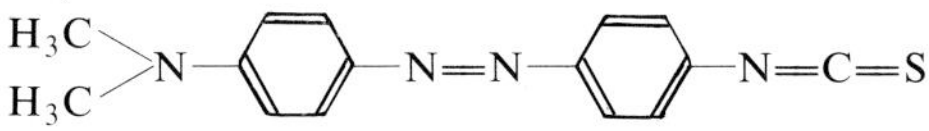	(Chang et al., 1976; § 2.7.2)
Fluorodinitrobenzene, (Fdnb)	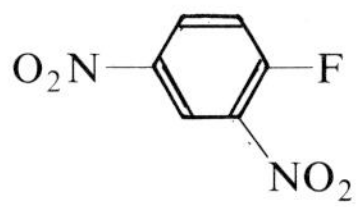	(Sanger, 1945)
Cyanate, OCN^-		(Stark, 1972a; Stark and Smyth, 1963)
Leucine aminopeptidase (EC 3.4.11.1)		(Light, 1972)
Aminopeptidase M (EC 3.4.11.2)		(Light, 1972)
Co (III) chelate		(Bentley, 1976)

The amino-terminal determination may reveal more than one residue, which could indicate the presence of an impurity in the protein, or the occurrence of partial proteolytic cleavage close to the amino terminus. Unexpected results may occasionally be obtained. For example, the apparent observation of two amino-terminal residues in ribosomal protein S11 from *Escherichia coli* (Chen and Chen-Schmeisser, 1977), which was shown to be due to an unexpected cleavage of the *N*-phenylthiocarbamyl-*N*-methyl-alanyl-lysine bond at slightly alkaline pH (Chang, 1978).

Many proteins have blocked amino-terminal residues, commonly through acetylation or cyclization of amino-terminal glutaminyl residues to pyrrolidonecarboxylyl (pyroglutamyl) residues. Some amino-terminal blocking groups found in proteins are listed in Table 2.4. The determination of such groups is generally made by the study of small peptides obtained after digestion with proteases, and isolated along with other peptides required for the determination of the complete sequence. Alternatively, methods devised for the selective isolation of peptides lacking a free amino-group may be used (§ 4.16). The identification of blocking groups is described in Chapter 7.

2.7.1. Determination of the N-terminal residue using dansyl chloride

Dansyl chloride (1-dimethylaminonaphthalene-5-sulphonyl chloride, Dns-Cl) reacts with amino groups in proteins under alkaline conditions to give sulphonamide derivatives which are resistant to hydrolysis in 6 M HCl at 110 °C (except for Dns-Pro, which is about 70% destroyed after overnight hydrolysis). The thiol group of cysteine residues, the imidazole group of histidine residues and the hydroxyl group of tyrosine residues also react, but the cysteine and histidine side-chain derivatives are unstable under the acid hydrolysis conditions. The principle of the method is to label the amino groups in the protein with dansyl chloride, followed by hydrolysis of the peptide bonds and identification of the α-*N*-Dns-amino acid derived from the amino terminus:

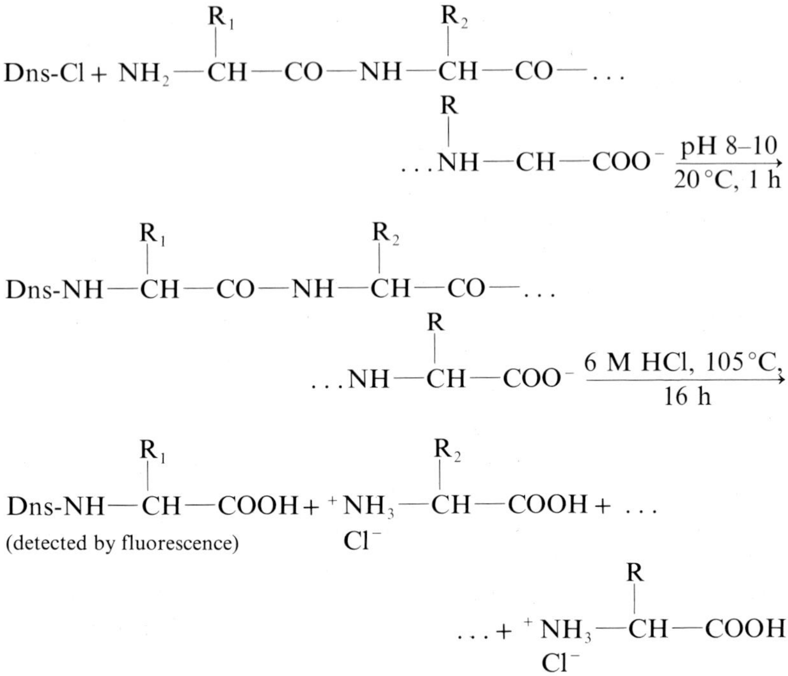

The excess reagent is hydrolyzed to dansic acid (Dns-OH), which is also fluorescent.

The following method is slightly modified from that described by Gray (1972).

The protein should be prepared as for amino acid analysis, free from salts and ammonia or other nucleophiles, and with cysteine residues converted to *S*-carboxymethylcysteine or some other stable derivative. 1–2 nmol (i.e. 50–100 μg of a protein of molecular weight 50,000) of the protein is dissolved in 20 μl of 1% (w/v) sodium dodecyl sulphate, with heating to 100 °C if necessary. The solution is placed in a borosilicate glass test tube, 4 mm i.d. × 50 mm. 20 μl of pure *N*-ethylmorpholine is added, and the contents of the tube are mixed. A freshly prepared solution (30 μl) of dansyl chloride in

TABLE 2.4
Examples of amino-terminal blocking groups

Acetyl, 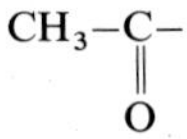	Many proteins, including horse heart cytochrome *c* (Margoliash and Smith, 1961), *Neurospora* NADP-specific glutamate dehydrogenase (Morris and Dell, 1975), rabbit sarcoplasmic reticulum Ca^{2+}-ATPase (Tong, 1977), *Neurospora* tyrosinase (Nau et al., 1977)
Formyl, 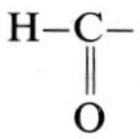	Nascent proteins in procaryotic systems (Webster et al., 1966; Osborn et al., 1970), with *N*-formyl methionine initiation. Mellitin (Kreil and Kreil-Kiss, 1967)
Pyrrolidone carboxylyl, 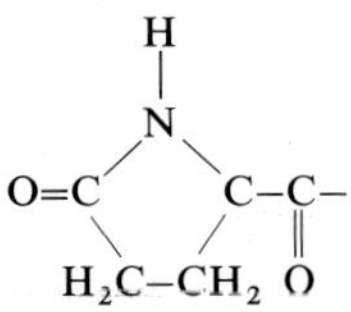	Many proteins, including some immunoglobulin chains (Gally and Edelman, 1972), and hog amylase (Kluh, 1979)
N-Trimethylalanyl, 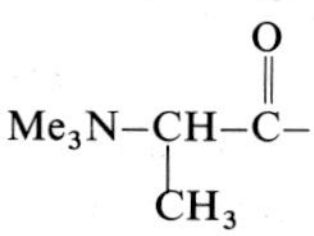	Ribosomal protein L-11 from *Escherichia coli* (Dognin and Wittmann-Liebold, 1977; Lederer et al., 1977)
Ketoamine group derived from glucose	Haemoglobin $A1_c$ (Bunn et al., 1978)
Fatty acyl	Murein lipoprotein from *E. coli* (Hantke and Braun, 1973)

pure dry dimethylformamide (25 mg/ml) is added, and the contents of the tube are mixed thoroughly. The reaction is allowed to proceed for 1 h at about 20 °C. Some proteins may precipitate upon addition of the dansyl chloride solution or during the reaction, but this may not prevent the reaction from occurring. Modified conditions, such as dilution of the protein solution, may help to keep the protein in solution.

The solution of dansylated protein is cooled on ice, and acetone (analytical grade) (total 250 μl) is added, with mixing. The dansylated protein forms a flocculent precipitate. The suspension is kept at 0 °C for 10 min, and the protein is collected by brief centrifugation. The supernatant is carefully removed with a drawn-out Pasteur pipette and discarded. The precipitate is suspended in 80% (v/v) acetone, and collected by centrifugation; the supernatant is discarded. The precipitate is suspended in 50 μl of 0.1 M HCl at 0 °C and 200 μl of cold acetone is added, with rapid mixing. The precipitate is once more collected by centrifugation, the slurry is distributed around the bottom of the tube by vortexing, and is dried thoroughly in vacuo. 25 μl of 6 M HCl is added. The tube is sealed under vacuum, and placed in an oven at 105 °C for 16 h. If N-terminal proline is suspected, the hydrolysis may be limited to 8 h. The tube is broken open and the acid removed in a vacuum desiccator over NaOH pellets; the contents of the tube must be thoroughly dry.

The residue is dissolved in 2 μl of 50% aqueous pyridine, and the dansyl amino acids are identified by chromatography on 5 cm × 5 cm double-sided polyamide layers as described in section 6.2.1. With larger proteins, the quantity of material (mainly unlabelled amino acids) in the hydrolyzate may interfere with the chromatography, and extraction of the dansyl amino acids with wet ethyl acetate (10 μl) may be preferred. Dns-His, Dns-Arg and Dns-$CySO_3H$ are not effectively extracted, however; but identification of these derivatives is not easy in any case (the former two having properties similar to those of ε-*N*-Dns-Lys, and the latter chromatographing close to dansic acid). Tryptophan and its derivatives are destroyed during hydrolysis in HCl; hydrolysis in vacuo with mercaptoethane sulphonic acid, followed by extraction of the non-polar dansyl amino acids into ethyl acetate may be used for the detection of N-terminal tryptophan (Giglio, 1977).

2.7.2. Determination of the N-terminal residue using 4-dimethylaminoazobenzene-4′-isothiocyanate (Dabitc)

The original method of Chang et al. (1976) used a triethylamine–acetic acid buffer, pH 10.1, to dissolve the protein, and the reagent was added in acetone. The following improved procedure is essentially that described by Chang et al. (1978) without the addition of phenylisothiocyanate.

The reduced and carboxymethylated protein (or other suitable derivative) (2–5 nmol), free from salts, especially amines, is dissolved in 20 μl of 50% (v/v) aqueous pyridine in a small glass test tube (4 mm i.d. × 50 mm). A freshly prepared solution of Dabitc in pyridine (10 μl of 2.82 mg/ml) is added. The tube is flushed with oxygen-free N_2, sealed with parafilm, and incubated at 55 °C for 20 min. Heptane/ethyl acetate (2 : 1, v/v; 150 μl) is added, and the phases are thoroughly mixed by vortexing for 10 s. The phases are separated by brief centrifugation. The upper, organic phase is removed with a fine Pasteur pipette and discarded. The extraction is repeated with a further 150 μl of the heptane/ethyl acetate mixture. The aqueous phase is dried thoroughly in vacuo over P_2O_5 and NaOH, care being required to avoid too sudden application of the vacuum which may lead to ejection of the sample from the tube. The dry residue is treated with 50% (v/v) aqueous trifluoroacetic acid (about 20 μl). The tube is flushed briefly with N_2, sealed with parafilm, and incubated at 80 °C for 10 min. The trifluoroacetic acid is removed in vacuo over NaOH pellets. The residue is extracted with wet ethyl acetate (20 μl), and 1 μl is applied to a polyamide thin-layer sheet (3 cm × 3 cm) for identification of the 4-*N*,*N*-dimethylaminoazobenzene-4′-thiohydantoin derivative of the N-terminal amino acid, as described later (§ 6.5.2).

If the protein is not soluble in 50% pyridine, different conditions for the coupling with Dabitc may be used. For example, the dodecyl sulphate-containing *N*-ethylmorpholine solution described above for the dansylation of proteins (§ 2.7.1) is satisfactory; the dimethylaminoazobenzenethiocarbamyl protein is similarly precipitated with cold acetone, but not washed with acidic acetone. The dried derivative is then treated with 50% trifluoroacetic acid solution as above for the cleavage and conversion reactions.

The chemical reactions are described in Chapter 6.

2.8. Determination of the carboxy-terminal residue

The methods available for the determination of the C-terminal residues in proteins are less satisfactory than those for the determination of the N-terminal residue. Some examples of methods which have been described are listed in Table 2.5. Carboxypeptidase

TABLE 2.5
Some methods for the determination of carboxy-terminal residues in proteins

Carboxypeptidase digestion	§§ 2.8.1–2.8.4
Hydrazinolysis	Schroeder (1972a); Narita et al. (1975)
Tritiation	Narita et al. (1975)
Hydantoin formation, using acetic anhydride and ammonium thiocyanate followed by treatment with acetohydroxamic acid	Cromwell and Stark (1969), Stark (1972b)

digestion is often the most convenient method, and experimental details for use with various enzymes are given below. On the whole, the C-terminal structure of a protein is most reliably determined from the study of small peptides. A tryptic peptide lacking lysine or arginine, a CNBr peptide lacking homoserine, or a staphylococcal (strain V8) protease peptide lacking glutamic acid or aspartic acid most probably originate from the C-terminus of the protein.

Occasionally, the terminal α-carboxyl group may be modified as, for example, in the peptide hormone oxytocin (Du Vigneaud et al., 1953), but such modification is much rarer than that of terminal α-amino groups.

2.8.1. *Carboxypeptidase A* (EC 3.4.12.2) (Ambler, 1972a)

Carboxypeptidase A releases non-polar residues rapidly from the C-terminus of polypeptides. Histidine, glutamine, threonine and homoserine are also released rapidly, while asparagine, serine and methionine sulphone are released slowly. Lysine is released slowly, but more rapidly at high pH. Glycine and acidic residues are released very slowly, and proline and arginine are not released. The nature of the penultimate residue affects the rate of release of the terminal residue. Thus the interpretation of the results is difficult if the terminal residue is released slowly and the penultimate residue is released rapidly. A more serious problem is the possible contamination of the carboxypeptidase with endopeptidases. Endopeptidase-free carboxypeptidase A (usually treated with diisopropylfluorophosphate) is required.

Carboxypeptidase A is insoluble at low ionic strengths, and is stored as a suspension at 4 °C. Before use, the enzyme is suspended in water and centrifuged. The supernatant, containing contaminating amino acids, is discarded, and the pellet is dissolved in a small volume of 2 M NH_4HCO_3. This enzyme solution is added to a solution of the denatured substrate protein (10–100 nmol) in 0.2–1 ml of 0.2 M *N*-ethylmorpholine acetate buffer, pH 8.5, to give a molar ratio of enzyme to substrate of about 1 : 100. If the substrate is insoluble in this buffer, the digestion may be performed in the presence of sodium dodecyl sulphate (1%, w/v) or 6 M urea. An internal standard, norleucine, is included (1 mol/mol of protein). The solution is incubated at 25 °C. Aliquots containing 2–20 nmol of the protein are removed after 0.5, 1, 4 and 16 h and added to an equal volume of 1 M acetic acid to stop the digestion. Any precipitated protein is removed by centrifugation and the supernatants are dried in vacuo. The residues are dissolved in the amino acid analyzer sample buffer and the amino acid contents determined relative to the value of norleucine. An enzyme blank is carried through the same procedure.

The number of moles of each amino acid released per mole of

protein is plotted as a function of time. In favourable cases, the order of release of the amino acids may be unambiguous, and the C-terminal sequence of several residues may be determined. If a large number of residues are released even after a short digestion time, the experiment should be repeated using lower concentrations of enzyme and removal of aliquots after 5, 10, 20 and 60 min, or other suitable times. Conversely, if only small amounts of amino acids are released, an enzyme/substrate ratio of 1:20, and digestion at 37 °C for longer times, up to 24 h, may be tried. In the latter case, interference in the results may arise from trace endoprotease activity, and a significant enzyme blank correction may be required. A better course is the use of a different carboxypeptidase, or the use of carboxypeptidases A and B together.

2.8.2. Carboxypeptidase B (EC 3.4.12.3) (Ambler, 1972a; Folk, 1970)

Carboxypeptidase B releases lysine and arginine from the C-termini of peptides and proteins. The enzyme is not very stable, and is stored frozen at −20 °C. Precautions concerning its use, and suitable digestion conditions, are as for carboxypeptidase A.

2.8.3. Carboxypeptidase C (EC 3.4.12.1) (Tschesche and Kupfer, 1972)

Carboxypeptidase C from orange leaves releases all the common amino acids from the C-termini of proteins, but at different rates. Conditions used with ribonuclease were as follows. Ribonuclease (1.08 μmol; 15.2 mg) was dissolved in 0.5 ml of 50 mM sodium citrate buffer, pH 5.3. An internal standard of 1 μmol norleucine in 0.5 ml of the same buffer was added. Carboxypeptidase C, at a weight ratio of enzyme: substrate of 1 : 100, was added in 0.5 ml of the buffer, and the solution was incubated at 30 °C. Aliquots of 0.15 ml were withdrawn after 2, 6, 12 and 20 min, and added to 0.6 ml of 0.2 N HCl, resulting in a pH of 2.5–3 and stopping the digestion. The samples were frozen. The solutions were then heated rapidly to 70 °C for 1 min, centrifuged to remove denatured protein, acidified to pH 1–1.5 and immediately applied to the amino acid analyzer. The norleucine (0.1 μmol in each aliquot) was used to normalize the results. Valine, serine, alanine and phenylalanine were released rapidly whith C-terminal, while aspartic acid was released more slowly. A peptide with C-terminal —Gly—Cys—Pro—Lys was digested much more

|

CH_2COOH

slowly by carboxypeptidase C.

Further details of the use of carboxypeptidase C, including the determination and inhibition of endoprotease activity, are given by Tschesche (1977).

2.8.4. Carboxypeptidase Y (Hayashi et al., 1973; Hayashi, 1977)

Carboxypeptidase Y from bakers' yeast also has a broad specificity, but hydrophobic amino acids are released faster, and charged amino acids more slowly, than others. Glycine in the penultimate position slows the release of the terminal amino acid. Dipeptides are not hydrolyzed. The enzyme is stable as a suspension in saturated $(NH_4)_2SO_4$ solution at $-20\,°C$, and in solution at $25\,°C$ for at least 8 h in the pH range 5.5–8.0. It is inhibited by metal ions such as Cu^{2+} or Hg^{2+}. The protein substrate is dialyzed against 10 mM sodium phosphate buffer, pH 7.0, or, if necessary for subsequent electrophoretic analysis, against the volatile buffer, 0.1 M pyridine–acetic acid, pH 5.5. Preparations of the carboxypeptidase may be contaminated with yeast proteinase A, which may be inhibited by the addition of pepstatin (Sigma Chemical Co.), and EDTA may be added to inhibit aminopeptidase contamination. Carboxypeptidase Y is added to the substrate at a weight ratio of 1 : 100, and digestion is performed at $25\,°C$, with removal of aliquots for acidification and amino acid analysis as described for carboxypeptidase A.

Further details of the use of carboxypeptidase Y, including several examples, are given by Hayashi (1977), Martin et al. (1977) and Lee and Riordan (1978).

With all the methods employing carboxypeptidases, samples (1–5 μg) of the protein substrate and of the final digestion product may be subjected to SDS-polyacrylamide gel electrophoresis; significant diminution in the intensity of staining of the protein band, and particularly the presence of additional bands of lower molecular weight, indicate that endoproteolysis is occurring, and the carboxypeptidase digestion results should be treated with caution. Examples of the incorrect determination of C-terminal sequences are frequent in the literature, as noted by Rask et al. (1979) among others.

Specific cleavage of the protein

3.1. General approach to cleavage

For the determination of the complete sequence by manual methods, polypeptides of less than about 400 residues may most conveniently be digested into relatively small peptides with enzymes such as trypsin, staphylococcal protease or chymotrypsin, or degraded with cyanogen bromide. The sequences of these small peptides may often be determined completely by degradation from the N-terminus, and the complete sequence of the protein deduced using overlapping sequences in two or more sets of peptides. The amino acid analysis will indicate which cleavage methods are likely to be of use. For example, a typical content of 5 mol% of each of the basic residues, lysine and arginine, will suggest that trypsin will be suitable, yielding peptides of average size 10 residues, while the absence of methionine will show that treatment with cyanogen bromide will be fruitless.

Cleavage by these methods of larger polypeptides gives rise to complex mixtures of fragments, and the complete separation of these in satisfactory yields becomes a formidable task. The identification of overlapping sequences also becomes much more difficult with increasing size of the polypeptide. It is therefore preferable to cleave larger polypeptide chains at a small number of points and to treat the large fragments produced as equivalent to whole proteins for further sequence analysis. A few large fragments may often be placed in order by amino- or carboxy-terminal analysis. The isolation of large fragments is even more desirable when automated sequencing machines are to be used; unfortunately, it is often not easy.

Methods for cleavage at a small number of specific points are discussed first (§ 3.2), followed by other enzymic (§ 3.3) and chemical (§ 3.4) methods. The quantity of protein required for digestion is discussed in the following chapter.

3.2. Selective cleavage of the polypeptide chain to produce large fragments

3.2.1. Limited proteolysis

In general, proteins in their native, typically globular, conformations are resistant to proteolysis. However, many proteins possess one or more peptide bonds which are unusually sensitive to particular proteases. Such sensitive regions of the polypeptide chain may link globular domains, or may simply lie on the surface of a tightly folded molecule. Naturally occurring specific cleavage of polypeptide chains is very common; examples include the conversion of zymogens into active enzymes, such as pepsinogen to pepsin and trypsinogen to trypsin, of prohormones, such as proinsulin, to the active hormones, and of fibrinogen to fibrin. Occasionally the amino- or carboxy-terminal regions of the polypeptide chain may be relatively freely accessible, and limited proteolysis may merely result in shortening of the polypeptide by a few residues. More useful, from the protein chemist's viewpoint, are those points of cleavage which lie closer to the middle of the polypeptide chain.

There are no generally applicable conditions which may be recommended for such cleavage; a particular protein may be sensitive to several proteases, or only one, and homologous proteins from different sources may behave quite differently. Relatively slight changes in such parameters as pH, ionic strength and composition, temperature, or the presence of cofactors, substrates, allosteric effectors, chaotropic agents or glycerol may lead to large changes in the rates of digestion. Preliminary experiments should be performed on a small scale, using a wide variety of conditions, with analysis of the products of digestion by polyacrylamide gel electrophoresis in the presence of dodecyl sulphate. An indication of the susceptibility of a protein to selective proteolysis may often be gained from the observation of cleavage of the protein during its purification.

The usefulness of limited proteolysis for analysis of primary structures is affected by the yields of fragments obtained. If a specific cleavage occurs in low yield, or if isolation of pure fragments is difficult, it may be more efficient to concentrate upon the isolation of smaller peptides from the whole protein, and to use the small amounts of large fragments for peptide mapping and aminoterminal analysis only.

During treatment with a protease, partial cleavage may occur at alternative peptide bonds, which may be close to each other in the peptide chain. Several very similar fragments, which differ by only a few residues, and which are not readily separated, may thus be formed. If the differences are at the N-terminus of a fragment, direct sequential degradation may not be possible. Another problem encountered during limited proteolysis is that very small peptides may also be produced, and these may be lost during procedures designed for the isolation of large fragments.

In the absence of a standard set of conditions for limited proteolysis, several examples, illustrating the variety of conditions which may be used, are given here; further examples are given in Table 3.1.

3.2.1.1. Ribonuclease A (Richards and Vithayathil, 1959) Subtilisin cleaves ribonuclease A into a peptide of 20 residues (the 'S-peptide') and a large polypeptide (the 'S-protein'). Ribonuclease A (730 mg) was dissolved in 5 ml of 0.1 M KCl at 3 °C and the pH was adjusted to 8.0 with NaOH. Subtilisin (1 mg, in 0.2 ml H_2O) was added, and 0.1 M NaOH was added to maintain pH 8.0, using a pH-stat. After 3 h the uptake of NaOH became slow; a total of 0.9 ml of NaOH solution was added, corresponding to the cleavage of 1–2 peptide bonds. There was no change in the ribonuclease activity. The protein was purified by ion-exchange chromatography on a column of Amberlite IRC-50 in 0.2 M sodium phosphate buffer, pH 6.35. A new N-terminal residue, serine, was observed.

TABLE 3.1
Examples of limited proteolysis suitable for use in sequence analysis

Bromelain was used to cleave influenza virus haemagglutinin (Band and Skehel, 1972).
Papain has been used to cleave immunoglobulin G (§ 3.2.1.3), myosin (§ 3.2.1.5) and murine histocompatibility antigens (Shimada and Nathenson, 1969)
Pepsin cleavage of immunoglobulins was described by Parr et al., 1976
Plasmin cleaved the α-chain of human fibrinogen, giving large fragments (Doolittle et al., 1977)
Rennin cleaved the κ-casein molecule into a large peptide and para-κ-casein at a single Phe-Met bond (Jollès et al., 1973)
Subtilisin cleaves ribonuclease (§ 3.2.1.1) and phosphorylase (§ 3.2.1.2) at a few bonds. Aspartokinase I-homoserine dehydrogenase I of *E. coli* K12 yielded an active homoserine dehydrogenase fragment (Briley et al., 1978), while aspartokinase II-homoserine dehydrogenase II gave two large fragments and many small peptides (Dautry-Varsat and Cohen, 1977) after treatment with subtilisin
Thrombin cleaved troponin C at a single Arg-Ala bond, giving two large fragments (Leavis et al., 1978)
Trypsin splits rabbit sarcoplasmic reticulum Ca^{2+}-ATPase at two points (§ 3.2.1.4)

For the separation of the fragments, a 1% solution of the salt-free protein in water at 0 °C was treated with one fifth of the volume of 20% (w/v) trichloroacetic acid at 0 °C. Precipitation occurred as the solution was allowed to warm slowly to room temperature. After 1 h the precipitated S-protein was collected by centrifugation, dissolved in water and reprecipitated with trichloroacetic acid. The S-peptide was recovered from the supernatants after extraction of the trichloroacetic acid with ether.

Gross and Witkop (1967) showed that there was some heterogeneity in the products, arising from cleavage at Ser^{21}-Ser^{22} as well as at Ala^{20}-Ser^{21}.

3.2.1.2. Glycogen phosphorylase (Koide et al., 1978) Rabbit muscle phosphorylase *b* (1 g) in 100 ml H_2O was dialyzed for 16 h at 5 °C against 6 litres of 50 mM Tris–acetic acid buffer, pH 8.5, 10 mM in 2-mercaptoethanol. The solution was diluted to 210 ml. AMP (306 mg) and subtilisin (2 mg) in 210 ml of the buffer without mercaptoethanol were added. After 12 h at 5 °C the protein was precipitated with 620 ml of 5% trichloroacetic acid. The cleaved protein was reduced and carboxymethylated, and separated into two large fragments by gel filtration on a column of Sephadex G-150 in 9% formic acid, 7 M urea. More than one peptide bond in one region of the molecule was cleaved.

3.2.1.3. Immunoglobulin G (Porter, 1959) γ-Globulin (150 mg) and mercuripapain (the complex of papain with mercuric chloride) (1.5 mg) were dissolved in 10 ml of a buffer containing 0.1 M sodium phosphate, pH 7.0, 10 mM cysteine and 2 mM EDTA. The solution was incubated for 16 h at 37 °C in the presence of a little toluene, followed by dialysis against several changes of water over 48 h, with vigorous stirring. This procedure removed the cysteine and EDTA and facilitated oxidation, inactivating the protease. The cleaved globulin was dialyzed against 0.01 M sodium acetate buffer, pH 5.5, and fractionated by chromatography on a column (30 cm × 2.4 cm) of carboxymethylcellulose in 200 ml of 0.01 M sodium acetate buffer, pH 5.5, followed by a linear gradient prepared from 1000 ml each of this buffer and of 0.9 M sodium acetate buffer, pH 5.5. The eluate was analyzed for absorbance at 280 nm, and three fractions were obtained.

Many alternative procedures are available for the limited cleavage of antibody molecules, including the use of pepsin (Parr et al., 1976). The subject is reviewed by Nisonoff et al. (1975).

3.2.1.4. Sarcoplasmic reticulum ATPase (Thorley-Lawson and Green, 1975) Rabbit skeletal sarcoplasmic reticulum vesicles were suspended at 5.3 mg/ml in 0.1 M KCl, 20 mM Tris–maleate, pH 7.0, 4 mM $CaCl_2$ and 1 M sucrose at 20 °C. Trypsin was added to 0.066 mg/ml. After 3 min the digestion was stopped by the addition of five volumes of a solution of soya bean trypsin inhibitor (2 mg/mg of trypsin) in 0.3 M sucrose, 10 mM Tris–HCl buffer, pH 7.3. The vesicles were collected by centrifugation for 15 min at

165,000 × *g*, and the cleaved preparation was purified by extraction with Triton X-100. Two large fragments, of mol. wt. 60,000 and 55,000, were produced from the ATPase molecule of mol. wt. 115,000. Full enzymic activity was retained.

Cleavage at a second point, converting the larger fragment into two smaller polypeptides, mol. wt. 33,000 and 24,000, resulted from incubation with 1 mg/ml trypsin for 20 min under the same conditions.

The fragments were separated by gel filtration on a column of Sephadex G-150, superfine (1.6 cm × 90 cm) in 50 mM sodium phosphate, pH 7.3, containing 0.1% sodium dodecyl sulphate, 0.1 mM dithiothreitol and 0.002% (w/v) hibitane.

3.2.1.5. Myosin The heavy chains of skeletal muscle myosin may be cleaved into various large fragments by trypsin (Lowey and Cohen, 1962) or by papain (Lowey et al., 1969). For tryptic cleavage, myosin was dissolved at 15 mg/ml in 0.5 M KCl, 0.03 M phosphate buffer, pH 6.2. To 10 ml of this solution was added 1 ml of 0.05% (w/v) trypsin. After stirring for 50 min at 20 °C, 1 ml of 0.1% (w/v) soybean trypsin inhibitor was added to stop the digestion. The solution was dialyzed against a 10-fold volume of 7 mM neutral phosphate buffer. The light meromyosin fragment precipitated, while the heavy meromyosin fragment remained soluble. The fragments were separated by centrifugation, and each was purified by fractional precipitation.

Similar fragments could be obtained by digestion with papain, insolubilized by reaction with diazotized *p*-aminobenzylcellulose, but the whole myosin rod could be recovered, and there was less digestion to low molecular weight peptides than there was with trypsin (Lowey et al., 1969).

In some cases, protease treatment is used during the isolation of proteins, especially membrane proteins, which may be difficult to purify intact. Examples include cytochrome b_5 (Nóbrega and Ozols, 1971) and the major murine histocompatibility antigens (Shimada and Nathenson, 1969). For complete sequence analysis, alternative purification procedures are therefore required, e.g. for H-2K and H-2D antigens, the use of non-ionic detergent (Freed et al., 1979).

3.2.2. Selective chemical cleavage for the production of large fragments

Although several of the methods discussed in § 3.4 may be expected to yield large fragments, depending upon the distribution of particular residues along the peptide chain, this section is concerned only with methods likely to cleave only one or two bonds in a typical protein. Many of the chemical cleavage methods require conditions incompatible with retention of the native conformation of the protein, but under mild conditions selective cleavage of those methionyl peptide bonds exposed at the surface of the protein may be achieved using CNBr. For example, bovine serum albumin may be cleaved with CNBr to give large fragments (Goossens et al., 1973). Moreover, S-cyanylation (see § 3.4.2) may be restricted to one or two reactive cysteine residues in a native protein; denaturation of the cyanylated protein may then lead to specific cleavage at these cysteine residues only (Stark, 1977). Cyanylation with the 6-thiocyano analogue of ATP has been observed to lead to specific cleavage of some proteins (Yount, 1975).

Two methods for the cleavage of denatured proteins into large fragments are hydroxylaminolysis of Asn–Gly bonds and mild acid hydrolysis of Asp-Pro bonds.

3.2.2.1. Cleavage at Asn-Gly bonds with hydroxylamine Studies on collagen chains showed that Asn-Gly bonds are far more susceptible to cleavage by hydroxylamine than are other peptide bonds (Butler, 1969; Bornstein, 1969). Bornstein (1970) proposed that a cyclic imide is formed by Asn-Gly sequences, and the imide is cleaved by hydroxylamine; the asparaginyl residue is converted to a mixture of α- and β-aspartyl hydroxamates. This reaction pathway is consistent with other observations, including the greatly reduced yield at Asn-Gly bonds during Edman degradation of peptides (§ 6.3.2.3) and the prevention of cleavage by pre-treatment with alkali (Bornstein, 1970). The cleavage reaction is discussed in detail by

Bornstein and Balian (1977). Yields of cleavage reported in the literature vary widely, but in favourable cases a yield of about 70% is obtained. Prior exposure of the protein to acidic conditions (e.g. anhydrous trifluoroacetic acid) should lead to higher yields. Small extents of cleavage at other asparaginyl peptide bonds may occur. The following procedure follows that of Bornstein and Balian (1977).

A solution, 6 M in guanidine, 2 M in hydroxylamine and 0.2 M in K_2CO_3 is prepared as follows. Guanidinium chloride (23 g) and hydroxylamine hydrochloride (5.5 g) are placed in a 100 ml beaker. 4.5 M LiOH solution is added slowly with vigorous stirring in an ice bath. The pH is maintained at 7–8 until all the solute has dissolved (total volume about 35 ml). K_2CO_3 (138 mg) is added, the pH of the solution is adjusted to 9.0 and the volume made up to 40 ml with water. A suitable volume of the solution is warmed to 45 °C, and the protein is added to give a concentration of 1–5 mg/ml. The pH is adjusted to 9.0 if necessary, and the solution is incubated at 45 °C for 4 h. The reaction is terminated by acidification with formic acid to pH 3, and the mixture is desalted on a column of Sephadex G-25 in 9% formic acid.

3.2.2.2. Cleavage at Asp-Pro bonds in dilute acid Although partial acid hydrolysis was extensively used in early sequence analysis, the relative lack of specificity of most procedures limits the usefulness of this method. However, Piskiewicz et al. (1970) observed that Asp-Pro bonds were readily cleaved under mild acidic conditions. This is a consequence of the relatively high basicity of the prolyl nitrogen atom coupled with neighbouring group catalysis by the aspartyl β-carboxyl group.

A variety of conditions may be used for the specific cleavage, including 10% acetic acid, adjusted to pH 2.5 with pyridine, with or without the presence of 7 M guanidinium chloride, and 70% formic acid. Several examples are given by Landon (1977). Temperatures around 40 °C and reaction times from 24 to 96 h have been used. Yields are typically between 30 and 80%. The use of 70% (v/v) formic acid at 40 °C for 48 h may be recommended; most proteins are denatured and soluble under these conditions, and the peptides may be recovered readily by dilution with 20 volumes of water followed by freeze-drying. A preliminary experiment using these conditions, with analysis of the products by SDS polyacrylamide gel electrophoresis, may indicate whether the method is likely to be of use. Small scale experiments using a variety of conditions may then be used to discover the optimum conditions for the particular protein being investigated.

A disadvantage of this cleavage method is that some loss of amide groups from asparagine and glutamine residues may occur, and extended reaction times may lead to non-specific cleavage of the peptide chain.

3.3. Enzymic cleavage of denatured polypeptides

There are many methods for the cleavage of denatured protein molecules at specific peptide bonds. Cleavage at arginine residues with trypsin (after modification of the lysine residues) or with an arginine-specific protease, at glutamic acid residues with staphylococcal protease or at methionine residues with cyanogen bromide usually yield a small number of relatively large peptides. Cleavage at cysteine residues via cyanylation, or at tryptophan residues with mild oxidizing reagents, may also yield a few fragments, but these methods have not in general proved to be so useful.

For cleavage of proteins into small peptides, of average length about 10 residues (depending on the content of lysine and arginine residues in the protein), the most specific method is cleavage with trypsin. Other proteases which are frequently used are chymotrypsin, pepsin, thermolysin, papain and subtilisin, in approximate order of decreasing specificity. The latter proteases are best reserved for secondary digestion of larger peptides derived by cyanogen bromide or tryptic treatment of the protein, since their low specificity leads to extremely complex mixtures of peptides, in widely varying yields, from whole proteins. However, specific isolation of arginine- or methionine-containing peptides after digestion with one of these proteases may be useful for the determination of alignments of tryptic or cyanogen bromide peptides.

Several other proteases have been described for applications in sequence analysis, and some examples are given in Table 3.2. It is likely that more proteases will be discovered and brought into routine use in the future. Experimental details for digestion with several enzymes are given in the following sections.

TABLE 3.2

Further examples of proteases used in sequence analysis

Armillaria mellea protease (Doonan et al., 1974).
The specificity is directed to bonds at the N-terminal side of lysine residues, but some cleavage occurs at the C-terminal side of arginine residues.

Clostripain (Mitchell, 1977). Specific for arginine residues.

Elastase (EC 3.4.21.11) (Shotton, 1970). Short peptides suitable for mass spectrometric sequence analysis were produced from dihydrofolate reductase (Morris et al., 1974).

Post-proline cleaving enzyme from lamb kidney (Koida and Walter, 1976). The enzyme was used in the determination of the sequence of troponin C from frog skeletal muscle (Van Eerd et al., 1978).

Protease II from *Myxobacter*, strain AL-1 (Wingard et al., 1972). The enzyme, with specificity for peptide bonds on the N-terminal side of lysine residues, was used during the determination of the sequence of dihydrofolate reductase from *Lactobacillus casei* (Freisheim et al., 1978)

Mouse submaxillary gland protease. Highly specific for the C-terminal side of arginine residues, although some arginine peptide bonds are cleaved slowly (Boseman et al., 1976; Schenkein et al., 1977).

Thrombin (EC 3.4.21.5) (Lundblad et al., 1976). Specific for certain -Arg-Gly- bonds in fibrinogen, but cleavage at other arginyl peptide bonds sometimes occurs, depending upon the neighbouring sequence.

3.3.1. Preparation of the protein for digestion

For rapid and complete digestion, the protein should be denatured and soluble, or at least in the form of a flocculent precipitate allowing access of the protease. Often the procedures described in Chapter 2, such as reduction and carboxymethylation, lead to a suitable denatured product. Freeze-drying may lead to insolubility, as may denaturation by heat. Many proteases remain active in the presence of chaotropic agents, such as urea, and dilution of a solution of the protein in 8 M urea into three volumes of the protease solution in water may give efficient digestion. Prior denaturation with acid, including precipitation with trichloroacetic acid or acidic acetone may be satisfactory. A finely divided substrate, easily digested by proteases, may sometimes be obtained by homogenization or ultrasonication. Solutions of denaturing detergents, such as sodium dodecyl sulphate, in which some proteases remain active, may be used, but the detergent could interfere with the subsequent isolation of peptides. Even after denaturation, part of the polypeptide chain may re-form secondary structure which inhibits cleavage.

It is possible that peptides produced during the course of digestion may have a high affinity for the active site of the protease, acting like small polypeptide protease inhibitors. This product inhibition could reduce the rate of digestion of a particular protein with a particular protease; the effect would be reduced at lower substrate concentrations. Thus the standard conditions described below may require modification for some proteins. Under some circumstances, especially at very high protein concentrations, some proteases catalyse the formation of peptide bonds; this is unlikely to occur significantly if the substrate concentration is about 10 mg/ml or below.

3.3.2. Trypsin (EC 3.4.21.4)

Trypsin catalyses the hydrolysis of lysyl and arginyl peptide bonds, except where the following residue is proline, although very slow cleavage of Lys-Pro or Arg-Pro bonds may be observed. The rates of cleavage vary considerably. Arginyl peptide bonds are generally cleaved slightly faster than lysyl peptide bonds, and the presence of acidic residues on either side of the basic residue reduces the rate significantly. *S*-(2-Aminoethyl)cysteinyl peptide bonds are cleaved at about one tenth the rate of lysyl peptide bonds. Trypsin has low exopeptidase, especially aminopeptidase, activity, and adjacent lysine or arginine residues usually provide alternative points of cleavage, with low yields of free lysine or arginine. Commercially available trypsin usually contains some contaminating chymotrypsin, and this should be inhibited with L-1-chloro-3-tosylamido-4-phenylbutan-2-one (also named '*N*-tosyl-L-phenylalanyl chloromethyl ketone' and abbreviated 'TPCK') (Kostka and Carpenter, 1964) or with other specific inhibitor. TPCK-treated trypsin is commercially available. Even pure trypsin has some chymotryptic-like activity (Inagami and Sturtevant, 1960) which is contributed by ψ-trypsin, an autolysis product of trypsin (Keil-Dlouhá et al., 1971) which may be present in commercial preparations. Prolonged digestion with trypsin should thus be avoided. Low concentrations of Ca^{2+} retard autolysis of trypsin and may contribute to retention of the specificity as well as stabilizing the enzyme in the presence of 2 M urea. Trypsin stock solutions in 0.1 mM HCl may be stored frozen at −20 °C for months without loss of activity.

It is useful to follow the time course of digestion of large quantities (more than 50 mg) of protein by the use of a pH-stat (§ 3.3.2.1). With smaller quantities of protein this is not practicable, and a volatile buffer is often used (§ 3.3.2.2). If the denatured protein is insoluble at pH 8, it may be dissolved in 8 M urea and diluted into a solution of trypsin, to give a final concentration of 2 M urea. Alternatively it may be dissolved at pH 10.5 or at acid pH, and the pH rapidly adjusted to pH 8 in the presence of trypsin. The finely divided precipitate which may form will usually be digested readily.

3.3.2.1. Use of a pH-stat to follow tryptic digestion The denatured, salt-free protein, dissolved or finely suspended in 0.1 mM $CaCl_2$ at about 10 mg/ml is rapidly stirred in a thermostated vessel at 37 °C, and a slow stream of N_2 is introduced to exclude atmospheric CO_2. The pH is adjusted to 8.3, and NaOH solution (0.10 M) is introduced from a micrometer syringe to maintain this pH constant. When the steady state has been attained, a solution of trypsin (1% by weight of the protein) at pH 8.3 is added. The addition of NaOH solution is recorded as a function of time using the pH-stat. After about 1 h, when further addition is usually slow, a further 1% by weight of trypsin is added. Unless significant uptake of NaOH continues, the digestion is stopped after a further 30 min by freeze-drying, acidification, or the addition of trypsin inhibitor (from soya bean) or N^{α}-tosyl-L-lysyl chloromethyl ketone (TLCK; 3-tosylamido-7-amino-1-chloroheptan-2-one) in a slight molar excess over the protease. Unless the inhibitors are used, digestion with proteases may continue after freeze-drying if the peptide mixture is redissolved at a suitable pH. The number of peptide bonds cleaved may be estimated approximately from the uptake of NaOH, with a correction for the proportion of peptide α-amino groups protonated using the formula: $\log([-NH_2]/[-NH_3^+]) = pH - 7.8$, where the average p$K_a$ of peptide α-NH_2 groups is 7.8 (Steinhardt and Beychok, 1964).

3.3.2.2. Tryptic digestion in volatile buffers The denatured protein is dissolved or finely suspended in 0.1 M NH_4HCO_3, 0.1 mM $CaCl_2$, at about 5–10 mg/ml. The pH is checked, and adjusted to 8–8.5 with NH_3 if necessary. Trypsin (1% by weight of the protein) is added as a solution (10 mg/ml) in 0.1 mM HCl. The solution or suspension is stirred at 37 °C for 1 h. A further addition of trypsin (1% by weight of the protein) is added, and after a further 30 min the digestion is stopped as described above. An alternative volatile buffer which is often used is *N*-ethylmorpholine acetate, pH 8–8.5. Thiodiglycol (0.01% by vol) may be included in the mixture as an antioxidant.

If the protein contains a number of slowly cleaved bonds, longer digestion times, up to 24 h, may be used, but non-specific cleavage of some tyrosyl, phenylalanyl or tryptophanyl peptide bonds may be expected to occur.

3.3.3. Cleavage by trypsin at arginine residues after modification of lysine residues

Many chemical modifications of lysine residues prevent subsequent cleavage of lysyl peptide bonds by trypsin, restricting cleavage to arginine residues. Reaction with dicarboxylic anhydrides, particularly succinic, maleic and citraconic anhydrides, has been most useful; other methods, such as carbamylation or trifluoroacetylation, are less suitable. Details of succinylation and maleylation of amino groups are given by Glazer et al. (1975) and Butler and Hartley (1972). Citraconylation may be performed similarly (Dixon and Perham, 1968). It is essential that the unsaturated anhydrides, if used, should be pure and free from polymeric material.

Briefly, 50 mol of the dicarboxylic anhydride per mol of amino groups is added in small aliquots to a rapidly stirred solution of the protein in 6 M guanidinium chloride, and the pH is maintained at pH 8–8.5 by the addition of 1 M NaOH. A small extent of acylation of the hydroxyl groups of serine and threonine residues occurs; the *O*-acyl groups may be removed by treatment with 1 M NH_2OH at pH 10 for 1 h at room temperature. Cysteine residues should be modified before reaction with the unsaturated (maleic or citraconic) anhydrides, to avoid alkylation of the thiol groups. The α-amino group of the protein is modified as well as the ϵ-NH_2 groups of lysine residues.

The *N*-3-carboxypropionyl groups introduced by reaction with succinic anhydride are only removed using conditions of hydrolysis which split peptide bonds, but the unsaturated groups introduced by reaction with maleic or citraconic anhydride are removed by treatment with dilute acid (48 h at pH 3.5 at 37 °C for maleyl groups; 2 h at pH 2 at 20 °C for citraconyl groups). Unless there is a need for removal of the blocking groups (for subsequent tryptic cleavage at lysine residues of the purified peptides), succinylation is to be preferred, since problems arising from the still reactive ethylenic groups are avoided. Such problems may also be avoided through the use of exo-*cis*-3,6-endoxo-Δ^4-hexahydrophthalic anhydride (Riley and Perham, 1970), or phthalic anhydride (Pechère and Bertrand, 1977). Phthaloyl groups are removed under acidic conditions (pH 3.5 at 50 °C for 48 h).

Guanidinium chloride and reaction by-products are removed by dialysis at pH 8–8.5, and the modified protein is digested with trypsin as described in § 3.3.2. Proteins modified by dicarboxylic anhydrides are generally soluble at pH 8, and the polypeptide chains are unfolded, allowing ready digestion. In addition, long polyanionic peptides are generally soluble at pH 8 and may be purified by ion-exchange chromatography in volatile and UV-transparent NH_4HCO_3 solutions.

It is possible that arginine-specific proteases (Table 3.2) will replace trypsin for this purpose in the future.

3.3.4. Cleavage by trypsin at lysine residues after modification of arginine residues

The specific modification of arginine residues in proteins is less readily accomplished than that of lysine residues. Reagents such as butan-2,3-dione and phenylglyoxal also react with amino groups. However, in the presence of borate reaction of α-diketones with the guanidino group is favoured, and cyclohexan-1,2-dione in borate buffer may be used for the selective blocking of arginine residues during sequence analysis (Patthy and Smith, 1975). Tris or other amine buffers should not be used, and clearly guanidinium chloride may not be used, for the reaction with cyclohexanedione. The product of the reaction with arginine residues is stable at alkaline pH only in the presence of borate, and the isolation of peptides after tryptic digestion, particularly those containing several arginine residues, may be easier after regeneration of the arginine residues by hydroxylaminolysis, unless acidic conditions are used throughout the isolation procedures. The following method follows that of Patthy and Smith (1975).

The reduced, carboxymethylated protein is dissolved at 10–20 mg/ml in 0.25 M sodium borate buffer, pH 9.0, and treated with cyclohexan-1,2-dione (0.15 M) at 35 °C for 2 h in a sealed vial. An equal volume of 30% acetic acid is then added, and the protein is dialyzed against 5% acetic acid to remove excess reagents and salts, followed by freeze-drying. For digestion with trypsin, the modified protein is dissolved in 0.1 M sodium borate buffer, pH 8.0, and trypsin added as described above (§ 3.3.2.2).

If regeneration of arginine residues is desired, the trypsin should be irreversibly inhibited, and the peptides separated from borate by gel filtration on a short column of Sephadex G-10 in 0.1 M NaCl. The peptides are incubated in 0.5 M hydroxylamine, 1 mM EDTA, pH 7.0 in an evacuated, sealed tube under nitrogen for 6–7 h at 37 °C (Smith, 1977). The peptides, with regenerated arginine residues, are again isolated by gel filtration.

Use of lysine-specific proteases is likely to be more efficient (Table 3.2).

3.3.5. Staphylococcal protease

An extracellular protease from *Staphylococcus aureus* (strain V8) cleaves specifically glutamyl peptide bonds and sometimes aspartyl peptide bonds (Drapeau et al., 1972). The enzyme showed a double maximum in the rate vs. pH profile with haemoglobin, at pH 4 and pH 8. Examples of the use of the enzyme have been given by Austen and Smith (1976), Wootton et al. (1975) and in many recent papers on sequence analysis. In general a high specificity for cleavage at the C-terminal side of glutamic acid residues, with cleavage at a few aspartic acid residues, has been reported. The rate of digestion is low if the glutamic acid is within three residues from either the N- or the C-terminus of a peptide, and Glu-Pro bonds are not cleaved (Austen and Smith, 1976). Non-specific digestion has sometimes been observed. For example, cleavage on the C-terminal side of some Gly and Ala residues in *E. coli* ribosomal proteins L11 (Dognin and Wittmann-Liebold, 1977) and S 5 (Allen, G., unpublished data) was noted.

For cleavage at glutamic acid residues, the denatured protein is dissolved or finely suspended at about 10 mg/ml in 0.1 M NH_4HCO_3. Staphylococcal protease (2% by weight of the protein) is added, and the mixture is stirred at 37 °C for 2 h. Much longer digestion times have been used in some cases, e.g. 48 h at pH 4 (Heiland et al., 1976).

For extending the cleavage to aspartyl peptide bonds, the digestion is performed in 50–100 mM sodium phosphate buffer, pH 7.8 (Drapeau, 1977).

Solutions of the enzyme in distilled water may be frozen and thawed without loss of activity, and the enzyme is active in 4 M urea or 0.2% (w/v) sodium dodecyl sulphate solutions (Drapeau, 1977).

3.3.6. α-Chymotrypsin (EC 3.4.21.1)

α-Chymotrypsin cleaves peptides and proteins at the carboxyl side of tryptophan, tyrosine, phenylalanine, leucine and methionine residues, with occasional cleavage at other sites. If the following residue is proline, cleavage does not occur, and the presence of acidic residues on either side of the susceptible residue reduces the rate of cleavage, while neighbouring basic residues enhance the rate. The action of chymotrypsin is difficult to predict accurately, since it depends upon factors outside the immediate environment of the bond cleaved (Kaspar, 1970). The commercially available protease may contain some tryptic activity, which may be inhibited with tosyl lysyl chloromethyl ketone (TLCK) or small amounts of trypsin inhibitor.

Conditions of digestion are the same as those used with trypsin (§ 3.3.2). After short digestion times cleavage at only a few particularly susceptible bonds, giving relatively large fragments, may have occurred, while longer digestion, up to 24 h, may be used if smaller fragments are required. A few drops of toluene or a crystal of thymol may be added to inhibit microbial contamination during long periods of digestion. Properties of the enzyme are reviewed by Wilcox (1970).

3.3.7. Pepsin (EC 3.4.23.1)

The best small substrates for pepsin are those containing two adjacent hydrophobic residues (especially phenylalanine), cleavage occurring between them (Fruton, 1974), but the enzyme hydrolyzes proteins at a wide range of peptide bonds, especially those involving aromatic or leucine residues. Prolyl peptide bonds are not cleaved. The extended sequence around the bond to be cleaved exerts a strong effect upon the rate of hydrolysis. It is therefore not possible to predict with certainty the points of peptic cleavage (Fruton, 1970; Sampath-Kumar and Fruton, 1974). As with chymotrypsin, limitation of cleavage to a few especially susceptible bonds may be achieved using short digestion times or low concentrations of the protease. The use of pepsin has two distinct advantages: the enzyme is active at low pH, so that prior denaturation of the substrate is rarely required, and at low pH disulphide interchange reactions are very slow. Pepsin has thus been used for the determination of cystine cross-links in proteins (§ 7.7.1). The relative non-specificity of pepsin militates against its more frequent use in sequence analysis.

Several acidic media have been used for peptic digestion; the following method has the advantage that proteins are usually soluble in strong formic acid solutions.

The protein is dissolved at 50 mg/ml in 99% formic acid, and diluted by slow addition, with rapid stirring, to twenty volumes of a solution of pepsin (1% by weight of the substrate) in 1 mM HCl at 25 °C. After the addition the solution is incubated, with stirring, at 25 °C for 2 h. The digestion may be stopped by raising the pH to above 5, or by freeze-drying. The digestion may sometimes be limited to a few specific points by digestion with less pepsin for short times (e.g. 0.1% by weight for 15 min), but the usefulness of such a procedure for the generation of large peptides from a particular protein cannot be predicted.

Further details of the enzyme are given by Wyle (1970).

3.3.8. Thermolysin (EC 3.4.24.4)

The thermostable endopeptidase from *Bacillus thermoproteolyticus* catalyzes the hydrolysis of peptide bonds involving the amino groups of the hydrophobic amino acids leucine, isoleucine, methionine, phenylalanine, tryptophan and valine. Other sites, at the N-terminal side of alanine, tyrosine, threonine and occasionally other residues may also be cleaved, but usually more slowly. Peptide bonds involving the carboxyl groups of proline residues are susceptible, but the presence of a proline residue at the C-terminal side of a hydrophobic residue prevents cleavage at the N-terminal side of the latter (Matsubara and Feder, 1971; Matsubara, 1970). The enzyme requires zinc, which is tightly bound, for its activity, and Ca^{2+}

enhances its thermostability. Thermolysin is inhibited by EDTA. The thermostability of the enzyme allows the use of high temperatures (e.g. 60 °C) which may help to denature the substrate.

Because of the large number of sites of cleavage in a typical protein, thermolysin is not recommended for the initial cleavage of proteins; it is, however, valuable for the subfragmentation of purified peptides. A particular advantage is that the N-terminal residues of the peptides released usually give strong fluorescence with fluorescamine, and may readily be detected on peptide maps.

For digestion, the protein or peptide is dissolved or suspended at about 10 mg/ml in 0.1 M NH_4HCO_3, 1 mM $CaCl_2$. Thermolysin (1–2% by weight of the substrate) is added, and the mixture is stirred at 45 °C for 1 h. Digestion is stopped by freeze-drying.

The protease is active in 8 M urea or 1% sodium dodecyl sulphate solution at room temperature, and these agents may be used to dissolve substrates if they are otherwise resistant to digestion. Some applications of thermolysin are given by Heinrikson (1977).

3.3.9. Subtilisin (EC 3.4.21.14)

Subtilisin is a non-specific protease, catalyzing the hydrolysis of a wide variety of peptide bonds, and is not useful for the cleavage of denatured proteins. However, it may prove useful for the cleavage of peptides from which other proteases have not yielded suitable subfragments for the completion of their sequence determinations. The conditions of digestion are as for trypsin. The enzyme is stable in 1% sodium dodecyl sulphate solution.

3.3.10. Papain (EC 3.4.22.2)

Papain is another non-specific protease, but the sequence around a site of potential cleavage exerts a considerable influence upon the rate of cleavage. In particular, the bond X-Y in the sequence -Phe-X-Y- is particularly susceptible in peptide substrates (Berger and Schechter, 1970). Papain is a thiol protease, and is inhibited by oxidizing reagents, heavy metal ions and alkylating reagents such as iodoacetic acid; the presence of complexing and reducing reagents is necessary for full activity. The enzyme is, however, active in cyanate-free 8 M urea solution.

The protein or peptide is dissolved in 0.1 M NH_4HCO_3, 1 mM 2-mercaptoethanol, 0.1 mM EDTA, and papain solution (previously activated by incubation in this solution for 30 min at 20 °C) is added, to give an enzyme to substrate ratio of 1 : 50 by weight. The solution is incubated with stirring at 37 °C for 2 h. Digestion may be stopped by the addition of iodoacetic acid (1 mM).

With some proteins, the specific formation of large fragments may be achieved by cleavage of particularly susceptible bonds with low papain concentrations for short periods, but limited digestion of denatured proteins has rarely been of use in sequence analysis. As with the other non-specific proteases, papain is best reserved for the subfragmentation of peptides of up to 100 residues in length.

3.4. Specific cleavage of proteins by chemical methods

Although several methods have been proposed for the cleavage of polypeptide chains into fragments suitable for sequence analysis, many of these are of low specificity or employ conditions which degrade the peptide side chains. Into the latter category fall partial hydrolysis with acid (§ 3.4.6) (except the mild conditions for cleavage of Asp-Pro bonds) and oxidative cleavage with *N*-bromosuccinimide (Ramachandran and Witkop, 1967). Thus only a few chemical cleavage methods are useful in sequence analysis. Two methods have been discussed already (§ 3.2.2). By far the most important chemical cleavage method is that of methionyl peptide bonds with cyanogen bromide. More recently introduced methods are cleavage at cysteine residues via cyanylation with 2-nitro-5-thiocyanobenzoate, and cleavage at tryptophan residues with 2-(2-nitrophenylsulphenyl)-3-methyl-3-bromoindolenine (BNPS-skatole), *o*-iodosobenzoic acid or dimethylsulphoxide/HCl/HBr (§ 3.4.2–3.4.5).

3.4.1. Cleavage with cyanogen bromide

Since its introduction by Gross and Witkop (1961), the cleavage at methionine residues with CNBr under acidic conditions has been used frequently. The reaction has been discussed in detail by Gross (1967), and is outlined below:

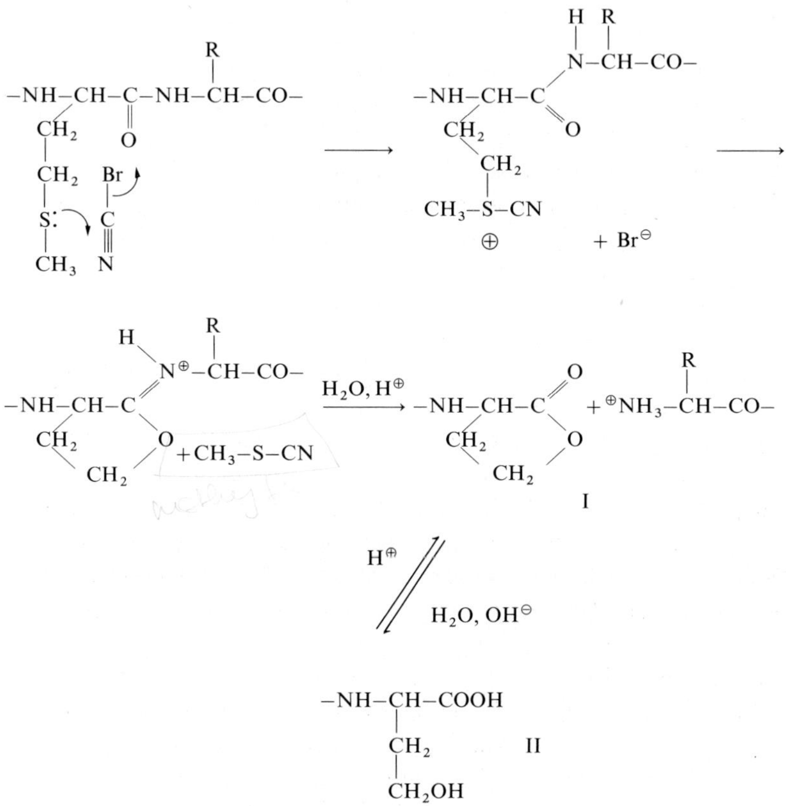

Methionine residues are converted to a mixture of C-terminal homoserine residues (II) and homoserine lactone residues (I), which are interconvertible (Ambler, 1965).

Cleavage of -Met-Thr- bonds, and to a lesser extent of -Met-Ser- bonds, often occurs in low yield, and it has been suggested that the β-hydroxyl groups of these residues take part in the reaction (Schroeder et al., 1969), probably via the intermediates III and IV:

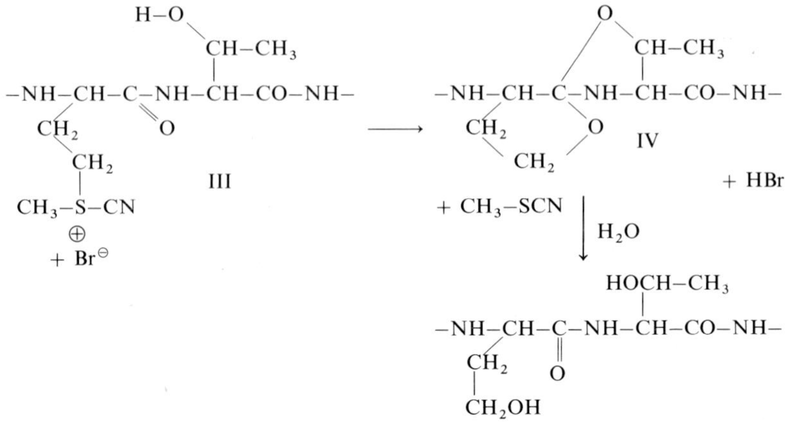

The methionine residue is converted to a homoserine residue (V) without peptide bond cleavage. Lack of cleavage of a methionine-cystine bond has also been reported (Doyen and Lapresle, 1979).

The reaction is usually performed for about 24 h at 20 °C in 70% (v/v) formic acid in water, although other conditions, such as the use of aqueous trifluoroacetic acid, have been used with success. Side reactions such as partial cleavage of Asp-Pro bonds, partial loss of side-chain amide groups, partial cyclization or rearrangement of Asn-Gly sequences to the imide or β-aspartyl peptide, and cyclization of released N-terminal glutaminyl residues to pyrrolidone carboxylyl residues (and analogous reactions with carboxymethyl cysteine, etc.) may be expected to occur. Additionally, unless pure fresh CNBr is used and precautions against autoxidation are taken, oxidative degradation of tryptophan residues may occur. Cleavage of tryptophanyl peptide bonds has been observed (Blumenthal et al., 1975). The cleavage reaction may be performed without prior reduction of cystine residues, which are unaffected by CNBr in acid. Oxidation of methionine residues prevents the reaction with CNBr from taking place, but the methionine sulphoxide residues which may be produced by autoxidation during the isolation of polypeptides may be reduced by thiols (Jori et al., 1968; Neumann, 1972; Westhead, 1972).

The following method should prove satisfactory for most proteins. Note that CNBr is toxic (Appendix 3).

The protein (previously treated with 5% 2-mercaptoethanol at pH 8 for 24 h if autoxidation has occurred, and separated from the thiol by dialysis) is dissolved at about 5 mg/ml in 70% (v/v) aqueous formic acid at room temperature. A small amount (about 5 mol/mol of methionine residues) of tryptamine or tryptophan may be included as a scavenging reagent to protect tryptophan residues in the protein. A 50-fold molar excess over methionine residues (typically a weight equal to that of the protein) of CNBr, which should be absolutely colourless, in a small volume of 70% formic acid, is added, with stirring, and the mixture is incubated in the dark under oxygen-free N_2 at 20–25 °C for 16–24 h. The mixture is diluted with 15 vol of water and freeze-dried. For complete removal of the acid and by-products, the freeze-drying is repeated after further addition of water.

Alternative conditions which have been used with integral membrane proteins are discussed in Chapter 9.

Cleavage of the peptide chain at *S*-methylcysteine residues has been described (Gross and Morell, 1974), but the reaction is much slower, and proceeds by a different mechanism.

3.4.2. Cleavage at cysteine residues after cyanylation

Cleavage of peptide chains under alkaline conditions after cyanolysis of disulphide bonds was observed by Catsimpoolas and Wood (1966). However, because of several side reactions, such as the reversibility of the cyanolysis reaction and the elimination of thiocyanate, the yields were low. A close study of the reaction mechanism (Jacobsen et al., 1973; Degani and Patchornik, 1974) led to the development of reaction conditions leading to high cleavage yields (80–90%). For the attainment of high yields, the reaction must be performed on the soluble, denatured protein.

Cysteine residues are converted to S-cyanocysteine residues by reaction with 2-nitro-5-thiocyanobenzoic acid (I) at pH 8.

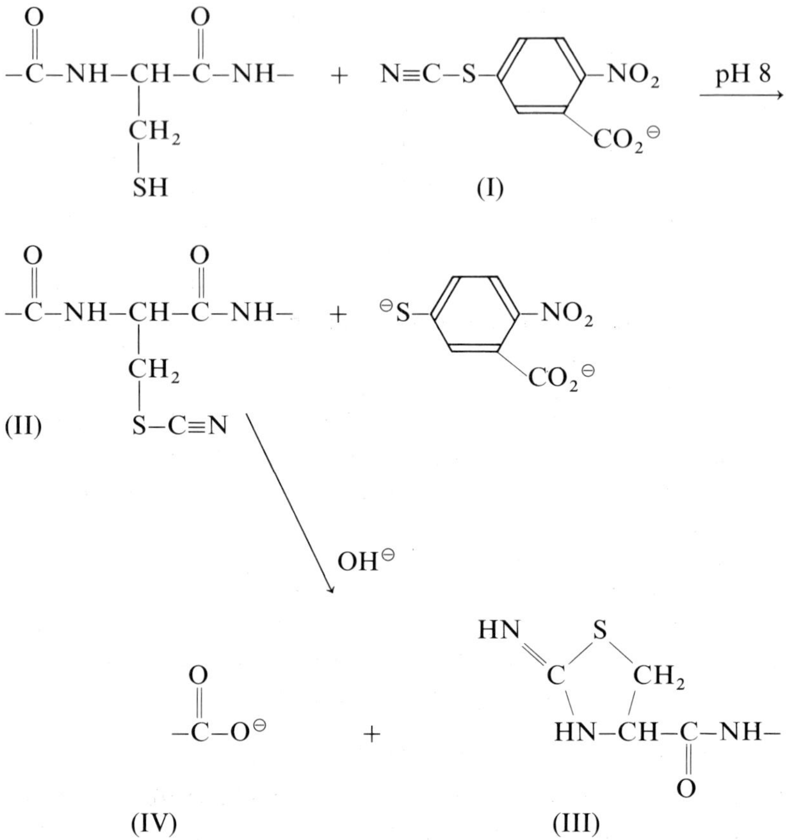

A large excess of (I), and a low concentration of total thiol groups, must be used, to avoid the side reaction of displacement of CN^- from S-cyanocysteine residues by unreacted cysteine thiol groups. The excess reagents and by-products are removed by gel filtration in acidic solution, and the S-cyanylated protein (II) is incubated at pH 9 to effect cleavage of the peptide bond, with the formation of an amino-terminal iminothiazolidine-4-carboxyl residue (III) from the S-cyano-cysteine residue, and liberation of the α-carboxyl group of the preceding residue (IV).

Since (III) is not degraded by the phenylisothiocyanate method, the polypeptides produced (except for that derived from the N-terminus of the protein) cannot be subjected directly to automated or manual sequence determination. Removal of the blocking group with a large excess of Raney nickel has been reported (Schaffer and Stark, 1976) but it remains to be seen if this will prove useful for application to routine sequence analysis. For the determination of the sequences of blocked peptides, subfragmentation with various proteases is generally required. Use of $^{14}CN^-$ in the synthesis of reagent (I) (Degani and Patchornik, 1971) leads to incorporation of the radioactive label in the iminothiazolidine derivative, and this may be helpful in the isolation of the blocked peptides.

Selectivity between cysteine and cystine residues may be achieved by carrying out the reaction without prior reduction of the protein.

The following reaction conditions are slightly modified from those of Stark (1977).

The protein is dissolved at 1–2 mg/ml in 6 M guanidinium chloride, 1 mM EDTA, 1 mM dithiothreitol, 0.2 M Tris–acetate buffer, pH 8.0, and incubated at 37 °C for 15 min. If disulphide bonds are present in the protein, the dithiothreitol concentration is raised to 5 mM, and the solution is incubated for 2 h. If cleavage at cysteine, but not cystine, residues is desired, no dithiothreitol is added, and the solution is immediately treated with the cyanylation reagent. A 10-fold molar excess of 2-nitro-5-thiocyanobenzoic acid (I) over total thiol groups is added, and the pH is returned to 8.0 with NaOH. After 15 min at 37 °C, the mixture is cooled to 4 °C and acidified to pH 4 or below with acetic acid. The cyanylated protein is separated from small molecules by dialysis or gel filtration into 50% (v/v) acetic acid, and recovered by freeze-drying. For cleavage, the cyanylated protein is dissolved at about 5 mg/ml in 6 M guanidinium chloride, 0.1 M sodium borate buffer, pH 9.0. The pH is adjusted to 9.0 with NaOH if necessary, and the solution is incubated at 37 °C for 12–16 h. The peptides are recovered by gel filtration on a column of Sephadex G-10 in 50% acetic acid or other suitable solvent, such as 0.1 M NH_3.

3.4.3. Cleavage at tryptophan residues by BNPS-skatole

Oxidative cleavage of peptide bonds on the carboxyl side of tryptophan, tyrosine and histidine residues with reagents such as bromine or *N*-bromosuccinimide is not of great value in sequence determination, since many side reactions occur and the yields of cleavage are low. However, the milder reagent BNPS-skatole, the product of oxidation by *N*-bromosuccinimide of 2-(2-nitrophenylsulphenyl)-3-methylindole, is more specific for cleavage at tryptophan residues (Omenn et al., 1970; Fontana, 1972). Side reactions, such as oxidation of tyrosine and histidine residues, occur at concentrations of the reagent required for high cleavage yields, however, and in practice cleavage yields are typically only 15–60% (Omenn et al., 1970; Fontana, 1972). Since tryptophan residues are usually sparse in proteins, such low yields may be acceptable for the generation of large fragments suitable for automated sequence analysis. If more than three or four tryptophan residues are present, however, the reaction product will be a complex mixture of many polypeptides in varying amounts, and the isolation of these peptide fragments may present considerable difficulty, depending upon the distribution of the tryptophan residues along the chain. Pilot experiments, with analysis of the reaction products by polyacrylamide gel electrophoresis, will indicate if this cleavage method is likely to be of value for a particular protein.

The following reaction details are taken from Fontana (1972).

BNPS-Skatole is unstable at room temperature, but may be stored dry at −20 °C for months. The pure compound is yellow and crystalline. The preparation of the reagent from 2-(2-nitrophenylsulphenyl)-3-methylindole is described by Omenn et al. (1970).

The protein is dissolved in 50% (v/v) acetic acid at about 10 mg/ml. A freshly prepared solution in acetic acid of BNPS-skatole (100-fold molar excess over tryptophan residues) is added, and the reaction is allowed to proceed at room temperature with stirring for 48 h in the dark. The peptides are separated from excess reagents and by-products on a column of Sephadex G-25 in 50% acetic acid. Tyrosine may be added to the protein as a scavenger to minimize reaction of tyrosine residues in the protein.

3.4.4. *Cleavage at tryptophan residues with* o-*iodosobenzoic acid*

This mild and selective reagent for the cleavage of tryptophanyl peptide bonds in high yields (70–100%) has recently been described by Mahoney and Hermodson (1979). The reduced and S-alkylated protein is dissolved in 80% acetic acid, 4 M guanidinium chloride, and *o*-iodosobenzoic acid (2 g/g of protein) is added. The reaction is allowed to proceed for 24 h at room temperature, and peptides are isolated by gel filtration in 50% acetic acid.

Several other reagents for cleavage at tryptophan residues have been described. For example, the use of very high concentrations of CNBr (Ozols and Gerard, 1977a) or a mixture of dimethyl-sulphoxide, HCl and HBr (Savige and Fontana, 1977).

Other workers have noted lower selectivity with iodosobenzoic acid than that reported by Mahoney and Hermodson (1979); for example, modification and cleavage of peptides containing tyrosine or histidine has been observed (Wachter, 1980). Fontana (1980) has concluded that similar pathways of oxidative halogenation are involved in all these cleavage methods. The products are C-terminal spirolactones:

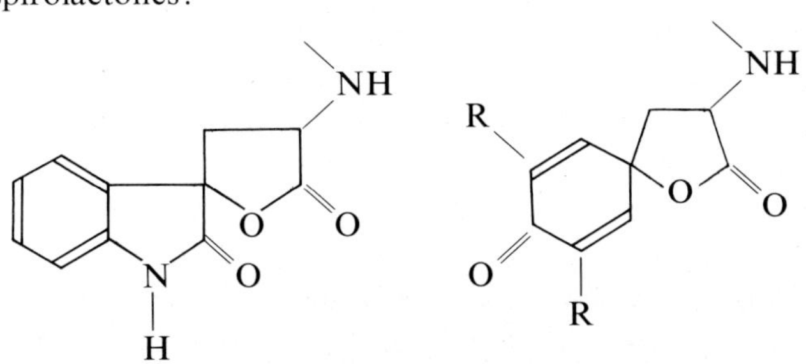

(R = Cl, Br or I)

derived respectively from tryptophan and tyrosine residues. The use of dimethyl sulphoxide/HCl/HBr has the advantage of ready availability of reagents, and appears at least as selective as other reagents. Methionine is converted to the sulphoxide, and cysteine to cystine.

3.4.5. Cleavage with dimethylsulphoxide/HCl/HBr

The protein (100 mg) is dissolved at room temperature in a mixture of glacial acetic acid (3.6 ml), 12 N HCl (1.8 ml), dimethylsulphoxide (0.15 ml) and phenol (100 mg). After 30 min, 48% HBr (0.6 ml; the acid must be colourless) and dimethylsulphoxide (0.15 ml) are added, and incubation at 20–25 °C is continued for a further 30 min. At this stage a cleavage yield of about 60% may be obtained. Some oxidized but uncleaved tryptophan residues are present, and the cleavage yield may be increased by treatment with aqueous acid. The reaction mixture is diluted with 6 ml H_2O, and the pH is adjusted to 2.0 with pyridine. The mixture is incubated at 60 °C for 15 h. The mixture is concentrated by rotary evaporation and the peptide fragments are separated by gel filtration in 10% formic acid.

3.4.6. Partial acid hydrolysis

Although partial acid hydrolysis is rarely used any longer, there may be occasions, for example in the study of blocked peptides, where this method may be suitable. A measure of specificity may be attained through the use of one or other of the following conditions.

(a) The peptide is incubated in 11 M HCl for 4 days at 37 °C. The acid is removed in vacuo over NaOH pellets. Cleavage at the N-terminal side of serine and threonine residues is faster than that of other peptide bonds (Sanger and Tuppy, 1951).

(b) The peptide is treated with 0.03 M HCl at 105 °C for 6–18 h. Under these conditions, cleavage on either side of aspartic acid residues is faster than that of other peptide bonds (Tsung and Fraenkel-Conrat, 1965).

Both procedures lead to the hydrolysis of side-chain amide groups, and at least partial destruction of tryptophan residues may be expected. The major problem with the use of these methods for the cleavage of peptides of more than about twenty residues is likely to be the isolation of the peptide fragments from the complex mixtures produced.

Inglis (1980) has re-investigated the use of dilute acid hydrolysis, and observed yields of about 60% for cleavage on the carboxyl side of aspartic acid residues after treatment of peptides at 108 °C with 0.0125 M HCl (pH 2) for 2 h in vacuo. Only small extents of side reactions were observed, although C-terminal asparagine residues may be deamidated. This method of partial hydrolysis is thus of comparable value to those for cleavage at tryptophan residues.

Separation and purification of peptides

4.1. Outline of methods used

The cleavage of a protein by enzymic or chemical methods results in a mixture of a few or many peptide fragments, which must be separated from each other for determination of their structures. The purification of peptides is the most time-consuming stage in the determination of the amino acid sequence of a protein, but the time required is now significantly lower than it was. Improvements in the efficiency of the Edman degradation allow the data required to be obtained from fewer peptides, and the trend towards higher sensitivity in amino acid analysis and in sequence determination allows the preparative use of thin-layer chromatography and electrophoresis, and smaller chromatographic columns, which are faster than earlier methods.

Separation methods for peptides are based on one or more of the following physical chemical characteristics: molecular size, charge, polarity, solubility and specific covalent or non-covalent interactions. The charge on a peptide may be altered by a change in the pH.

Successive application of gel filtration (based on molecular size), ion-exchange chromatography (based mainly on the charge), thin-layer electrophoresis (based on both size and charge) and thin-layer chromatography (based on polarity) will usually be sufficient to resolve the most complex mixtures encountered, although a pair of peptides with very similar properties, differing, for example, in the substitution of a leucine residue for an isoleucine residue, may require additional steps for resolution.

A summary of the methods used for the fractionation of peptides is given in Table 4.1. Many of these methods may also be used as analytical procedures for monitoring the progress of purification of peptides; for example, thin-layer chromatography on small aliquots of fractions obtained by column chromatography. The choice of preparative procedures will be influenced by the availability of compatible analytical procedures. The detection of peptides is discussed in Chapter 5. Descriptions of the principal techniques for the separation of peptides are presented in the following sections. An idealized scheme for the separation of tryptic peptides from a soluble polypeptide of molecular weight about 50,000 is shown in Fig. 4.1.

The discussion in this chapter is of methods appropriate for typical globular proteins; specialized classes of proteins, such as collagen, with unusual distributions of amino acid residues, will require a modified approach. The separation of peptides from integral membrane proteins presents serious problems, owing to the great tendency of such peptides to aggregate or be insoluble in the usual media for resolution of peptide mixtures. The isolation of such peptides is therefore considered separately, in Chapter 9.

One point of general importance in peptide fractionation is the susceptibility of peptides to proteolysis by contaminating proteases or microorganisms. Cleanliness is essential. Any materials suspected of microbial contamination should be discarded, and exposure of peptides to warm or wet conditions should be minimized. Peptide mixtures should be stored at or below −20°C, preferably in the dry state. Particular difficulty may be experienced in laboratories which have previously been used for the cultivation of micro-organisms or work on large quantities of proteases; complete redecoration of all surfaces may be necessary. The purity of all components of the solutions used must be ensured to avoid reaction of impurities, such as aldehydes or peroxides, with peptides.

TABLE 4.1
Methods for the fractionation of peptide mixtures

Technique	Examples of materials used	Properties of peptide molecules exploited
Centrifugation	Dilute buffers, trichloroacetic acid	Solubility
Gel filtration	Sephadex, BioGel P, Spheron P.	Size
Ion-exchange chromatography	Derivatives of cellulose or Sephadex (DEAE, CM, QAE, SP, phospho-); polystyrene derivatives	Charge, with some influence of polarity
Paper electrophoresis	Whatman Nos. 1, 3MM; volatile buffers	Charge and size
Paper chromatography	Whatman Nos. 1, 3MM aqueous organic solvents	Polarity
Thin-layer electrophoresis	Cellulose, silica gel; volatile buffers	Charge and size
Thin-layer chromatography	Cellulose, silica gel; aqueous-organic solvents	Polarity
Polyacrylamide gel electrophoresis	(a) In urea solution	Charge and size
	(b) In dodecyl sulphate solution	Size
High-performance liquid chromatography (Hplc)	Reversed-phase columns; ion-pairing reagents; aqueous-organic solvents	Polarity
Gas chromatography	Coated packings; combination with mass spectrometry	Volatility of derivatives
Counter-current extraction	Two immiscible solvents	Polarity; sometimes specific interactions
Affinity chromatography	Antibody-Sepharose	Specific interactions
Covalent chromatography or irreversible binding	Thiol-Sepharose derivatives; aminopolystyrene	Disulphide bond formation; reactivity of homoserine lactone

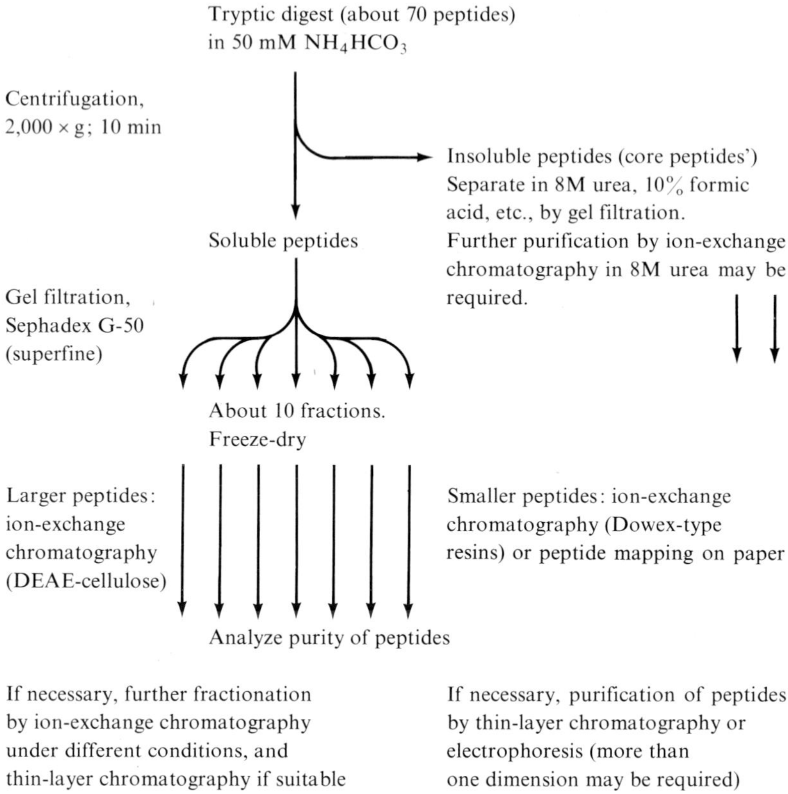

Fig. 1. Outline of idealized fractionation scheme for tryptic peptides from a typical protein of molecular weight 50,000.

4.2. Amounts of digested protein required

To decide the quantity of protein required for each digestion, the yield of each fractionation step must be considered. Handling and transfer of peptides, especially in small amounts, should be minimized. Often yields of 50% or less are obtained after each step (allowing for the material used in analysis); thus a final yield of only about 6% is to be expected after four steps of purification. If analytical methods at the 10 nmol level are available for amino acid analysis and sequence determination, a minimal initial amount of protein required would thus be 300 nmol. Since cleavage methods give yields less than 100%, this quantity must be increased. A reasonable minimum quantity of protein of molecular weight 50,000 is about 500 nmol (25 mg) for each digest. A protein of molecular weight 15,000 may be expected to yield pure tryptic peptides after only three fractionation steps, and about 200 nmol (3 mg) would be required to obtain the sequences of these peptides. These amounts may be reduced by a factor of three or so if techniques of higher sensitivity are used.

If more material is available, it is recommended that larger quantities of protein are used for each digest, to provide reserve material if some analytical procedures are unsuccessful, and to allow for subfragmentation of peptides which can not be completely sequenced by the Edman degradation. In addition, the determination of the structures of modified residues may require substantially more material than the routine determination of the sequences of peptides which contain no modified residues.

4.3. Preliminary study of the peptides in a digest

Before the bulk of the digest is subjected to fractionation, a useful preliminary is the analysis of a small portion of the mixture by SDS-polyacrylamide gel electrophoresis or by thin-layer peptide mapping. The former technique is useful for the analysis of larger peptides obtained either by limited proteolysis, by CNBr cleavage or by another method of narrow specificity. The distribution of sizes and amounts of peptides determined by gel electrophoresis will indicate which fractionation methods should be employed. Iso-electric focusing or electrophoresis in urea solutions in polyacrylamide gels may also be useful, indicating which methods of ion-exchange chromatography might be successful in separating the peptides. Many techniques for electrophoresis in polyacrylamide gels have been described, and some examples are given below. Thin-layer peptide mapping is suitable for the analysis of mixtures of peptides containing smaller peptides, of less than about 20 residues. Such mixtures are present in tryptic, chymotryptic, peptic and similar digests of proteins. If ninhydrin is used as the detection reagent, about 2 nmol of the digest is required. Fluorometric methods are capable of higher sensitivity, but the variety of colours obtained with ninhydrin reagents gives additional information. Other reagents are described in Chapter 5. Either electrophoresis or chromatography may be used in each dimension, but the most effective separation of peptides is usually obtained by electrophoresis in one dimension followed by chromatography in the perpendicular direction. Many different buffers may be used for electrophoresis, but volatile buffers containing no primary or secondary amines are clearly preferable, and pyridine acetate buffers are particularly suitable. Similarly, many chromatographic solvents have been used; those based on butanol are most common.

Other methods of peptide mapping are discussed in Chapter 10. Of these, the recently introduced high-performance liquid chromatographic methods are sensitive and rapid, but some limitations, such as incomplete elution of peptides, have yet to be overcome.

4.3.1. Analytical SDS-polyacrylamide gel electrophoresis (Laemmli, 1970)

The following stock solutions are prepared:

(a) Acrylamide 30 g, *N,N′-bis*-methylene acrylamide 0.8 g, made up to 100 ml with water. Filter and store in the dark at 4 °C. Use within one month.
(b) Tris buffer, pH 8.8. Tris 121 g, conc. HCl 18.8 ml per litre.
(c) Tris buffer, pH 6.8. Tris 121 g, conc. HCl 83.2 ml per litre.
(d) SDS 10 g in 100 ml.
(e) Running buffer, pH 8.3. Concentrated stock solution. Tris 30.3 g, glycine 144.2 g, SDS 10 g, in 1 litre.
(f) Sample cocktail. 1.88 ml of solution (c), 6.0 ml of (d), 3.0 ml glycerol, 231 mg dithiothreitol and 2.0 mg bromophenol blue, made up to 15 ml with water. Store as 1 ml aliquots at −20 °C.

(g) Coomassie brilliant blue R-250, 1 g; methanol, 450 ml; glacial acetic acid, 100 ml; and water, 450 ml. Filter.
(h) Destain solution. Acetic acid, 70 ml; methanol, 200 ml; water, 730 ml.
(i) Drying reagent. Acetic acid, 100 ml; glycerol, 10 ml; water, 890 ml.
(j) Ammonium persulphate, 1 g in 10 ml. Freshly prepared.

Gel electrophoresis may be performed in tubes or in thin slabs, the latter method being generally preferred. Several types of apparatus are available. For full discussion of the various types of equipment the monograph by Gordon (1975) in this series should be consulted. Usually the gel is formed between two glass plates separated by spacers made of acrylic plastic, 1–2 mm thick. To prevent leakage of the gel solution, the sides and bottom of the assembly are sealed with agar, 15% polyacrylamide gel or thin silicone rubber tubing. Silicone grease may also be used but is less satisfactory.

The separating gel solution of the desired concentration is prepared:

	5%	7.5%	10%	12.5%	15%
(a)	5.0	7.5	10	12.5	15 ml
(b)	11.2	11.2	11.2	11.2	11.2 ml
H_2O	13.7	11.2	8.7	6.2	3.7 ml

The solution is mixed and degassed (O_2 inhibits polymerization), 0.3 ml of solution (d), 0.1 ml of (j) and 20 μl of TEMED (*N*,*N*,*N*′,*N*′-tetramethylethylenediamine) are added and the solution is mixed again and poured into the slab assembly. The gel should be set within 30 min at 20 °C. A flat surface of the gel may be obtained by overlayering the solution with *n*-butanol.

A 'comb' for the formation of sample slots is inserted between the glass plates, and the stacking gel (1.67 ml of (a), 1.25 ml of (c), 7.03 ml H_2O, 0.1 ml of (d), 50 μl of (j) and 10 μl of TEMED) is poured to fill the remaining space. The stacking gel should be completely set in 60 min.

The sample (2–50 μg, depending upon the number of different components in the mixture, in 5–25 μl of a solution of low ionic strength and low buffering capacity) is mixed with an equal volume of solution (f), and the mixture is heated at 100 °C for 2–5 min. The running buffer is prepared by diluting solution (e) 10-fold. The bottom spacer of the gel is removed, and the electrophoresis apparatus is assembled, with running buffer in anode and cathode compartments. The samples are loaded into the slots beneath the less dense buffer solution. Markers consisting of 2 μg each of proteins of suitable molecular weight are also applied; these also serve to aid the identification of the positions of sample slots after staining the gel.

The power supply is connected (-ve terminal at the top) and electrophoresis is performed until the blue marker dye reaches the bottom of the gel (about 3 h at 100 V and 20 mA for a slab 150 mm wide, 100 mm high, 1.5 mm thick; the exact values being dependent upon the gel concentration and temperature).

The power supply is disconnected, the gel carefully removed from the apparatus, and immersed in solution (g) overnight. The stain solution is replaced by solution (h), which is renewed after 3 h. A permanent record may be made by photography, or the gel may be soaked in solution (i), and dried on filter paper in vacuo with application of heat. The dried gel may be placed next to an X-ray film for radio-autography if the samples are labelled with ^{14}C, ^{35}S or ^{32}P isotopes. Fluorography may be performed after soaking the wet gel with dimethylsulphoxide, then with a solution of PPO (2,5-diphenyloxazole) in dimethylsulphoxide followed by washing with water and drying. Alternatively, a proprietary reagent (‘EN^{3}HANCE’ from New England Nuclear) may be used. Polypeptides labelled with ^3H are detected by radioautography at −70 °C after this treatment, and the sensitivity for ^{14}C and ^{35}S isotopes is increased.

4.3.2. Analytical polyacrylamide gel electrophoresis in the presence of urea and SDS

In the presence of urea, the range of molecular weights of polypeptides separable by SDS-gel electrophoresis is extended, and even small peptides of 10–15 residues may be separated; detection may prove difficult, however, since small peptides are not generally fixed and stained as efficiently as are proteins.

Swank and Munkres (1971) described a system, with the presence of 8 M urea and a Tris–phosphate buffer, suitable for the separation of peptides of molecular weights 1,200–10,000. Peptides did not migrate strictly according to their molecular weights.

4.3.3. Analytical isoelectric focusing of peptides in polyacrylamide gels

Isoelectric focusing has rarely been used for the separation of peptides, because peptides often have no clearly defined isoionic points, and the specific staining of peptides in the presence of ampholines is difficult to achieve. One procedure, using 0.7 mm thick gels, containing 8 M urea, with detection by staining with Coomassie blue G-250 in 1 N H_2SO_4–12% trichloroacetic acid solution, has been described recently (Righetti and Chillemi, 1978), but a low sensitivity was observed.

4.3.4. Analytical thin-layer peptide mapping

Many systems have been described for peptide mapping; the following details follow those of Heiland et al. (1976). Details of apparatus for thin-layer electrophoresis are given in § 4.10.1.

A sample of the digest (2–5 nmol) is dissolved in 5 µl of 20% pyridine and applied to the origin (see Fig. 4.2) on a 20 cm × 20 cm plastic-backed cellulose thin-layer plate (Macherey and Nagel, Polygram CEL 300) as a spot about 10 mm diameter. An internal marker (1 µl of a solution containing about 0.5 mg/ml of ε-dinitrophenyllysine and 0.2 mg/ml of xylene cyanol FF) may also be applied at the origin to aid comparison of maps. The solvent is evaporated in a stream of warm air.

The thin layer is placed on the cooling plate of the electrophoresis chamber, and electrophoresis buffer (pyridine/acetic acid/acetone/water; 20 : 40 : 150 : 790, by vol.), pH 4.4, is applied, as described in § 4.10.1, such that the sample is concentrated to a narrow band by the capillary flow of the buffer solution. Any surplus fluid is removed,

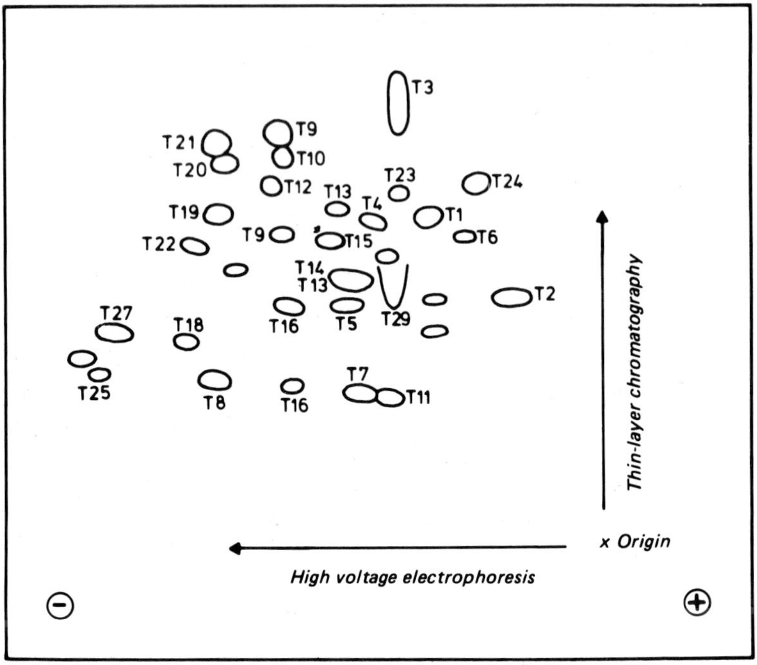

Fig. 4.2. Thin-layer peptide map of the tryptic peptides of *Escherichia coli* ribosomal protein L10. The map was prepared essentially as described in the text, with electrophoresis at pH 4.4 followed by chromatography in butanol/acetic acid/water/pyridine (15 : 3 : 12 : 10, by vol.), on cellulose. Reproduced with permission from Heiland, I., Brauer, D. and Wittmann-Liebold, B. (1976) Hoppe Seyler's Z. Physiol. Chem., *357*, 1751–1770.

the wicks are placed in position, and electrophoresis is performed for 90 min at 500 V (about 20 mA) at 15–20 °C.

The solvent is evaporated in a stream of warm air. The thin layer is allowed to equilibrate with atmospheric humidity for 30 min, then placed in the chromatography tank containing *n*-butanol/acetic acid/water/pyridine (15 : 3 : 12 : 10, by vol.), so that chromatography is performed at right angles to the direction of electrophoresis. When the solvent front reaches the top of the thin-layer plate (about 6 h), the solvents are evaporated in a stream of warm air, and peptides are detected by one or more of the methods described in Chapter 5. Ninhydrin/Cd reagent is recommended for general use.

4.4. *Separation of peptides on the basis of solubility*

The digest of a protein may not be completely soluble under the conditions of digestion, and it has often been found that a useful preliminary fractionation procedure is the separation of soluble from insoluble peptides. Adjustment of the pH of the digest may also result in the specific precipitation of some peptides. In general, relatively large, non-polar peptides tend to be insoluble in water, especially at their iso-ionic points. Such peptides may originate from the hydrophobic interiors ('cores') of globular proteins.

Standard conditions are not available for the optimal separation by precipitation of one or a few peptides. Care should always be taken to avoid non-specific precipitation, and to wash the precipitate thoroughly, since peptides which would normally be soluble may be adsorbed to the precipitate. The precipitation of peptides during enzymic digestion may prevent expected cleavage of some peptide bonds, and alternative conditions for the digestion, such as the inclusion of 2 M urea, should be considered. A better separation of peptides insoluble in water from other peptides may often be achieved by gel filtration under conditions where all peptides are soluble, but some may be aggregated. For example, the insoluble 'core' tryptic peptide from papain, which formed a gel in water, could be separated and purified by gel filtration in 50% acetic acid, where it was still aggregated (Husain and Lowe, 1969).

The isolation of pure peptides from the precipitate may be difficult, but if only a few insoluble peptides are produced, gel filtration in urea or formic acid solutions may suffice.

The separation by fractional precipitation of fragments produced by limited proteolysis of ribonuclease A and myosin have been described in Chapter 3. Separation of insoluble peptides from integral membrane proteins, such as glycophorin, is discussed in Chapter 9. For the complete separation of peptides from actins,

Vandekerkhove and Weber (1978a) digested the performic acid-oxidized proteins with trypsin, and suspended the freeze-dried digests in pH 6.5 pyridine/acetic acid electrophoresis buffer (10 mg in 15 ml). The insoluble peptides were collected by centrifugation at 7000 × *g* for 10 min. The pellet was resuspended in the buffer, and the insoluble peptides again collected by centrifugation. The pellet was further digested, with thermolysin.

4.5. Gel filtration

The technique of gel filtration, otherwise known as gel permeation chromatography, has been described in detail in another volume in this series (Fischer, 1980), and only brief information of practical importance is included here.

The separation of peptides according to molecular size is a useful early step in the fractionation of complex mixtures, since different methods are appropriate for the subsequent purification of peptides of different size classes. Two types of gel filtration materials have been used widely, consisting of cross-linked dextran (Sephadex) or polyacrylamide (BioGel P), each available in a variety of pore sizes. Alternative media are commercially available, including porous glass (and coated derivatives suitable for high-pressure liquid chromatography, § 4.12), agarose beads and synthetic polymers.

The choice of the pore size of the gel filtration medium will be governed by the range of sizes of peptides in the mixture to be separated. It should be noted that extended peptide chains, particularly when complexed with dodecyl sulphate, are eluted from gel filtration columns at volumes comparable with those of globular proteins of three or four times the molecular weight. Suitable media for the separation of peptides from different digests of proteins are summarized in Table 4.2.

The resolution obtained from columns packed with small beads (superfine grades of Sephadex or -400 mesh BioGel) is significantly greater than that using larger beads. The flow-rates obtained are lower, but this is rarely a disadvantage, as columns may be left

TABLE 4.2
Gel filtration media suitable for the separation of peptides

Bio-Gel[a]	Sephadex or Sepharose[a]	Useful range of molecular weights (extended peptide chains)[b]	Typical digests of proteins
A-15 m	Sepharose 4B	$20–300 \times 10^3$	Partial proteolysis
A-5 m	Sepharose 6B	$10–100 \times 10^3$	Partial proteolysis
A-0.5 m	Sephadex G-200	$10–50 \times 10^3$	Partial proteolysis
P-300		$10–50 \times 10^3$	Partial proteolysis
P-200	Sephadex G-150	$5–30 \times 10^3$	Limited chemical cleavage
P-150	Sephadex G-100	$2–20 \times 10^3$	CNBr digest
P-100		$2–20 \times 10^3$	
P-60	Sephadex G-75	$1–15 \times 10^3$	CNBr, cleavage at Arg residues
P-30	Sephadex G-50	500–8,000	Tryptic, chymotryptic, staphylococcal protease, peptic
P-10		500–6,000	
P-6	Sephadex G-25	200–5,000	Peptic, thermolytic, subtilisin
P-2	Sephadex G-10	100–1,500	Desalting

[a] Bio-Gel is a registered trade mark of Bio-Rad Laboratories, and Sephadex and Sepharose of Pharmacia Fine Chemicals; for further information on these materials, the manufacturers' catalogues should be consulted.
[b] Approximate values: for globular polypeptides the useful range is for molecular weights about 4-fold greater, and for small peptides hydrophobic interactions significantly affect the elution volumes. In mixed aqueous-organic solvents, the separating range may be changed.

unattended if a reliable fraction collector is used. The small bead sizes are thus recommended, except in those cases where speed is important, such as in the isolation of unstable chemically modified peptides.

4.5.1. Chromatographic columns

Long columns are required for high resolution; a suitable length is 2 m. The diameter of the column is chosen according to the amount of sample to be applied. Columns 10 mm in diameter are suitable for up to 20 mg of a mixture of peptides, 20 mm diameter for 20–100 mg, and 30 mm for 100–500 mg. The limitation of sample weight depends upon the volume to be applied, which should, for maximum resolution, be less than 1% of the total column volume, and the viscosity of the sample, which limits the concentration of the sample to about 20 mg/ml, depending on the peptides present. The viscosity of a solution of large polypeptides may be very high, limiting the concentration in the sample to perhaps less than 5 mg/ml, especially if dodecyl sulphate solution is used. Considerably larger sample volumes may be applied if only a crude separation into broad size classes is desired, or for desalting (up to 20% of the column volume for desalting large peptides on Sephadex G-25).

Columns designed for gel filtration are commercially available, and these may be fitted with water-jackets for precise thermal control. However, satisfactory columns may readily be made from precision-bore glass tubing, and used at room temperature in a draught-free area. The most critical part of the column is the bed support. The volume of the outlet tubing should be minimized to prevent re-mixing of resolved components. The tubing should not release plasticisers into the buffers, since these would give artefactual absorbance peaks; polyethylene or polytetrafluoroethylene tubing is usually satisfactory. A suitable design for a column and arrangement for chromatography are shown in Fig. 4.3.

4.5.2. Eluents for gel filtration of peptides

The choice of solvents for the gel filtration of peptides is made after the consideration of several factors, including solubility of the peptides, freedom from chemical reaction with the peptide or the gel, ease of detection of peptides in the fractions collected, and ease of recovery of peptides from these fractions. The ideal liquid phase

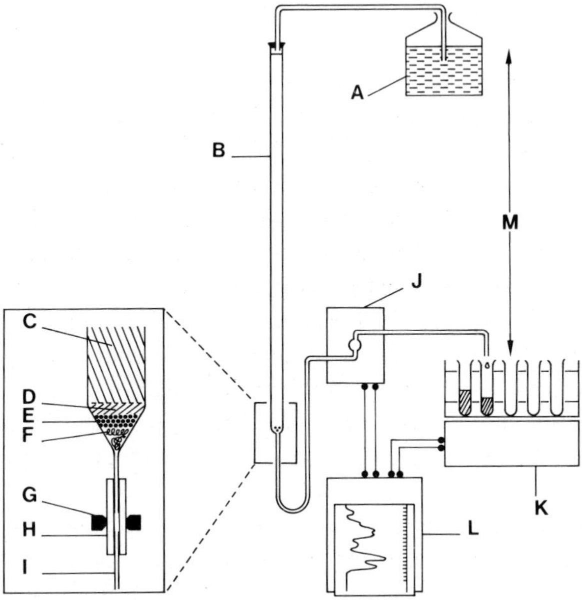

Fig. 4.3. Assembly for gel filtration. (A) Eluent reservoir, (B) column, (C) gel bed (e.g. Sephadex G-75, superfine), (D) Sephadex G-25 (medium), (E) acid-washed sand or small glass beads, (F) loose cotton-wool or glass-wool plug, (G) clamp for shutting off the flow, (H) silicone rubber connection, (I) Narrow-bore polyethylene outlet tubing, (J) spectrophotometer flow-cell, (K) fraction collector, (L) chart recorder, (M) hydrostatic head. For further details Fischer (1980) should be consulted.

would thus be an excellent solvent for peptides, close to neutral pH, with high optical transmission down to 215 nm, and readily volatile. No single solution meets all these criteria. Dilute (0.05–0.1 M) NH_4HCO_3 solution is often suitable. In distilled water, non-specific interactions between peptides and the support, particularly if the peptides contain tryptophan or other hydrophobic groups, such as iodotyrosine residues, are stronger, and slight ion-exchange properties of gels, which are suppressed in 50 mM salt solutions, are significant. Thus in water alone, peaks tend to be broader, and may trail, and elution volumes for hydrophobic peptides are greater. The latter property may be utilized in the selective purification of tryptophan-containing peptides.

Dilute NH_3 solution may be used, but there is some danger that alkali-catalyzed reactions, such as the deamidation of asparagine residues or the destruction of methionine sulphone residues, may occur. Solutions of organic acids, such as 5% formic acid or 30% acetic acid are good solvents for many peptides, but the far UV-absorbance is high and detection of peptides by absorbance is less sensitive than in UV-transparent solvents. Deamidation of asparagine and glutamine residues, a lower stability of tryptophan residues, cyclization and blockage to Edman degradation of N-terminal glutamine and *S*-carboxymethylcysteine residues, and cleavage of some sensitive peptide bonds may occur under these acidic conditions. Buffer solutions such as *N*-ethylmorpholinium formate, pH 3, may be useful.

In general, basic peptides, obtained from basic proteins such as histones or ribosomal proteins, are more soluble in acidic solutions, while acidic peptides, such as those derived from succinylated proteins, are more soluble under alkaline conditions.

For peptides insoluble in any of these solvents, recourse may be made to other solvents which are less satisfactory in some respects. 8 M urea or 6 M guanidinium chloride dissolve most peptides, but removal of these compounds from the separated peptides requires more work and may not be easy to achieve. Solutions of pyridine, such as 20% aqueous pyridine or pyridine–acetic acid buffers, are good solvents for many peptides, but have high UV-absorbance and an obnoxious smell. 70% pyridine/30% 1 M NH_3 has been used (Casey and Rees, 1979) on Sephadex G-25, but probably the separation of peptides is not purely according to size in this solvent. Solutions of sodium dodecyl sulphate are useful for the separation of large polypeptides (mol. wt. > 20,000) on columns of Sephadex G-200 or agarose beads. Smaller peptides are not effectively separated, since the micellar weight (about 20,000, depending upon temperature, ionic strength and composition) of the detergent swamps small differences in the size of peptides complexed with dodecyl sulphate.

Recoveries of peptides from gel filtration columns are usually high, but some peptides may adsorb to the stationary phase, particularly if small amounts of peptides are chromatographed on large columns. Freshly packed columns of Sephadex may contain some actively absorbing sites, and pretreatment of the column by chromatography of a digest of serum albumin or other protein may be used to block these. Adsorption effects are less pronounced if solvents containing urea, pyridine, etc. are used.

The addition of a small amount (0.01% by volume) of thiodiglycol to elution buffers, to inhibit autoxidation of methionine, derivatives of cysteine, or tryptophan, is recommended (Harris, 1967). The use of buffers supporting microbial growth should be avoided wherever possible.

4.5.3. Packing of gel filtration columns

Gel filtration columns may be used many times and it is worthwhile to take every precaution in the packing of the columns. For detailed advice on the packing of columns Fischer (1980) should be consulted. The gel beads should be completely equilibrated with the elution buffer. Fines should be removed by several cycles of suspension in about ten settled bed volumes of the buffer followed by decantation of the supernatant after settling of the beads. Stirring of the suspension should be gentle, to avoid destruction of the beads.

Packing of beads of low porosity presents few problems, but the higher porosity gels, of lower mechanical strength (such as Sephadex G-150 and G-200 and BioGel P-150), require more careful handling. A layer, about 5 mm deep, of more rigid beads, such as Sephadex G-25 (medium) at the base of the column (Fig. 4.3) helps to prevent blockage of the column by flattening of less rigid beads or intercalation of fine beads into the supporting nylon mesh or glass beads.

Packing is aided by the use of an extension tube of the same diameter as the column, but a powder funnel is satisfactory. A slurry prepared from equal volumes of the buffer and of the settled gel is degassed under vacuum and carefully poured into the column, which must be exactly vertical. A steady flow of buffer is maintained during the packing. For the less rigid gels, the hydrostatic head during packing should not exceed about 30 cm, the outlet tube being raised to within this distance below the top of the slurry. When the packing is complete, the buffer is allowed to pass through the column until the bed height is constant. A disk of filter paper may be placed on the top of the bed to prevent disturbance during application of the sample.

4.5.4. Loading the sample, and elution

The mixture of peptides is dissolved in the elution buffer, subject to the conditions outlined above (§ 4.5.1). It is often observed that peptides which were soluble in, for example, dilute NH_4HCO_3 solution do not re-dissolve in the same solvent after freeze-drying. In such cases the peptides may be dissolved in 8 M urea solution for application to the column; once separation of the peptides has begun, the peptides generally remain in solution. Only small tryptophan-containing peptides are likely to emerge with urea from columns of Sephadex G-50, and these may be separated from the urea by the use of ion-exchange resins. Dense sample solutions should not be used with gels of high porosity, such as Sephadex G-200, since irregular flow, or even flotation of the gel beads above the sample solution, may result.

The peptide solution is carefully layered onto the gel bed and allowed to soak in (see also Fischer, 1980). The flask is rinsed with a small volume of the buffer, and the washings used to rinse the walls of the column. The buffer is then placed carefully above the gel bed, and the tubing from the reservoir containing de-gassed buffer is connected. A coloured marker, ε-*N*-dinitrophenyllysine, may be added to the column immediately after the sample. This compound is slightly adsorbed to Sephadex, and is eluted after the total column volume. It is unlikely that any peptides would be eluted so late from the column, unless hydrophobic prosthetic groups were present, and the yellow peak serves as a measure of the resolution of the column and a marker for the comparison of column runs.

Fractions of about 1/150 of the column volume are collected, at a flow-rate appropriate to the particular column packing. With most peptides, and particularly large peptides, higher resolution is obtained at low flow rates. Columns of high porosity gel beads may be eluted at 1–2 cm/h, while the less porous gels, for the separation of smaller peptides, may be eluted at 3 cm/h (i.e. 3 ml/h for a column 1 cm^2 in cross-section). These flow rates may be achieved by adjusting the hydrostatic pressure. With gels of low mechanical strength the head should never exceed about 30 cm. If preferred, the flow rate may be controlled by a peristaltic pump, preferably fitted at the inlet side. A reliable fraction collector is essential, preferably with capacity for at least 150 tubes. Time or volume may be used to signal fraction changes. Sophisticated equipment is commercially available, but gravity feed (made constant by means of a Mariotte flask) and timed collection of fractions is probably the most reliable combination. A solenoid valve may be used to shut off the flow if the electricity supply should fail; a peristaltic pump acts as a similar safeguard.

4.5.5. Recovery of peptides from fractions

Methods for the detection of peptides in column eluates are discussed in the next chapter; measurement of absorbance at 215 nm and 280 nm, scintillation counting if radioactive label has been incorporated, and, for small peptides, thin-layer chromatography of samples followed by detection with ninhydrin are usually the best methods. SDS-Polyacrylamide gel electrophoresis (§§ 4.3.1, 4.3.2) followed by staining with Coomassie blue is valuable for the detection of large polypeptides.

Fractions are combined according to the distribution of peptides. The resolution of long columns allows up to twenty sets of fractions to be collected; narrower selection of fractions to be combined results in greater overlap of peptides between adjacent sets, while broader cuts of the eluate lead to inefficient use of the available resolution. However, if a method of high resolution, such as ion-exchange chromatography on Dowex-type resins, is to be applied subsequently it may be more efficient to sacrifice some of the resolution obtained by gel filtration for a gain in simplicity in later separation steps.

The best method for the recovery of peptides is removal of the solvent by rotary evaporation or (preferably) freeze-drying. Freeze-drying from 50 mM NH_4HCO_3 is convenient, as the salt forms a stable matrix to which peptides are adsorbed during the freeze-drying process; the residual salt then sublimes steadily, leaving a fine mesh of peptide; complete removal of NH_4HCO_3 may be obtained after a further 24 h in a good vacuum dessicator over P_2O_5 and NaOH at room temperature. Freeze-drying from solutions containing no salt may lead to loss of peptide blown away in the stream of vapour. The vessel should be equipped with a cover of filter paper (e.g. Whatman No. 4) to prevent loss of material, even if NH_4HCO_3 is present. Ammonium acetate is less volatile than ammonium bicarbonate, but it may be removed by sublimation in a desiccator over P_2O_5 and NaOH pellets under high vacuum at 50 °C for 24 h.

Non-volatile components, such as urea, may be removed from large peptides by dialysis, or by gel filtration on short columns of Sephadex G-10 (or equivalent) in 0.1 M NH_3 or 5% formic acid. The removal of sodium dodecyl sulphate is less easy; the bulk of the detergent may be removed from large peptides (mol.wt. > 10,000) by dialysis against water, but a complex of the polypeptide and residual dodecyl sulphate may precipitate. Precipitation of polypeptides from the solution with 80% acetone and 0.1 M HCl at −10 °C may be successful, but some peptides remain soluble under these conditions. Alternatively, dialysis against 80% acetone may be used (Rice, 1974). The method of Weber and Kuter (1971; § 9.1) is effective, but time-consuming.

4.5.6. Examples of the separation of peptides by gel filtration

Some examples of the separation of subunits of proteins have been given in Chapter 2, and of large peptides derived by limited proteolysis in Chapter 3. Two examples of separation of mixtures of smaller peptides are presented in Figs. 4.4 and 4.5.

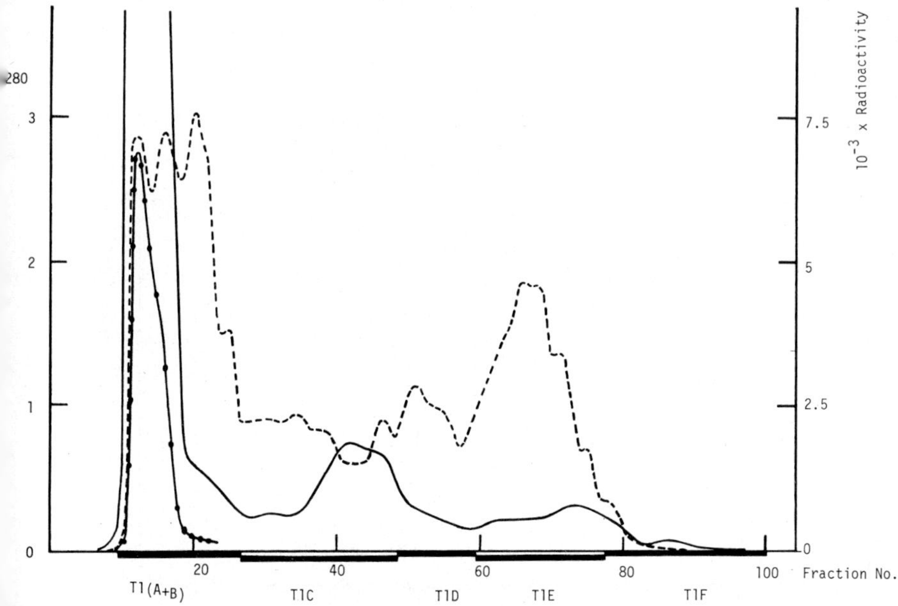

Fig. 4.4. Separation of peptides in the tryptic digest of succinylated, [^{14}C]carboxymethylated Ca^{2+}-ATPase of rabbit sarcoplasmic reticulum on a column (1.8 cm × 140 cm) of Sephadex G-50 (superfine). The digest (160 mg; approximately 1.4 μmol) was applied to the column in 20 ml of 50 mM NH_4HCO_3, and the column was eluted with 50 mM NH_4HCO_3/0.01% (v/v) thiodiglycol. Fractions of 3.5 ml were collected at 15 ml/h, and were assayed by absorbance and by liquid-scintillation counting of samples. (———), $A_{280}^{10\,mm}$; (—•—•—), $A_{280}^{2\,mm}$; (– – – –), radioactivity (10^{-3} × cpm in 10 μl). Fractions were combined for further separation of peptides as indicated by the bars at the base of the figure, labelled T1 (A + B), T1C, T1D, T1E and T1F. Reproduced with permission, from Allen, G. (1978), in FEBS 1977, Vol. *45*, Membrane Proteins (Nicholls, P., Møller, J.V., Jorgensen, P.L. and Moody, A.J., eds.), pp. 159–168, Pergamon Press, Oxford.

4.6. Ion-exchange chromatography

For the purposes of peptide separation it is convenient to classify ion-exchange materials into two groups, those based on hydrophilic matrices such as cellulose or cross-linked dextrans, and those based on cross-linked organic polymers, particularly polystyrene (Dowex-

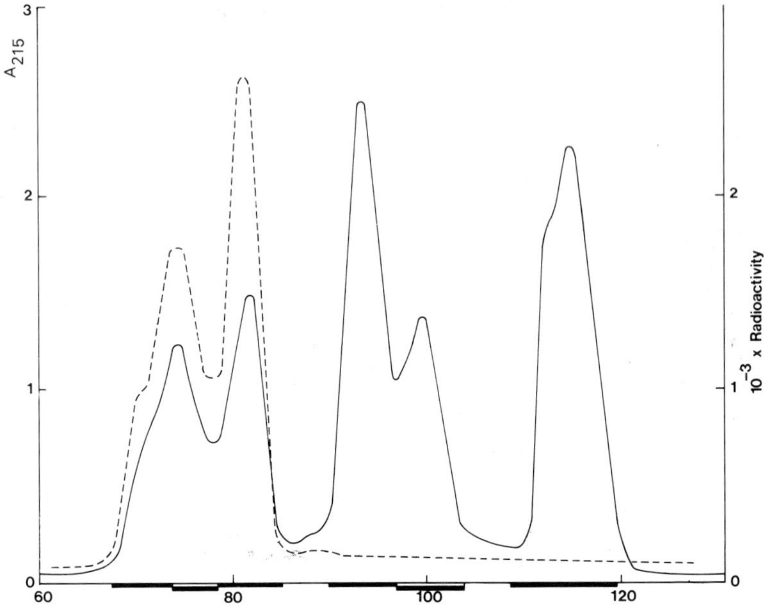

Fig. 4.5. Separation of peptides by gel filtration. A 31-residue peptide with a blocked amino-terminus was isolated from the tryptic digest of carboxymethylated, succinylated Ca^{2+}-ATPase of rabbit sarcoplasmic reticulum. The peptide (600 nmol; 2.0 mg) was digested with thermolysin and the digest was separated on a column (10 mm × 1430 mm) of Sephadex G-50 (superfine) in 50 mM NH_4HCO_3/0.01% (v/v) thiodiglycol. Fractions of 0.8 ml were collected at a flow rate of 5 ml/h. (———), $A_{215}^{10\,mm}$; (– – – –), radioactivity (10^{-3} × cpm in 20 μl). Fractions were combined for further purification on thin layers as indicated by the bars at the base of the figure. Reproduced with permission, from Allen, G. (1980) Biochem. J., *189*, 545–563.

type resins). The former group is useful for the separation of larger peptides (mol.wt. >1,000), while the latter group is capable of very high resolution of mixtures of peptides of lower molecular weight (< 1,500). Other types of ion-exchangers are commercially available, but their use in peptide fractionation for sequence analysis has been limited. As with other chromatographic materials, manufacturers regularly introduce advanced materials, and it is likely that higher resolution will be obtainable in the future. For the separation of peptides, ion-exchangers have been used almost exclusively in columns. A summary of suitable ion-exchangers is given in Table 4.3.

TABLE 4.3
Examples of ion-exchange materials used for the separation of peptides

Cation-exchangers	Type of peptides separated
Carboxymethyl-cellulose (CM-cellulose)	Large, mainly basic peptides
Carboxymethyl-Sephadex C-25 (CM-Sephadex)	Large, mainly basic peptides
Phosphocellulose	Large and small peptides
Sulphopropyl-Sephadex C-25 (SP-Sephadex)	Large and small peptides
Dowex 50W-X2, 50W-X4, Bio-Rad AG 50W-X2, Aminex 50W-X4, Dowex M71, etc.	Small peptides
Anion-exchangers	**Type of peptides separated**
Diethylaminoethyl-cellulose (DEAE-cellulose)	Large, mainly acidic, peptides
Diethylaminoethyl-Sephadex A-25 (DEAE-Sephadex)	Large, mainly acidic, peptides
Diethyl-(2-hydroxypropyl)aminoethyl-Sephadex (QAE-Sephadex A-25)	Large and small peptides
Dowex 1-X2, Bio-Rad AG1-X2, Durrum DA-X2	Small peptides

4.6.1. Cellulosic ion-exchangers

An earlier volume in this series deals with the properties and uses of cellulosic ion-exchange materials (Peterson, 1970). Two types of cellulose derivative have been used widely for the separation of peptides, the cation-exchanger carboxymethylcellulose (CM-cellulose) and the anion-exchanger diethylaminoethylcellulose (DEAE-cellulose). The pre-swollen microgranular forms (Whatman CM-52 and DE-52) are suitable. A beaded form, DEAE-Sephacel, has recently been introduced by Pharmacia, and may prove useful. Phosphocellulose is less frequently used, but has advantages when more acidic conditions than can be used with CM-cellulose are required.

Peptides are usually chromatographed using gradients of increasing ionic strength at constant pH. As for gel filtration, buffers should ideally be volatile, good solvents for peptides and have low UV-absorbance. The buffering capacity should be as high as is consistent with binding of the peptides to the stationary phase at the beginning of the elution. Linear gradients of NH_4HCO_3, pH 8, and ammonium acetate buffers, pH 5–6, are commonly used for chromatography on DEAE-cellulose and CM-cellulose, respectively. Triethylammonium or *N*-ethylmorpholinium acetate buffers are useful alternatives. Urea may be incorporated into the buffers if the peptides are not soluble. With peptides of mol.wt. $> 3,000$, the presence of 8 M urea usually results in better resolution and higher recoveries of peptides, by suppressing non-specific interactions of peptides with the cellulose matrix and with each other. The urea must be free from cyanate.

The choice of ion-exchanger is influenced by the properties of the peptides to be separated. Thus, peptides from a basic protein are more likely to be retained on the cation-exchanger, CM-cellulose, and separated by application of the gradient, while DEAE-cellulose is more useful for separating peptides from an acidic protein, especially one modified by succinylation. On the whole, the longer tryptic peptides from typical proteins will be anionic at pH 8, since they should contain only a single basic residue, the C-terminal Lys or Arg (His side-chains are not protonated at pH 8), while the probability of the presence of more than one acidic residue is high in long peptides. DEAE-cellulose is thus preferable for long tryptic peptides. The converse situation arises with peptides derived by cleavage at acidic residues, for example with *Staphylococcus aureus* V8 protease. For peptides derived by CNBr cleavage, cation-exchange chromatography at slightly acidic pH may be preferred, since the homoserine–homoserine lactone interconversion (§ 3.4.1) is less likely to interfere with the isolation. Similar considerations apply in the case of peptides produced by cleavage at tryptophan residues (§ 3.4.4).

4.6.1.1. DEAE-cellulose Microgranular DEAE-cellulose (Whatman DE-52) is supplied ready swollen, and precycling through 1 M HCl and 1 M NaOH is not required. The material should be washed with a strong solution of the eluting buffer, e.g. 1 M NH_4HCO_3, to remove any contaminants and for rapid equilibration. If buffers other than bicarbonate are to be used, degassing at acid pH should be performed to remove adsorbed CO_2. Equilibration with the starting buffer may be achieved rapidly by suspension in water of the DEAE-cellulose treated with the strong buffer solution, decanting the supernatant, and repeating until the ionic strength of the supernatant is identical with that of the starting buffer, which should be of low ionic strength, e.g. 10 mM NH_4HCO_3.

Suitable sizes of columns and gradient volumes for various weights of peptide mixtures are given in Table 4.4. Details suitable for the separation of about 10 mg of a mixture of peptides of molecular weights 1,500–3,000 are given here.

A suitable column, 6 mm × 300 mm, may be prepared by

TABLE 4.4

Column sizes and volumes of buffer solutions suitable for anion-exchange chromatography of mixtures of peptides on DEAE-cellulose

Weight of peptide mixture (mg)	Column size (diameter × height, mm)	Volume of each buffer component for preparation of a linear gradient (ml)	Fraction size (ml)
0.1– 1	3 × 300	50	0.5
1 – 20	6 × 300	100	0.7
20 – 100	10 × 300	300	2.0
100 –1000	20 × 500	1000	8.0

drawing out the tip of a calibrated 5 ml pipette. A small plug of non-absorbent cotton wool is placed loosely in the bottom of the column, and a small amount of the starting buffer is carefully drawn upwards through the outlet tube, avoiding the trapping of air bubbles. A slurry of DEAE-cellulose, prepared by suspension of the settled material with about two volumes of 10 mM NH_4HCO_3, is transferred to the column, with the use of a wide Pasteur pipette. The column may be packed by allowing the buffer to flow from the outlet under 30 cm–50 cm hydrostatic pressure; alternatively a slight nitrogen pressure may be applied. The column should be vertical and in a place free from thermal gradients during packing and chromatography. Repeated additions of the slurry are made until the packed bed is within 2 cm of the top of the column. The starting buffer, 10 mM NH_4HCO_3, with the addition of 0.02% (v/v) thiodiglycol, is then allowed to flow through the column (about 2 column volumes).

A linear gradient for elution of the column may readily be formed by the use of two identical cylinders, such as two narrow beakers, joined by a connecting siphon. One of the vessels is equipped with an efficient magnetic or mechanical stirrer. Electronically controlled gradient producing devices consisting of two or more reservoirs and a system of valves are commercially available. The column may be eluted under about 50 cm hydrostatic pressure, or piston or peristaltic pumps may be used. The simple column may be replaced by a commercially produced one with a nylon-mesh bed support and a water jacket. Some types of gradient-producing device and column assemblies are illustrated in Fig. 4.6. The buffers should be degassed, to prevent the formation of bubbles, both in the siphon of the gradient device and in the column. Detailed instructions for the preparation of gradients will be found in Peterson (1970).

The sample, dissolved in the starting buffer, or in this buffer made 8 M in urea, is applied carefully to the top of the column. Rinsings from the sample flask are added, and the sides of the column are carefully rinsed as the sample flows into the bed. The inlet tube is connected, and 10 ml of 10 mM NH_4HCO_3/0.02% thiodiglycol is passed through the column. Fractions of 0.7 ml are collected. A gradient prepared from 100 ml each of 10 mM NH_4HCO_3/0.02% thiodiglycol and 1 M NH_4HCO_3 is then applied to the column. It is important to avoid a sudden rise in the ionic strength at the start of the gradient. The meniscus in the mixing vessel should thus be slightly higher than that in the reservoir when the connecting tube is opened. With the apparatus shown in Fig. 4.6b, the gradient is started by temporary suction at the top of the Y-piece. A slight movement of the dilute buffer into the concentrated buffer does not significantly affect the separation of peptides.

A flow-rate of about 5 ml/h is used, and this may be obtained by gravity feed with a head of about 50 cm, depending upon the tightness of the cotton-wool plug and the bore of the outlet tube, which should be narrow (0.5 mm) to avoid loss of resolution. If desired, a peristaltic pump may be inserted between the mixing vessel and the column. Gravity-fed flow tends to decrease as the gradient progresses.

The eluate is monitored by absorbance at 220 nm; the use of thiodiglycol in the starting buffer serves to minimize the increase in absorbance through the gradient.

Few peptides are retained on the column beyond 0.5 M NH_4HCO_3, but the gradient is continued to ensure complete elution and to provide an absorbance baseline. An alternative elution schedule may be used, with a shallow gradient to start with, followed by a steeper gradient to elute the more firmly bound peptides.

For the separation of larger peptides, the same gradient may be used, but with the inclusion of 8 M urea in the buffers. Recovery of peptides, normally achieved by freeze-drying, then requires an additional gel filtration step for the removal of urea from each set of combined fractions comprising peaks in the chromatogram. A column of Sephadex G-25 (2 cm × 30 cm) in 0.1 M NH_3 is suitable for this purpose.

A typical result of the separation of peptides on a column of DEAE-cellulose is shown in Fig. 4.7.

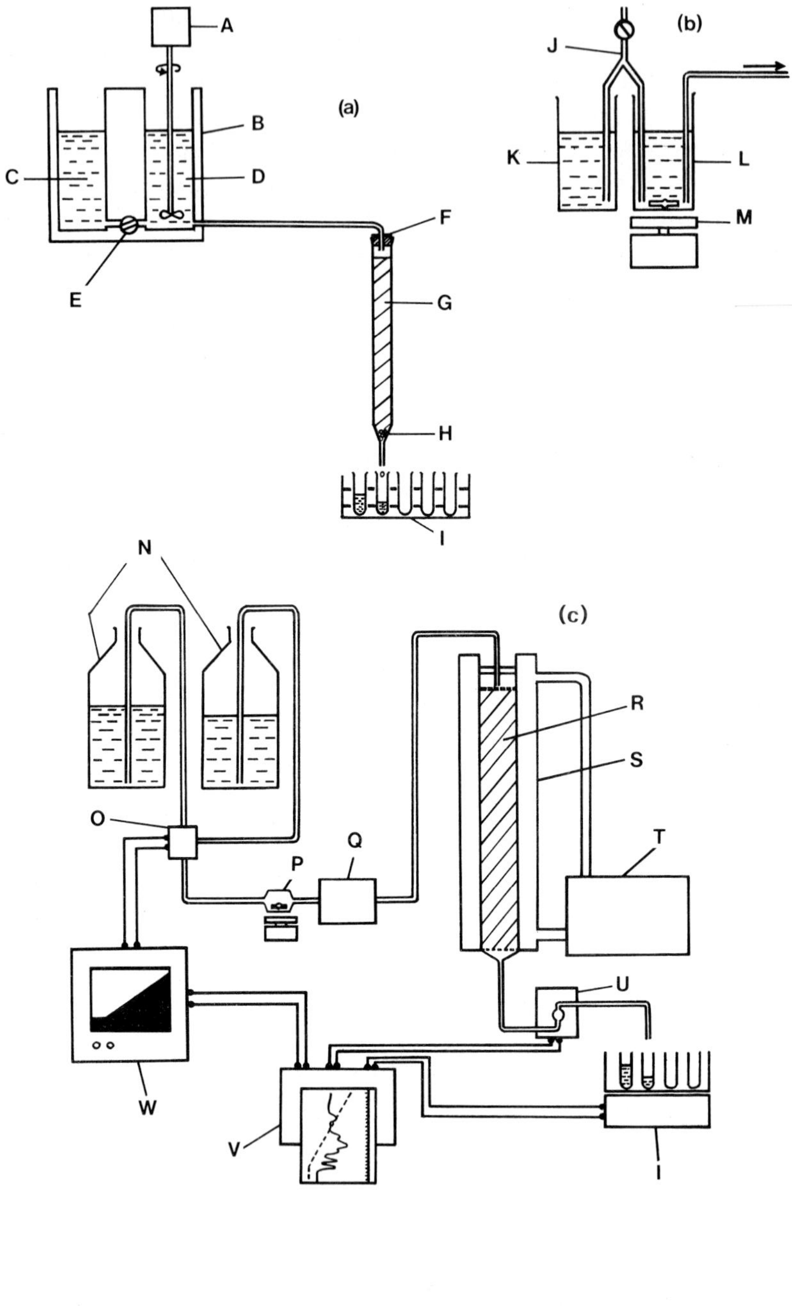
A
B
C
D
E
F
G
H
I
(a)
(b)
J
K
L
M
(c)
N
O
P
Q
R
S
T
U
V
W
I

4.6.1.2. CM-cellulose Details for the use of CM-cellulose are similar to those for DEAE-cellulose, with appropriate changes in buffer composition. Ammonium acetate buffers, pH 5–6 may be suitable; if urea is used for dissolving the peptides and achieving good resolution, a non-volatile salt gradient may be used instead, since a desalting step is in any case required for the recovery of each peptide. NaCl is suitable, allowing sensitive detection of peptides by UV-absorbance at 220 nm or below; a good buffering capacity is still required, however. An example of the separation of peptides by chromatography on CM-cellulose is given in Fig. 4.8, and the separation of domains of immunoglobulins has been given in the previous chapter (§ 3.2.1.3).

4.6.1.3. Phosphocellulose Phosphocellulose is a more strongly acidic cation-exchanger than CM-cellulose, and may be used at lower pH where peptides generally have a higher affinity for the charged matrix. Phosphocellulose is suitable for the separation of both large and small peptides, and high resolution may be obtained by elution with a non-linear gradient of ammonium acetate buffers, pH 4–6.3, followed by a wash at pH 9, or a gradient of pyridine–acetic acid buffers, pH 3.9–5.4 (Canfield and Anfinsen, 1963). These authors separated the chymotryptic peptides from 650 mg of reduced and carboxymethylated egg-white lysozyme on a 2.4 cm × 25 cm column

◁ Fig. 4.6. Apparatus for chromatography on columns of DEAE-cellulose. (a) Simple apparatus for the formation of linear gradients using gravity feed. (A) Stirring motor, (B) vessels machined from a perspex block, (C) buffer of high concentration, (D) mixing vessel, initially containing buffer of low concentration, (E) narrow-bore key, (F) silicone rubber stopper with hole for inlet tube, (G) DEAE-cellulose bed, (H) cotton-wool plug, (I) fraction collector. (b) Variant of (a), with mixing vessel (L) and reservoir (K) connected by a siphon (J). (M) Magnetic stirrer. (c) Sophisticated apparatus. (N) Buffer reservoirs, (O) proportional valve, controlled by an electronic gradient former (W), (P) mixing vessel, (Q) constant flow-rate pump, (R) DEAE-cellulose, supported on nylon mesh in a column with water jacket (S), (T) thermostated circulating water bath, (U) UV-absorbance flow cell, (V) multi-channel chart recorder.

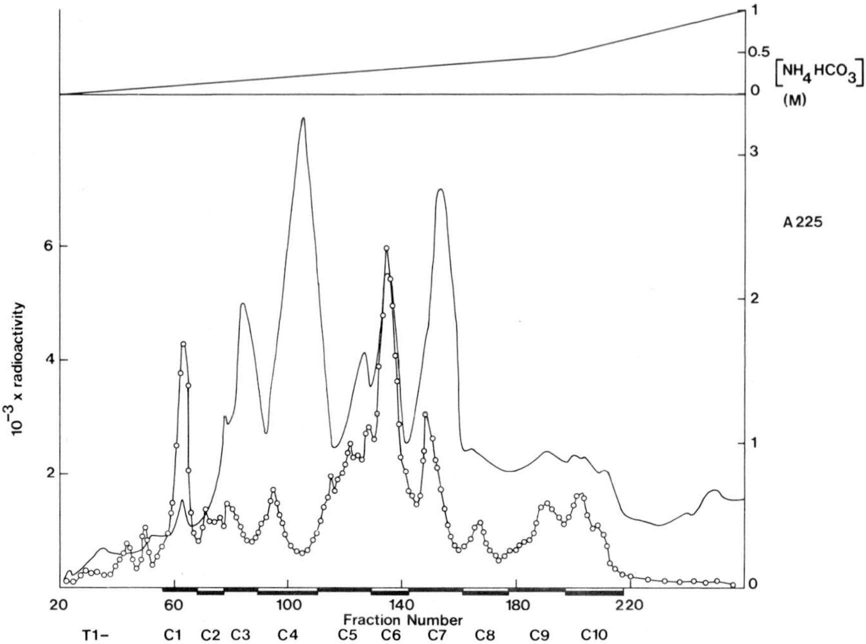

Fig. 4.7. Separation of peptides on DEAE-cellulose. A column of DEAE-cellulose (Whatman DE-52) (18 mm × 230 mm) was equilibrated with 10 mM NH_4HCO_3/0.01% (v/v) thiodiglycol. A peptide mixture (about 50 mg of peptides of 20–40 residues; fraction T1C in Fig. 4.4) was dissolved in 5 ml of this buffer made 8 M in urea, and applied to the column. The column was developed with 30 ml of 10 mM NH_4HCO_3/ 0.01% thiodiglycol, followed by two linear gradients of NH_4HCO_3 in 0.01% (v/v) thiodiglycol as indicated in the upper section of the figure (total volume 1.3 litre). Fractions of 5 ml were collected at 40 ml/h. $A_{225}^{10\,mm}$, ———; 10^{-3} × radioactivity (cpm in 0.05 ml), (○). Fractions were combined as indicated by the horizontal bars, and peptides were isolated from the combined fractions T1-C1 to T1-C10 by further steps of column chromatography. Reproduced with permission, from Allen, G. (1980) Biochem. J., *187*, 545–563.

of phosphocellulose at 4 °C. Fractions of 10 ml were collected at 60 ml/h, with a total 7.5 litres of eluate. Peterson et al. (1975) used gradients of KCl in 25 mM H_3PO_4, pH 2.5, to separate peptides from a dihydrofolate reductase protein, with detection of peptides

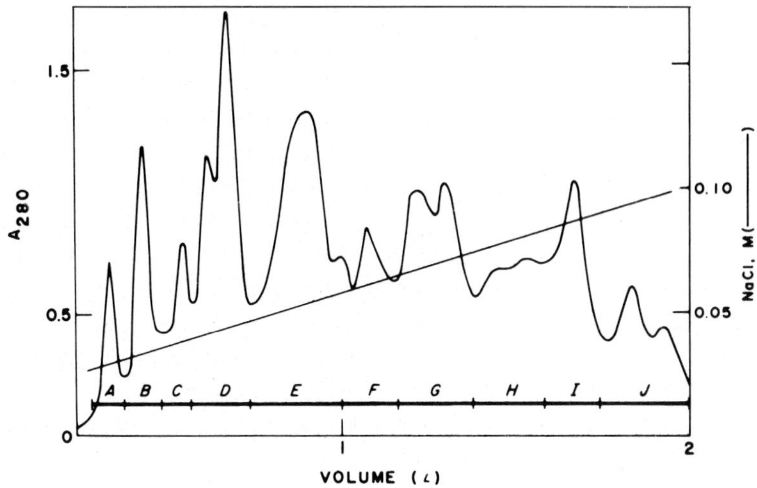

Fig. 4.8. Chromatography of peptides on CM-cellulose. Cyanogen bromide peptides of β-galactosidase were applied to a column (2.5 cm × 38 cm) of CM-cellulose in 0.02 M ammonium acetate, pH 5.0, and 8 M urea, and were eluted with a linear gradient of 0 to 0.15 M NaCl in the same buffer. The total volume was 3000 ml. The eluate was divided into 10 pools indicated by the letters A to J. Urea was removed from the pooled fractions by dialysis against 10% acetic acid using Spectrapor 3 dialysis membrane. This membrane, previously boiled in 10 mM EDTA, pH 8.0, and extensively washed, retained peptides as small as 7 residues. Reproduced with permission, from Fowler, A.V. (1978) J. Biol. Chem. *253*, 5499–5504.

by far UV-absorbance, but desalting steps were required for each peptide.

Phosphocellulose may be considered as an alternative to the cross-linked polystyrene derivatives, and although not giving such high resolution it possesses the advantage that extremely hydrophobic peptides which may absorb essentially irreversibly to the Dowex-type exchangers may be separated. However, this cellulose derivative has been used less frequently than the polystyrene derivatives.

4.6.2. Ion-exchange derivatives of cross-linked dextrans

Ion-exchange derivatives of Sephadex bearing strong or weak acidic or basic groups are available from Pharmacia. Carboxymethyl-

Sephadex C-25 and diethylaminoethyl-Sephadex A-25 may be used as alternatives to CM-cellulose and DEAE-cellulose, respectively, under similar conditions. The materials are subject to considerably greater volume changes with ionic strength and pH than are the cellulose derivatives, but the spherical form of the Sephadex beads may enhance the resolution obtainable. The more strongly acidic cation-exchanger, sulphopropyl-Sephadex C-25, is closer in its properties to phosphocellulose, but with a lower capacity, and may be used for the separation of both large and small peptides at low pH. Peptides may bind firmly to sulphopropyl-Sephadex at pH 3, and high salt concentrations may be required for elution of all peptides, e.g. 2 M Na^+ in phosphate buffers. An example of the use of sulphopropyl-Sephadex is the purification of the 31-residue phosphorylated active-site peptide from rabbit sarcoplasmic reticulum Ca^{2+}-ATPase (Allen and Green, 1976).

The strongly basic anion-exchanger, diethyl-(2-hydroxypropyl)-aminoethyl-Sephadex (QAE-Sephadex A-25) may be used at a higher pH than the DEAE derivatives (maximum pH about 9), and is more suitable than these for the separation of medium-sized neutral peptides. Volatile buffers, such as triethylamine–acetic acid or triethylamine bicarbonate, pH 9, may be used for the elution.

A booklet published by Pharmacia gives details of the use of the Sephadex ion-exchangers.

4.6.3. Ion-exchange derivatives of cross-linked polystyrene

Anion-exchange and cation-exchange resins of the Dowex type (Dowex-1 and Dowex-50, respectively) have long been used for the separation of peptides. Columns packed with these resins have high capacity and give high resolution. The high capacity was of great importance in earlier work when the methods of primary structural analysis of small peptides were far less sensitive than those now available. The tendency towards higher sensitivity has led to the use of very small columns of these ion-exchange resins, and the resolution and recovery of peptides have been improved through the use of smaller and more regularly sized analytical resin beads, as in the comparable separation of amino acids on the amino-acid analyzer.

The separation is based on both the charge and the polarity of peptides. Hydrophobic amino-acid residues, especially those with aromatic groups, interact strongly with the resins. This interaction, which leads to trailing of peaks and reduced yields, is suppressed by the use of organic bases, usually pyridine, in the elution buffers, or by the use of an organic co-solvent such as propanol. Elevated temperatures, up to 60 °C, are also used to increase the resolution. The accessibility to peptides of charged groups on these resins is a function of the degree of cross-linking of the polystyrene matrix. While Dowex X8 resins are suitable for amino-acid analysis, the more porous Dowex X2 types are more suitable for the chromatography of peptides; even the latter resins often do not resolve satisfactorily peptides of more than 20 residues. Thus the polystyrene ion-exchange resins are only suitable for the separation of small peptides, and complement the cellulosic derivatives.

A wide variety of elution gradients have een employed; the optimal conditions for the separation of peptides in a particular protein digest can only be determined from a number of trial separations. For practical purposes in primary structure determination, it is more efficient to employ standard conditions, which may be sub-optimal. If it should prove necessary to repeat the separation, the gradient may be modified, to be more shallow, for example, where many peptides are eluted close together. For the separation of peptides from a protein which is similar to one which has previously been investigated it will often be advantageous to use identical conditions to those used in the earlier work, since comparison of the elution positions of such peptides will be valuable. Allowance should be made for slight differences due to the use of different batches of resin.

A commonly encountered situation is that in which most peptides from a digest are suitably eluted by means of a shallow gradient, but a few peptides remain firmly bound to the resin and are only eluted at much higher ionic strengths or different pH values. This problem may be overcome by the use of a second, steep gradient after the completion of the shallow gradient, or a concave gradient may be constructed from suitable glassware or electronic or electro-mechanical gradient-producing apparatus (from Pye-Unicam, LKB, or other suppliers). Alternatively, the few firmly bound peptides may be eluted together in a wash of high ionic strength (and high pH for cation-exchangers or low pH for anion-exchangers), and separated by some other technique, such as paper chromatography.

Detection of peptides in the eluate by UV-absorbance is not usually feasible, since the pyridine-containing buffer solutions absorb strongly. Colorimetric, or the more sensitive fluorometric, methods have frequently been used. These are most satisfactory after alkaline hydrolysis of aliquots of each fraction collected. Details of suitable detection methods using ninhydrin, fluorescamine or *o*-phthalaldehyde are given in the following chapter. Liquid-scintillation counting of radio-labelled peptides is a sensitive quantitative technique. Thin-layer chromatography or electrophoresis of aliquots from each fraction gives information about the complexity of the mixture in each peak and an indication of which subsequent purification techniques are likely to be useful. Many methods for the detection of peptides on thin layers are available.

Detection of peptides in the eluate may be automated by means of a system similar to that in amino-acid analyzers. Buffers containing ammonia, primary or secondary amines may not be used with such systems. A small proportion of the eluate from the column is diverted from the stream passing to the fraction collector, mixed with NaOH solution and heated in a reaction coil for partial hydrolysis of the peptides. The solution is then neutralized and mixed with a stream of ninhydrin solution. After passage through a second reaction coil at 90 °C for 20 min, the absorbance at 570 nm is monitored using a flow-cell photometer and chart recorder. Accurate calibration is required to correlate the fractions collected with the absorbance detected, after a considerable time delay, from the same portion of the eluate. Fluorometric detection with fluorescamine or *o*-phthalaldehyde may be used similarly. Details of the construction of such peptide analyzers have been given by Hill and Delaney (1967),

Herman and Vanaman (1977) and Benson (1976). Since it is unlikely that frequent use of such ion-exchange chromatography would be made during the determination of the sequence of a single protein, the construction of an automated detection system will not often be worthwhile. However, for work which requires repeated identical chromatographic separations, for example in the comparison of highly homologous proteins, such a system may be useful for both preparative and analytical purposes.

Two simple but effective gradients for ion-exchange chromatography of small peptides on Dowex-type resins are described below. Full details of some alternative procedures are given by Schroeder (1972b), and some examples of the use of these resins are listed in Table 4.5.

TABLE 4.5

Some examples of the use of polystyrene ion-exchange resins for the separation of peptides

Margoliash and Smith (1962). Chymotryptic peptides from horse heart cytochrome *c* were separated on a 1-litre column of Dowex 50-X2 in a gradient of pyridine/acetic acid buffers, pH 3.1 to pH 5, followed by a wash with 3M trimethylamine.
Heiland et al. (1976). A micro-column of Dowex M71 cation-exchange resin was developed with three successive gradients of pyridine-formic acid buffers, pH 2.6–3.3, for the separation of tryptic peptides from ribosomal protein L10 of *Escherichia coli*.
Herman and Vanaman (1975, 1977) described the separation of peptides on small columns of small bead-size cation-exchange resins in pyridine-acetic acid gradients.
Benson (1977) described the separation of peptides on highly cross-linked cation-exchange resins using citrate buffers, similar to those used for amino acid analysis.
Van Eerd et al. (1978) purified peptides by anion-exchange chromatography on Durrum DA-X2 and cation-exchange chromatography on Technicon Chromobeads-P during the determination of the sequence of troponin C from frog skeletal muscle.

4.6.3.1. A standard procedure for chromatography on Dowex-50-type resins The following procedure has given excellent results for the separation of small (2–10 residues) tryptic peptides from a trypanosome variant glycoprotein. The column size is suitable for 100–500 nmol of digest. All buffer components must be highly purified.

Three grams of Aminex 50 W X4 resin (20–30 μm beads, Na^+ form, from BioRad) were converted to the pyridinium form by suspension in 50 ml of 1 M HCl, decanting the supernatant, washing with water in a Buchner funnel, and suspending in 4 M pyridine. The beads were then washed with water.

Three buffers were prepared:

(1) 0.05 M pyridine, adjusted to pH 2.8 with formic acid;

(2) 0.2 M pyridine, adjusted to pH 3.1 with acetic acid (1.61 ml pyridine and 27.9 ml acetic acid in 100 ml);

(3) 2.0 M pyridine, adjusted to pH 5.0 with acetic acid (16.1 ml pyridine and 14.32 ml acetic acid in 100 ml).

Each buffer was made 0.01% (v/v) in thiodiglycol, and the buffers were stored under N_2 in the dark at 4 °C.

The resin was equilibrated with buffer 1 by 3 cycles of suspension in 10 ml buffer and decantation of the supernatant after settling of the beads. The resin was suspended in 10 ml of buffer 1 and packed under pressure (1 atm. N_2) into a column, 2.5 mm i.d. × 250 mm (prepared from a graduated 1 ml pipette fitted with a water jacket at 45 °C), with a cotton-wool plug and a layer of fine washed sand as a bed support. 10 ml of buffer 1 was passed through the column.

The sample was dissolved in 0.3 ml of buffer 1 and 0.02 ml formic acid was added. The sample solution was applied to the column and carefully washed in, using a slight pressure of nitrogen. The column was developed at 45 °C with a linear gradient prepared from 10 ml each of buffers 1 and 2, followed by a gradient prepared from 50 ml each of buffers 2 and 3, using a gradient mixer of the type shown in Fig. 4.6a. Finally the column was washed with 10 ml of buffer 3 made 0.5 M in *N*,*N*-dimethylallylamine. Fractions of 0.5 ml were collected at a flow rate of 2.5 ml/h. A peristaltic pump was used on the inlet side of the column.

Peptides were detected in the eluate by thin-layer chromatography of samples from each fraction (§ 5.1.7). The resolution was high:

each peptide was eluted in a narrow peak, contained in two or three fractions. Almost all the peptides were eluted during the second gradient.

4.6.3.2. Chromatography on strongly cationic anion-exchange resins
Herman and Vanaman (1977) described the separation of peptides on a column of Durrum DA X8–11 anion-exchange resin (11 μm bead size). The column (3 mm × 210 mm) was equilibrated in 0.36 M pyridine (30 ml pyridine/litre), pH 9.1. The sample was applied in this solution, and the column was eluted with a linear gradient prepared from 30 ml of this solution and 30 ml of 2.0 M pyridine adjusted to pH 5.0 with acetic acid at 55 °C. With such small beads the back pressure is high, and columns, pumps and fittings similar to those used for amino acid analysis were required. Peptides were detected automatically in the eluate by a ninhydrin system.

4.7. Paper electrophoresis

Paper electrophoresis has been used frequently for the purification of peptides for sequence analysis. The increasing sensitivity of analytical techniques is leading to the displacement of paper methods by thin-layer methods, but paper methods are still valuable, especially when electrophoresis and chromatography are combined in preparative peptide mapping (§ 4.9). Both paper and thin-layer methods are limited to the separation of peptides of less than 20 residues, although some larger peptides may behave satisfactorily also.

The apparatus required for high-voltage electrophoresis has been described in detail in an earlier volume in this series (Brownlee, 1972). Two types of apparatus are illustrated in Fig. 4.9. Efficient cooling of the paper is required, and this may be achieved either by immersion of the paper in a tank containing a non-conducting inert solvent, in which peptides are not soluble, such as white spirit, and fitted with coils through which cooling water flows (Fig. 4.9a), or by close contact of the paper with insulated metal plates, also cooled by water (Fig. 4.9b). The advantages of the cooled plate

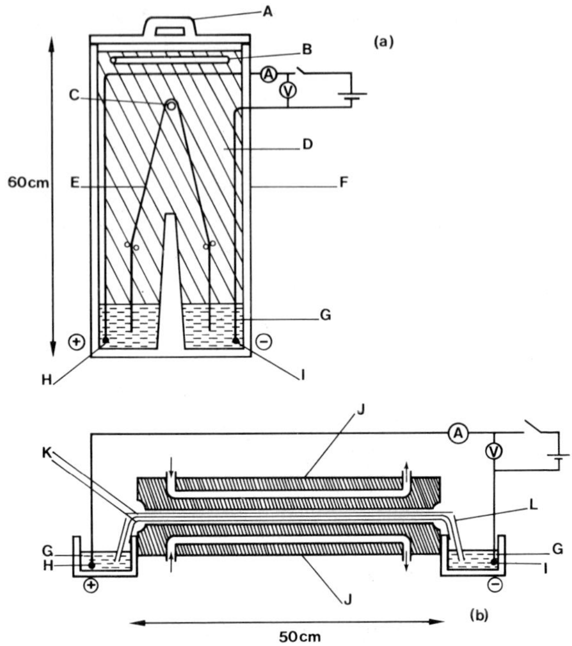

Pig. 4.9. Apparatus for paper electrophoresis. (a) Solvent tank design (cross section); (b) cooled plate design (cross section). (A) Lid, fitted with safety cut-out switch, (B) cooling coils, with tap water and a pressure switch safety device, (C) plastic support for paper, (D) inert solvent (white spirit), (E) paper, (F) glass or glass fibre tank, (G) electrophoresis buffer, (H) anode, (I) cathode, (J) metal cooling plates, with tap water cooling and safety cut-out device, (K) insulating polyethylene sheets sandwiching paper, (L) wicks, made from two thicknesses of Whatman No. 3MM paper. Designs after those of Michl (1951, 1952). In an alternative solvent tank apparatus, the cathode buffer compartment is higher than the anode compartment, so that the hydrostatic flow partially compensates for electroendosmosis (Ryle et al., 1955).

design are that a large volume of organic solvent, presenting serious fire risks, is not required, the electrode buffers are more readily exchanged, and there is no loss of sample or buffer components, such as pyridine, into the coolant. Disadvantages are that careful maintenance is required; small particles of hard materials may penetrate the insulating layers and metal parts may be corroded by acidic buffers. Assembly of the wicks connecting the paper to the buffer vessels is not quite so straightforward as is loading the papers into the solvent tanks.

The apparatus should be large enough to accommodate a full sheet (46 cm × 57 cm) of chromatography paper. A direct current power supply with an output of at least 200 mA at 3 kV is required. Such potentials are lethal, and reliable safety devices are needed, including the provision of an inert atmosphere in case of fire if organic solvents are used. Large quantities of solvents, including noxious materials such as pyridine, are also used, and efficient fume extraction is necessary. The laboratory should be fitted with a glass-topped bench for applying the samples and wetting the papers, drying ovens, fume-hood space for dipping and spraying papers with detection reagents, and a sewing machine with cross-stitch capability for joining excised strips of paper, loaded with peptides, to fresh sheets for further separation. The laboratory should preferably be ventilated with filtered air. These requirements make the setting-up of a laboratory for paper electrophoresis rather expensive.

For preparative purposes, volatile buffer solutions are required. Examples of commonly used solutions are given in Table 4.6. For some purposes, such as the separation of two similar peptides, one of which contains a histidine residue, alternative buffers, prepared

TABLE 4.6

Commonly used volatile buffer solutions for paper electrophoresis

Nominal pH	Composition (by volume)
1.9	Acetic acid/formic acid/water (8 : 2 : 90)
3.5	Pyridine/acetic acid/water (1 : 10 : 189)
4.4	Pyridine/acetic acid/water (2 : 4 : 94)
6.5	Pyridine/acetic acid/water (10 : 0.5 : 89.5)
8.9	1% (w/v) ammonium carbonate

To these buffers may be added 0.1% (v/v) thiodiglycol to protect against autoxidation of methionine and carboxymethyl cysteine residues. All buffer components must be highly purified.

from different proportions of acetic acid and pyridine, for example, may be used. It is necessary to reserve separate tanks for each buffer if the solvent-cooled apparatus is used. The buffers may be used for several weeks between replenishment, but products of electrode reactions and oxidation of pyridine gradually develop, and frequent renewal is recommended.

Peptides may migrate in the electric field either towards the anode or towards the cathode, depending upon the pH and the structures of the peptides. At pH 1.9, only peptides lacking a free amino group or containing cysteic acid residues migrate towards the anode; the sample is therefore applied close to the anode. At pH 6.5, the sample is applied close to the middle of the paper. Suitable positions for the application of samples for one-dimensional separations are shown in Fig. 4.10. These positions may be varied if the electrophoretic properties of the peptides are known, to allow the optimum separation. The line of application of the sample, points of application

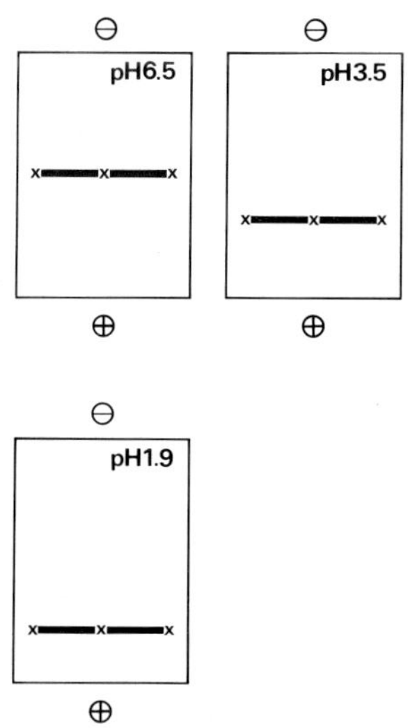

Fig. 4.10. Positions for the application of samples to a full sheet of paper for electrophoresis at pH 6.5, pH 3.5 and 1.9. ×, points of application of a marker mixture (ε-*N*-dinitrophenyllysine and xylene cyanol FF, and amino acids if desired).

of the markers, the anodic and cathodic ends of the paper and other information are marked on the paper with a soft graphite pencil.

The paper is placed on the clean glass surface with the origin line raised over two glass rods, as shown in Fig. 4.11. All handling of the paper should be done with the protection of plastic gloves. The sample is dissolved in the buffer (10–50 μl for each cm-width of Whatman 3MM paper). If the sample is not completely soluble, the insoluble material may remain at the origin or may partially dissolve during the electrophoresis, giving streaked-out bands of peptides. In either case, the insoluble material will interfere with the separation, and should be removed by centrifugation and purified separately by other means. The presence of salts in the sample is seriously detrimental; thorough removal of NH_4HCO_3 in vacuo over NaOH and P_2O_5 is required after freeze-drying from even dilute solutions of this salt. Small amounts of urea are, however, acceptable, although the urea will not be separated from uncharged peptides. About 10 nmol of each peptide in a mixture of up to 30 peptides may be applied per cm-width to Whatman No. 1 paper, and up to five times this amount to Whatman No. 3 MM paper. These two types of paper are recommended. Higher resolution is obtained at lower levels of loading, but recoveries are often lower.

The solution of the sample is applied from a small Pasteur pipette or a capillary pipette as a regular band, 5 mm–20 mm wide, along the origin line. Dye markers (ε-*N*-dinitrophenyllysine and xylene cyanol FF, and others if desired [Stevenson, 1971]) and amino acid markers (10 nmol each) are applied at the edges and in the middle of the origin line (Fig. 4.10). The sample and markers are dried with a stream of warm air. Hot air, or too thorough drying, should be avoided, since the sample may become tightly adsorbed to the paper by such treatment. The whole paper is wetted with the running buffer, applied carefully with a pipette to each side of the sample zone, so that the buffer front concentrates the sample evenly along the origin line (Fig. 4.11). Excess solvent is removed from the paper by blotting with clean sheets of chromatography paper, until no more liquid is removed and the paper has an evenly damp appearance.

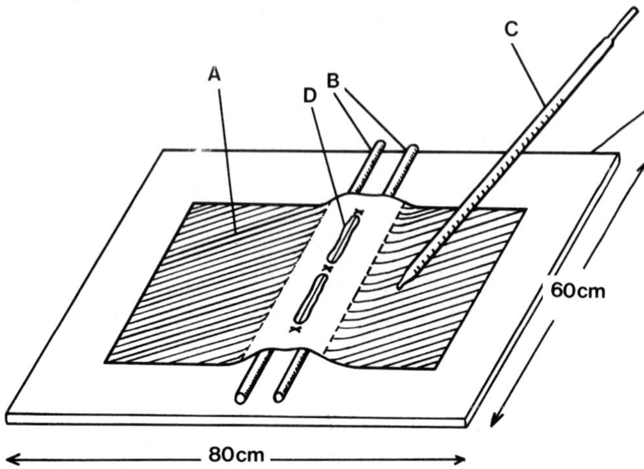

Fig. 4.11. Wetting the paper for electrophoresis. (A) Paper saturated with electrophoresis buffer, applied from a pipette (C) on each side of the sample zone (D), which is supported on two glass rods (B), 8 mm diameter on a glass plate (E).

Excessive wetness leads to broadening of the zones during electrophoresis. When wet, the thinner Whatman No. 1 paper is easily torn. Transfer of the paper to the electrophoresis apparatus is aided by the support of the glass rods.

Electrophoresis is performed for 40 min at 3 kV (60 V/cm) at about 25 °C. If the mobilities of the peptides are known from trial separations to be low, longer runs, e.g. 2 h, may be performed. The paper is then lifted from the apparatus with the support of glass rods, and clipped to a glass or stainless steel rod for drying in a cabinet at about 40 °C with a slow upward flow of clean air.

The zones of peptides are detected on the paper by one or more of the methods described in the following chapter. The mobility, relative to that of aspartic acid or some other marker is recorded. Chromatography paper bears a slight negative charge, due to the presence of some carboxyl groups on the cellulose matrix; this leads to the bulk movement of the neutral components of the buffer solution towards the cathode (electroendosmosis). The relative mobilities are determined not from the point of application of the sample, but from the position of a suitable uncharged marker, such as a neutral amino acid at pH 6.5.

The strips of paper containing each separated peptide are cut out (with pointed tips as shown in Fig. 4.12a), and the peptides are eluted with a suitable solvent, e.g. 10% acetic acid, 20% pyridine or 0.1 M NH_3, as shown in Fig. 4.12b. The square-cut end of the paper

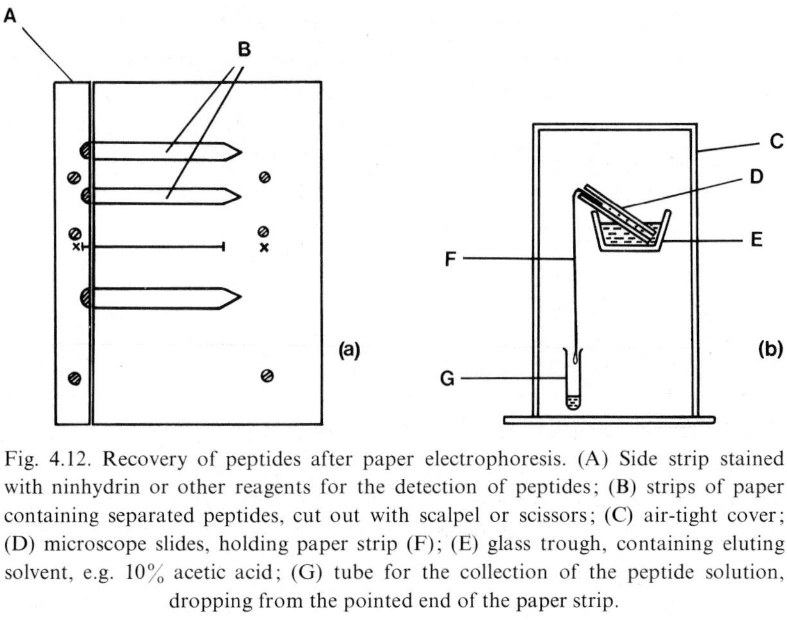

Fig. 4.12. Recovery of peptides after paper electrophoresis. (A) Side strip stained with ninhydrin or other reagents for the detection of peptides; (B) strips of paper containing separated peptides, cut out with scalpel or scissors; (C) air-tight cover; (D) microscope slides, holding paper strip (F); (E) glass trough, containing eluting solvent, e.g. 10% acetic acid; (G) tube for the collection of the peptide solution, dropping from the pointed end of the paper strip.

strip is held between two microscope slides dipping into a trough containing the eluting solvent, and the pointed end is inserted into the top of a rimmed collection tube, held in an adjustable stand. The assembly is covered to prevent evaporation. The solvent moves slowly through the pores of the paper by capillarity, and the peptide is often eluted in the first few drops from the tip. Larger volumes are usually collected, e.g. 1 ml from 10 cm^2 of Whatman No. 3 MM paper, to ensure high yields. If some peptides remain at the origin during electrophoresis, due to interaction with the cellulose, better recoveries may sometimes be obtained by elution with 50% pyridine.

The extraction solvent is removed in a vacuum desiccator over an appropriate drying reagent (NaOH pellets for acetic acid solution, P_2O_5 for pyridine or NH_3 solutions).

If the zones do not consist of pure peptides, a second method of separation, either electrophoresis at a different pH or chromatography, may be performed without first eluting the peptides. The strip of paper containing the peptides is cut out and sewn, using pure cotton thread, into position on the starting line for the second separation on a fresh sheet of paper. The area of paper below the sample on the second sheet is excised, allowing an overlap of about 2 mm–5 mm for the unrestricted migration of the peptides across the joins, and electrophoresis is performed after application of the buffer as described above, or chromatography as described below.

4.8. Paper chromatography

Although thin-layer chromatography has largely replaced paper chromatography, the latter technique is still of value, particularly for the purification of larger quantities of peptides. The development of the theory and technique of partition chromatography on paper (Consden et al., 1944) was historically of great importance. A still useful account of the method was given by Block et al. (1955). To a first approximation, the separation is based on the relative solubilities of peptides in two phases, a polar stationary phase consisting mainly of water bound to the cellulose fibres, and a less polar organic solvent phase, usually partially aqueous, which flows slowly through the capillary channels between the fibres. Hydrophobic peptides, which are more soluble in the moving phase, have relatively high R_F values, while polar, especially charged peptides have low R_F values. Changes in the acidity of the solvent change the relative rates of migration of peptides.

Although a large number of solvent systems have been used for paper chromatography (more than seventy are listed by Block et al.,

1955), only a few are suitable for routine separations of peptide mixtures (Table 4.7); *n*-butanol/acetic acid/water/pyridine (15 : 3 : 12 : 10, by vol.) and *n*-butanol/acetic acid/water (3 : 1 : 1, by vol.) are especially useful. Glycopeptides, which typically have very low R_F values, may be chromatographed in the more polar solvent *n*-propanol/water/NH_3 (sp.gr. 0.88) (30 : 10 : 1 by vol.).

TABLE 4.7
Some solvents for paper chromatography of peptides

1-butanol/acetic acid/pyridine/water (15 : 3 : 10 : 12, by vol.)	(Waley and Watson, 1953)
1-butanol/acetic acid/water (3 : 1 : 1, by vol.)	(Ambler, 1963)
1-butanol/acetic acid/water (4 : 1 : 1, by vol.)	(Block et al., 1955)
2-butanol/3% NH_3 (3 : 1, v/v)	(Block et al., 1955)
2,4,6-collidine/water (125 : 44, v/v), in an atmosphere of conc. NH_3	(Block et al., 1955)
m-Cresol, saturated with water, in an atmosphere of 0.03% NH_3	(Block et al., 1955)
Phenol, saturated with water, with 0.002% 8-hydroxyquinoline	(Block et al., 1955)
iso-amyl alcohol/pyridine/water (35 : 35 : 30, by vol.)	(Wittmann and Braunitzer, 1959)
1-butanol/pyridine/water (1 : 1 : 1 by vol.)	(Balian et al., 1971)

It is essential that pure solvents are used. Some observations of losses of aromatic amino acids after hydrolysis of peptides purified by paper chromatography have been reported (Furth et al., 1974), but impurities, such as aldehydes, may be responsible for such losses, since this is not a general phenomenon. In particular, mixtures of pyridine and butanol kept in air and light are likely to be degraded, and fresh solvents, stored in the dark, should be used. The addition of a trace of mercaptoethanol aids the isolation of peptides containing methionine or cysteine derivatives by preventing autoxidation.

The slow flow of moving phase is achieved by capillary action, either with an ascending or with a descending solvent front. Descending chromatography (Fig. 4.13) is usually preferred; equi-

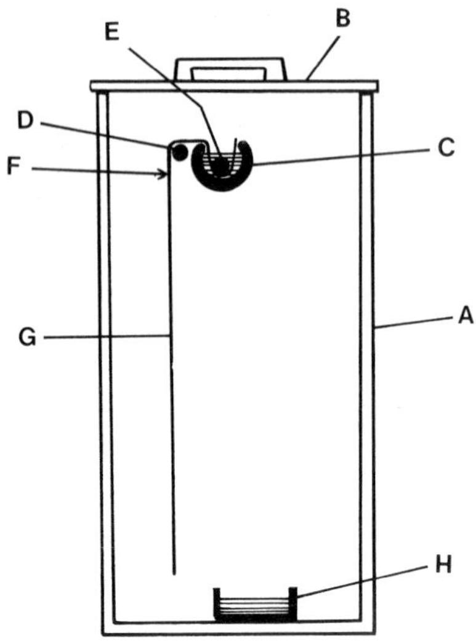

Fig. 4.13. Apparatus for descending paper chromatography (idealised cross-section). (A) Glass tank, 60 cm high × 60 cm wide × 30 cm deep (approx.), (B) closely fitting lid, (C) glass or stainless steel trough, containing solvent, (D) support rod of glass or stainless steel, (E) glass rod holding paper in trough, (F) position of application of sample, (G) paper, (H) dish containing chromatography solvent. Several troughs may be used in the same tank.

libration of the atmosphere in the tank and loading of the paper are more straightforward than with the ascending mode.

The paper should be neither too wet nor too dry before chromatography. Equilibration with atmospheric humidity typically in the range 40–70% is satisfactory. The vapour phase in the tank should be equilibrated with the liquid, using a container placed in the base of the tank, and it is essential to maintain a constant temperature, avoiding sunlight or draughts; even a slight temperature gradient across the tank will lead to uneven development.

The sample is applied evenly in a band across the paper, 5 mm–10 mm wide, and about 12 cm from one end of the paper. The edges of the paper (5 cm wide) are not used, since zone distortion may occur close to the edges. The loading is the same as that for paper electrophoresis. It is important that the sample be free from salt and soluble in the solvent (typically 20% (v/v) aqueous pyridine or 30% (v/v) acetic acid) used for the application to the paper. The solvent is removed in a stream of cool air. Suitable dye markers (e.g. xylene cyanol FF and ε-*N*-dinitrophenyllysine) may be applied at the edges and in the middle of the loading zone to aid in the comparison of chromatograms and to check on even development. The top of the paper is folded to fit into the solvent trough and over the supporting rod (Fig. 4.13), and the paper is held in position by a heavy glass rod in the trough. The solvent (about 150 ml for a full sheet, 46 cm × 57 cm, of Whatman No. 3MM paper) is poured carefully into the trough. As an aid to equilibration of the atmosphere in the tank, a sheet of paper lining the tank may be saturated with the solvent. The lid must be close fitting.

Chromatography is allowed to continue until the solvent front reaches the lower end of the paper (about 16 h with the butanol solvents, depending upon the temperature and type of paper). If the components to be separated are known to have low R_F values, the chromatography may be continued by allowing the solvent to drip from the end of the paper. To ensure an even flow of solvent across the width of the paper, the lower edge should be serrated with pinking shears. The paper is removed from the tank, supported on the glass or stainless steel rod, and dried in an upward flow of filtered air at 40 °C for 2 h. Some solvents, such as those containing phenol, are only removed effectively at a higher temperature.

The zones of peptides are located by methods described in Chapter 5, and the peptides are recovered from the zones as described above for paper electrophoresis.

4.9. Peptide mapping on paper

Good resolution of mixtures of up to fifty peptides may be obtained on a single sheet of paper by a combination of electrophoresis and chromatography in two dimensions, as described by Ingram (1956) and Katz et al. (1959). Alternatively, a narrow strip of paper may be used for the first dimension, followed by sewing the strip to a fresh sheet of paper for the second dimension; the join of the paper may distort the separation, however. More complex separation methods may be used; for example, the neutral peptides after separation by electrophoresis at pH 6.5 may be subjected to a further two-dimensional separation, while the acidic and basic peptides are separated by a second step alone (Svasti and Milstein, 1972; Chen and Krause, 1975).

Since up to 50 nmol of each peptide in a mixture may be separated with good resolution on a single sheet of Whatman No. 3MM paper, peptide mapping is a valuable preparative technique; larger quantities of peptides may be applied, but the resolution obtained is then lower. A peptide map should be performed with a small proportion of the material first, so that the distribution of the peptides may be assessed. The analytical peptide map may be stained with a variety of destructive reagents for the location of the peptides, while the subsequent preparative map may only be stained with non-destructive methods, such as dilute fluorescamine (see § 5.2.2.2). By comparison with the analytical map, peptides which would otherwise pass undetected may be located and recovered. The analytical peptide map may be made on thin layers of cellulose (§ 4.3.4), but a more precise comparison may be made if paper is used, although the latter method requires about 20 nmol of the mixture.

A useful combination for two-dimensional preparative peptide mapping is electrophoresis at pH 3.5 in the first dimension, followed by chromatography in *n*-butanol/acetic acid/water/pyridine (15 : 3 : 12 : 10, by vol.) in the perpendicular direction. The sample, about 50 nmol of each peptide in a salt-free mixture, in 10–25 μl of pH 3.5 electrophoresis buffer, is applied to the origin, about 10 cm from one edge, and 20 cm from the anodic end of a sheet of Whatman No. 3MM paper, and allowed to dry in cool air. The paper is wetted with the pH 3.5 buffer and blotted dry, as described above (§ 4.7), and electrophoresis is performed at 3 kV for 40 min. The paper is dried in warm air, and chromatography is performed in the second dimension as described above (§ 4.8).

Peptides are located on the paper with the dilute fluorescamine reagent (§ 5.2.2.2). The peptide-containing spots are excised, in the shape of a rectangle with a pointed end, and the peptides are eluted with 0.1 M NH_3 as described previously (§ 4.7). The NH_3 reacts with excess fluorescamine, preventing further reaction with the peptides.

The method may be modified for the separation of larger amounts, up to 500 nmol, of mixtures of fewer peptides, when maximal resolution is not required. The sample is applied to a line up to 8 cm long, as in one-dimensional electrophoresis, rather than to a spot. After the electrophoresis the partially separated zones of peptides are concentrated to spots by careful wetting of the paper on each side with the electrophoresis buffer and allowing the solvent fronts to merge. The paper is then dried again, and chromatography is performed in the second dimension.

4.10. Thin-layer methods for purification of peptides

The advantages of thin-layer methods over paper methods include a much simpler and more compact electrophoresis apparatus, lower voltage requirements, smaller amounts of solvents, easier drying after electrophoresis and chromatography, smaller amounts of detection reagents and small films for radioautography, and easier handling and storage of chromatograms. The main disadvantages are the limitation of loading to about 10 nmol/cm on 0.1 mm thick layers, and the high sensitivity to salt in the sample. As with paper methods, only small peptides of less than about 20 residues may, in general, be separated satisfactorily, larger peptides tending to streak out.

Two types of thin-layer materials have been used: cellulose and silica gel. Cellulose may contribute some impurities, especially after hydrolysis with HCl, while silica gel is generally completely free from amino acids or proteins. The properties of cellulose are very similar to those of paper, however, and the large amount of experience gained with paper may be used to advantage if this material is used. Suitable thin-layer plates are plastic-backed cellulose MN300 or

MN400 and silica gel G plates (Polygram series from Macherey and Nagel). These may be cut to size if a full 20 cm × 20 cm plate is not required.

4.10.1. Thin-layer electrophoresis

The principle is identical with that of paper electrophoresis, but since the layer is thinner and lower voltage gradients are used less heat is generated and there is less need for effective cooling. However, it is important that there should be no thermal gradients across the thin-layer, and the temperature must be controlled to prevent excessive evaporation. Suitable apparatus is commercially available, or may be constructed quite simply as described by Bates et al. (1975). A suitable apparatus is shown in outline in Fig. 4.14. It is important

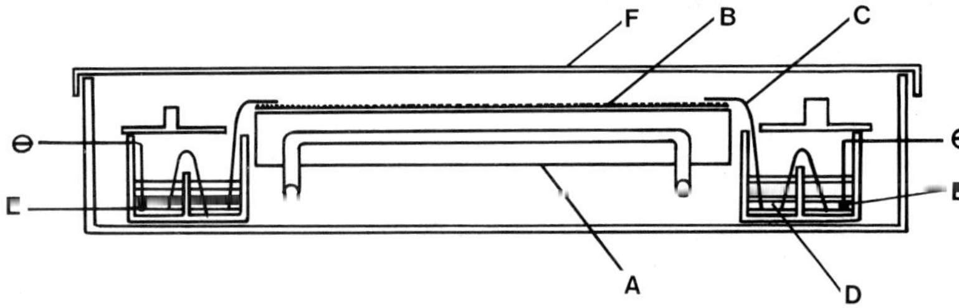

Fig. 4.14. Apparatus for thin-layer electrophoresis (cross-section). (A) Insulated platen, cooled by tap water, (B) plastic-backed thin-layer cellulose or silica gel plate, (C) wick, of Whatman No. 3MM paper, (D) electrophoresis buffer in electrode trough, partitioned to reduce the extent of diffusion of electrode reaction products to the wick, (E) platinum wire electrode, (F) closely fitting lid, fitted with a safety cut-out switch.

that the surface of the buffers in the electrode compartments and the parts of the wicks not in contact with the cooled plate are not exposed to the atmosphere above the thin-layer plate, since otherwise the solvent would evaporate from the relatively warm wicks and condense on the cooled thin-layer, causing flooding and smearing out of the zones of peptides.

Buffers similar to those used for paper electrophoresis may be employed, but the addition of 20% (v/v) acetone is recommended.

Electroendosmosis may be pronounced with some batches of cellulose plates, and this should be allowed for when positioning the sample application point or line.

The sample is dissolved in the electrophoresis buffer and applied from a capillary pipette to the lightly marked origin line, at about 5 μl/cm width. About 5 nmol of each peptide in a mixture is applied per cm on 0.1 mm thick cellulose plates, or double this amount on 0.25 mm thick silica gel plates. Dye markers are applied as for paper electrophoresis. Care is required to avoid damaging the thin layers; cellulose MN300 is particularly sensitive when wet. The solvent is allowed to evaporate in a stream of cool air. The thin-layer plate is placed on the cooled platen. A few drops of ethanol may be used to improve the thermal contact between the platen and the thin layer. The thin layer is wetted with the electrophoresis buffer on each side of the line of application of the sample, such that the solvent fronts concentrate the sample at a line where they meet. A simple method of wetting the layer is to place a piece of Whatman No. 3MM paper, saturated in the buffer, on the surface. Excess liquid is carefully blotted off with dry 3MM paper. The wicks, which may be made from the same paper, are connected. Capillary action is sufficient to hold them in place. The lid is fitted, and electrophoresis is carried out for a suitable time, depending upon the mobilities of the peptides to be separated; typically 90 min at 500 V at 20 °C for a 20 cm long plate. The power supply is disconnected and the thin layer plate is removed from the apparatus and dried in a stream of warm air.

Peptide zones are detected by one or more of the methods described in Chapter 5. Fluorescamine staining of the whole thin layer is useful, as any irregularities in migration may be detected. Strips may be cut from the edges and the middle of the plate and peptides detected with ninhydrin or other destructive reagents.

Peptides may be recovered from the zones by scraping off the cellulose or silica gel and eluting with a suitable solvent (0.1 M NH_3, 30% (v/v) acetic acid or 20% (v/v) pyridine). Filtration or centrifugation may be used to separate the peptide solution from the thin-layer material. The fibrous cellulose MN300 may most easily be removed from the plastic-backed layers with a small sharpened spatula: small scrolls of the support are formed, and these are transferred to small glass columns (3 mm i.d.) fitted with cotton-wool plugs (Fig. 4.15). The

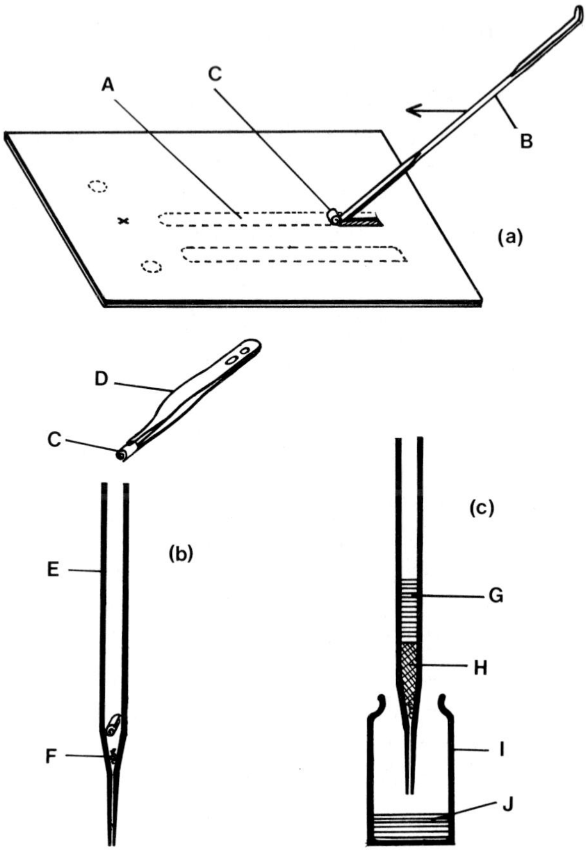

Fig. 4.15. Recovery of peptides from cellulose MN300 thin-layers. (a) Scraping off the thin-layer cellulose from the plastic backing: (A) Area of thin-layer bearing peptide, (B) small spatula with sharp chisel end, (C) 'scroll' of cellulose carrying peptide; (b) transfer of the scrolls to a small glass column: (D) lightweight forceps, (E) column (small Pasteur pipette), (F) small cotton-wool plug; (c) elution of the peptide from the cellulose: (G) solvent (e.g. 20% pyridine), (H) cellulose, stirred to release air bubbles, (I) collection vial, (J) solution of peptide.

eluting solvent is then applied to the column in several small aliquots, with stirring of the cellulose to ensure complete elution, and the peptide solution, filtered through the plug, is collected in small glass tubes. About 0.1 ml of solvent is sufficient for the elution of peptides in 1 cm^2 of cellulose.

Silica gel is readily collected using a 'vacuum-cleaner' technique (Janák, 1964). The apparatus is constructed as shown in Fig. 4.16. Application of a filter-pump vacuum while the tip of the tube is

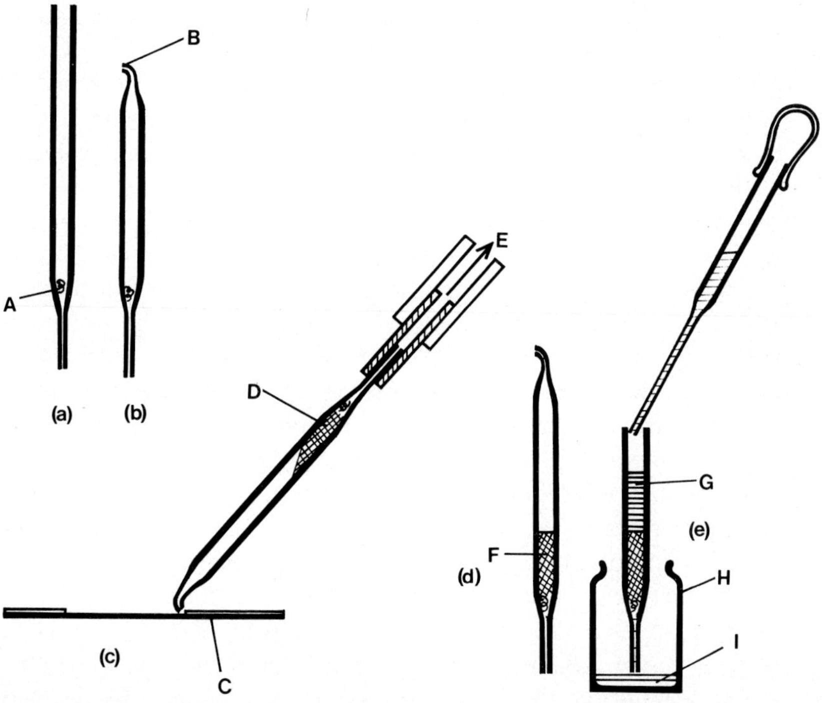

Fig. 4.16. Recovery of peptides from silica gel thin layers. (a) Small Pasteur pipette: (A) cotton-wool plug; (b) top of pipette drawn out and bent, and broken to form a sharp-edged tip (B); (c) removal of silica gel bearing peptide from the thin-layer plate: (C) thin-layer plate, (D) accumulation of silica gel particles, (E) to vacuum line; (d) tube placed vertically, containing a column of silica gel (F); (e) elution of peptides: (G) solvent (e.g. 20% pyridine), (H) collection vial, (I) solution of peptide.

scraped across the peptide zone causes the particles of silica gel to collect against the cotton-wool plug. The ends of the tube are broken off after scratching with a diamond pencil, forming a small column, from which the peptide is eluted, with stirring to release bubbles trapped in the silica gel.

4.10.2. Thin-layer chromatography

Chromatography on cellulose thin layers is similar to that on paper, and the same solvents (Table 4.7) may be used. The sample, which must be salt-free, is applied as a narrow band, 2–5 mm wide, about 2 cm from the lower edge of the plate. Dye markers (e.g. xylene cyanol FF and ϵ-*N*-dinitrophenyllysine) may be applied at the edges and in the middle of the sample line. The plate is placed in a glass tank with a closely fitting lid and containing the developing solvent, 5–10 mm deep. Ascending chromatography takes about 6 h with the butanol-containing solvents for a 20 cm high plate. Satisfactory resolution is often obtained with plates only 10 cm high. Similar precautions to those taken for paper chromatography (no thermal gradients, and preequilibration of the tank) are to be observed. When the solvent front reaches the top of the plate, the plate is removed from the tank and the solvents are evaporated in a stream of warm air, and peptides are detected and eluted as for thin-layer electrophoresis.

The process of chromatography on silica gel is somewhat different from that on cellulose or paper, adsorption effects being important as well as partition. Some solvents used on paper are incompatible with some types of silica gel; *n*-butanol/2-propanol/acetic acid/water/pyridine (15 : 10 : 3 : 12 : 10, by vol.) or chloroform/methanol/30% NH_3 (2 : 2 : 1, by vol.) (Baum et al., 1975) have been used.

The loading for thin-layer chromatography is the same as that for thin-layer electrophoresis, up to 10 nmol of each peptide in a mixture per cm width on 0.25 mm thick silica gel, although heavier loadings of simple mixtures may be used.

4.10.3. Preparative peptide mapping on thin layers

A combination of electrophoresis and chromatography in two dimensions, as with the paper methods, is a rapid and effective method for the resolution of a mixture of up to 50 peptides. For good resolution, the amount of sample applied is restricted to about 5 nmol of each peptide on the cellulose thin layers, so methods of high sensitivity are required for the determination of the sequences of peptides purified by this technique.

A variety of electrophoresis buffers and chromatographic solvents may be used. One system which has been used frequently in the sequence analysis of ribosomal proteins of *E. coli* consists of electrophoresis at pH 4.4 followed by chromatography in *n*-butanol/acetic acid/water/pyridine (15 : 3 : 12 : 10, by vol.) on cellulose MN300 (Heiland et al., 1976), as described for analytical purposes (§ 4.3.4). For the resolution of simple mixtures of peptides, for example those produced by further digestion of medium-sized peptides, where the maximum resolution is not required, the loading may be increased to 20 nmol, and the sample is applied as a band 3 cm wide.

Detection of peptides is by the methods mentioned for preparative maps on paper, and elution of the peptides is as described above. Dilute fluorescamine is a particularly valuable detection reagent (§ 5.2.2.2).

4.11. Preparative polyacrylamide gel electrophoresis

Electrophoresis in polyacrylamide gels is an important analytical technique (§ 4.3) but it has rarely been used preparatively for protein sequence analysis. There are several reasons for this: the high resolution obtained on the analytical scale is not achieved on the preparative (mg) scale; pre-running of the gel in the presence of thioglycollic acid or similar reducing agent is generally required, to eliminate oxidative damage caused by residual free radicals from the polymerization reaction; and recoveries of peptides have often been low. Koziarz et al. (1978) have described precautions which may be taken to protect tryptophan and terminal amino groups from degradative reactions.

Sequence information has been obtained using highly sensitive radiochemical methods from large peptides purified by polyacrylamide gel electrophoresis (Chapter 10). Sequence studies on mitochondrial proteins have successfully used a Tris–glycine system, with no special precautions, followed by detection of polypeptides by partial precipitation with potassium ions and electrophoretic elution from the gel slices (Walker, 1980).

Preparative polyacrylamide gel electrophoresis has been described in an earlier volume in the series (Gordon, 1975). Some recent articles describing preparative gel electrophoretic techniques are those of Ryan et al. (1976), Mardian and Isenberg (1978) and Koziarz et al. (1978).

4.12. High-performance liquid chromatography (Hplc)

The use of small ($\leqslant 10\,\mu$m diameter) regular, rigid spheres for column chromatography allows extremely high resolution to be obtained. High pressures, of the order of 200–500 atmospheres, are required for reasonable flow rates, and the equipment, principally the pumps, is expensive. For the separation of peptides, reverse-phase chromatography, in which the stationary phase consists of long-chain alkyl groups bonded to silica microspheres and the moving phase is a mixed organic/aqueous solvent, has been widely investigated in the last two or three years. However, the technique has not yet been developed to the stage of a routine preparative method, and may be best employed at present as a rapid and sensitive analytical tool.

Several problems have yet to be overcome before the method can be used routinely. Detection of the peptides in the eluate has usually been based on low-wavelength UV-absorbance (206–220 nm). Only a few solvents are sufficiently transparent at these wavelengths, and methanol/water or acetonitrile/water have been used most often. Very pure solvents are required to eliminate spurious peaks due to impurities. Charged compounds, including peptides, tend to give broad peaks, so a dilute buffer, such as ammonium acetate, pH 5 (10 mM NH_4^+) or an acidic 'ion-pairing reagent', such as acetic acid, trifluoroacetic acid, or phosphoric acid, may be added. Even with these modifications, recoveries of peptides other than small soluble peptides have not generally been high, and the addition of non-volatile compounds to the solvent makes recovery of peptides more difficult. Some recent accounts of the separation of peptides by Hplc are those of Hancock et al. (1978), O'Hare and Nice (1979), Schroeder et al. (1979) and Fullmer and Wasserman (1979).

An alternative detection procedure, using the reaction of fluorescamine with a portion of the eluate, has recently been described (Rubinstein et al., 1979). This system allows the use of better solvents for peptides, such as pyridine/acetic acid buffers (1 M pyridine). A gradient of *n*-propanol was found to be more effective than the methanol or acetonitrile gradients generally preferred because of their low viscosities and UV-absorbances. As much as 5 mg of a tryptic digest of a small protein was resolved in 160 min. Not all the peaks consisted of pure peptides, but the resolution was comparable with that obtained from columns of Dowex-type ion-exchange resins. This system looks promising for widespread use in the future.

Column packings for rapid gel permeation chromatography of large peptides and proteins have recently been described (e.g. TSK-GEL from Tokyo Soda Co., Rokushika et al., 1979). Resolution greater than that obtainable with Sephadex columns is achieved within less than one hour, through the use of porous silica microspheres with hydrophilic coatings.

It is likely that with further development of the technique Hplc will become of major importance for the purification of peptides for sequence analysis, as well as for the analysis of phenylthiohydantoin derivatives (§ 6.4.1.1).

4.13. Affinity chromatography

Several methods have been developed for the selective isolation of peptides containing particular functional groups, using receptors covalently coupled to an insoluble matrix, typically agarose (Sepharose) beads. The technique of affinity chromatography has been described in a recent volume in this series (Lowe, 1979) and comprehensively reviewed by Turková (1978).

Glycopeptides may be isolated by passage of a digest of a glycoprotein through a column of a lectin (e.g. concanavalin A or lentil lectin) with the correct specificity bound to Sepharose (Töpfer-Petersen et al., 1976). Glycopeptides containing the structural features recognised by the lectin remain bound to the column, while the other peptides may be washed off. The glycopeptides are subsequently eluted with a solution of a compound with a high affinity for the binding site of the lectin, such as α-methyl mannoside for concanavalin A. Yields may be low, since the binding may be essentially irreversible.

Biotinyl peptides may be isolated by an analogous method, using the egg-white protein, avidin, covalently bound to Sepharose (Buckley et al., 1979). A solution of biotin is used to displace the bound peptides, but the release may be very slow.

The specific chemical modification of a protein may be followed by the selective isolation of modified peptides by adsorption to a column containing receptors for the functional group introduced (e.g. anti-dinitrophenyl antibodies for dinitrophenylated proteins (Wilchek, 1974)). Antibodies with a specificity directed against a part of the primary structure may be used for the selective isolation of peptides bearing the antigenic site. Monoclonal antibodies are particularly suitable for such an approach.

These affinity methods have only rarely been used for the determination of primary structures.

4.14. Selective isolation of peptides by covalent binding to an insoluble matrix

Cysteine-containing peptides may be selectively isolated through covalent reaction with reactive disulphide derivatives of thiol-Sepharose (Egorov et al., 1975; Lin and Foster, 1975; Svenson et al., 1977). The thiol groups of cysteine residues form mixed disulphide bonds with the thiol-Sepharose as the mixture of peptides passes through a column of activated thiol-Sepharose. Peptides lacking cysteine are eluted from the column by washing. Cysteine-containing peptides are subsequently eluted with a solution of mercaptoethanol. A similar selective isolation of cysteine-containing peptides may be achieved on columns of mercurial gels (Turková et al., 1975).

Peptides containing methionine residues were selectively bound to an alkylating derivative of BioGel P-100 (Shechter et al., 1977). The reaction was reversed by thiolysis.

4.15. 'Diagonal' methods

It is possible to selectively alter, by chemical modification, the properties of peptides containing certain amino acid residues. If peptides are partially resolved by paper electrophoresis, treated so that the electrophoretic mobilities of a few peptides are altered, and subjected to the same electrophoretic separation in the second dimension, those peptides with altered mobilities are selectively separated from the bulk of the peptides which lie on a diagonal line across the paper. The most important of the 'diagonal' methods is that for the identification of cystine-containing peptides, which is described in Chapter 7. The method was applied to the purification of cysteine-containing peptides from myosin, using performic acid oxidation of the mixed disulphides formed with cystine (Weeds and Hartley, 1968). Naughton and Hagopian (1962) described a method for the selective isolation of C-terminal peptides. After electrophoretic separation of tryptic peptides from a protein, the paper was sprayed with a solution of carboxypeptidase B, which specifically removes lysine and arginine residues from the C-termini of all peptides except the C-terminal peptide (unless the protein has one of these residues at its C-terminus). After repeat electrophoresis in the perpendicular direction, all peptides migrate off the diagonal except for the C-terminal peptide. Other diagonal methods have been devised for lysine-containing and N-terminal peptides, using reaction of the protein with maleic or citraconic anhydride, separation of peptides in one dimension, hydrolysis in acid to remove the blocking groups, and repeating the separation in the second dimension (Butler and Hartley, 1972). Methionine-containing peptides may be selectively purified by alkylation with iodoacetamide after the first electrophoretic separation (Tang and Hartley, 1967).

These methods often give satisfactory results with small proteins, but the paper electrophoretic method is not so successfully applied to peptides from larger proteins. The principle of the diagonal methods may, however, be extended to other separation methods, including ion-exchange chromatography or gel filtration. With the latter techniques, considerably more work is required, since isolated fractions from the first separation must be independently subjected to the chemical reaction and the repeat of the separation method. In practice, therefore, this approach is limited to the selective purification of a small number of peptides which may be detected specifically in the eluate from the first chromatographic separation, such as cystine-containing peptides. Degen and Kyte (1978) recently described such a method for the purification of methionine-containing peptides, using reversible alkylation with I-[^{14}C]$CH_2 \cdot CONH_2$. The final purification of active-site peptides, containing radiolabelled aspartyl phosphate residues, from the sarcoplasmic reticulum Ca^{2+}-ATPase was by a diagonal method, based on the change in electrophoretic mobility at pH 1.9 of the peptides after hydrolytic removal of the phosphate moiety (Allen and Green, 1976).

4.16. Selective purification of blocked N-terminal peptides

For some purposes, such as locating regions in DNA sequences corresponding to the start of the expressed gene product, information on N-terminal amino-acid sequences is of particular interest. If the N-terminal residue is blocked to the Edman degradation, for example by acetylation or the presence of a pyrollidone carboxylic acid residue, the sequence can not be determined directly. Enzymic methods are available for the removal of pyrollidone carboxylic acid residues (§ 7.5.2), but the isolation of small peptides is in general required for the determination of the N-terminal sequences of blocked proteins.

If the total sequence is to be determined, such blocked peptides will normally be isolated along with all the other peptides in various digests, but if not, selective purification techniques may be used. In the simplest form, a digest of the protein with an enzyme such as thermolysin or pepsin (but not trypsin) is acidified to pH 2 and applied to a column of a strongly acidic cation-exchange resin (e.g. Dowex 50 X2, H^+ form) in water. All peptides containing a free amino-group or other basic group are bound to the column, while peptides lacking a basic group pass unretarded. If the blocked N-terminal peptide contains no basic residues (His, Lys or Arg) it should be found in the eluate. Several problems limit the use of this technique: large peptides, even those containing a basic group, may not be bound to the resin, since they do not penetrate the pores in the resin beads. Conversely, hydrophobic peptides, especially those containing tryptophan residues, may be bound even if they lack a free amino group. At low pH, peptides containing N-terminal glutamine residues derived from internal regions of the polypeptide may be converted to the blocked form, with N-terminal pyrrolidone carboxylic acid residues, and these peptides may be mistakenly identified as those derived from the N-terminus of the protein. The presence of a basic residue close to the N-terminus prevents application of the method. The method may be modified to circumvent some of these limitations, and a detailed description is given by Narita et al. (1975). The use of sulphoethyl-Sephadex (or sulphopropyl-Sephadex) may overcome some of the limitations inherent in the use of Dowex resins. Kluh (1979) used sulphoethyl-Sephadex for the selective isolation of a blocked N-terminal peptide from hog amylase.

Methods for the detection of peptides

Compatible methods for the detection of peptides in fractions eluted from columns or on paper or thin layers are required for all chromatographic methods. Quantitative determination is preferable, but in most cases semiquantitative methods suffice. Very many methods for detecting peptides have been described, but the majority of these are modifications of a few basic procedures. Sensitivity and rapidity are important attributes of those methods selected for discussion here. The methods fall into two categories: general methods for the detection of all, or almost all, peptides, and specific methods for the detection of peptides containing a particular amino acid residue. Methods in each category may be suitable for the detection of peptides either in solution or adsorbed to a support such as paper.

General methods for the detection of peptides are listed in Table 5.1, and methods for the detection of particular amino acid residues in Table 5.2.

5.1. General methods for the detection of peptides in column eluates

5.1.1. Low-wavelength UV absorbance

The peptide bond absorbs strongly ($\epsilon \sim 10^4$) in the region 200–210 nm, with significant absorbance extending to above 225 nm, and measurement of low-wavelength UV absorbance is a convenient and sensitive method for the detection of peptides. Other chromophoric groups, mainly the aromatic groups of tyrosine, phenylalanine,

TABLE 5.1
General methods for the detection of peptides

Method	Applicability	
	In solution	On paper or thin layers
Low-wavelength UV absorbance (206–230 nm)	+	–
Reaction with ninhydrin	+	+
Reaction with fluorescamine	+	+
Reaction with *o*-phthalaldehyde/mercaptoethanol	+	+
Alkaline hydrolysis, followed by reaction with ninhydrin, fluorescamine or *o*-phthalaldehyde	+	–
Acid hydrolysis and reaction with ninhydrin	–	+
Acid hydrolysis and amino-acid analysis	+	–
Acid hydrolysis and thin-layer electrophoresis at pH 2	+	–
Dansylation, followed by hydrolysis and thin-layer chromatography	+	–
Acid hydrolysis, dansylation and thin-layer chromatography	+	–
Transfer to thin-layers for spot tests, chromatography or electrophoresis	+	–
SDS-gel electrophoresis, acid/urea gel electrophoresis or isoelectric focussing, followed by staining with Coomassie blue	+	–
Chlorination, followed by reaction with *o*-tolidine or starch/KI	–	+

histidine and tryptophan residues, contribute to the absorbance, which is not, therefore, proportional to the concentration of peptide bonds in the solution. The relative contribution of aromatic groups is greater at higher wavelengths (220–230 nm).

Unfortunately, most of the solvents used for the separation of peptides also absorb strongly at wavelengths below 220 nm, and solutions of pyridine, commonly used for ion-exchange chromatography, absorb strongly at all wavelengths below 290 nm. The absorbance of the solvent may not be a serious problem in gel

TABLE 5.2
Methods for the detection of peptides containing particular residues

Method	Amino-acid residues detected	In solution	On paper
UV absorbance at 280 nm	Trp, Tyr, some modified residues	+	−
Fluorescence	Trp (weak), oxidized Trp, some modified residues	+	+
Liquid-scintillation counting	Radio-labelled residues, e.g. S-[^{14}C]carboxymethyl cysteine, [^{32}P]phosphoserine	+	+
Radioautography	Radio-labelled residues	−	+
Gamma-counting	Some radio-labelled residues (e.g. [^{125}I]iodoTyr	+	+
Reduction and reaction with DTNB	Cystine	+	+
Phenol/H_2SO_4	Glycopeptides	+	−
Iodoplatinate	Met, Cys	−	+
Phenanthrenequinone	Arg	−	+
p-Dimethylaminobenzaldehyde	Trp	−	+
Pauly test	His	−	+
Nitrosonaphthol	Tyr	−	+
Isatin	N-terminal Pro	−	+

filtration if stable detectors capable of the accurate determination of small increments of absorbance in the presence of a high background absorbance are used. The use of detectors operating at 206 nm (for which stable, intense light sources are available) may thus be suitable even if only 1% of the incident light is transmitted through the buffer solution. However, great care is required under such conditions to ensure baseline stability. Factors such as a change in temperature of one or two degrees or a change in the concentration of buffer components of 0.1% or less may lead to significant changes in the absorbance of the eluate, obscuring peaks due to peptides or producing misleading artefact peaks. The use of short path-length flow cells, e.g. 1 mm, while giving lower sensitivity, alleviates this problem. The alignment of the flow-cell and stray-light effects are also important at high buffer absorbances. More satisfactory results may often be obtained by using a longer wavelength, e.g. 215 or 220 nm with solutions of NH_4HCO_3 or NH_3, or 225 nm with dilute acetic or formic acids.

For ion-exchange chromatography using gradients of increasing buffer concentration, the use of wavelengths at which the buffer components absorb significantly is less satisfactory. With linear gradients, extrapolation of a gradually increasing background absorbance is often possible, and peaks may be detected after subtraction of this background. The baseline absorbance may be made almost constant by the addition of a neutral component, such as thiodiglycol, to the initial buffer, so that the absorbance of this buffer becomes close to that of the final buffer solution. This method is also advantageous when non-linear gradients are used.

Some commercially available photometers and flow-cells are listed in Appendix 1. A high-sensitivity in the presence of high background absorbance is attained by the Uvicord III and Uvicord S, while other instruments, with variable wavelength (e.g. Cecil), are more flexible. The flow cell should be of low volume, particularly for use with small columns, to avoid loss of resolution through re-mixing of the eluate. If the density of the eluate is not constant, or monotonically increasing, irregular flow and artefactual peaks due to refraction may be observed, particularly with flow-cells of larger volumes.

If no flow-cell and chart recorder are available, the absorbances of individual fractions may be determined with a spectrophotometer. The background scatter tends to be high at wavelengths where the buffers absorb, especially since considerable demands are made upon spectrophotometers operating at wavelengths below 220 nm. The instrument should be checked for accuracy at low wavelengths and high blank absorbances. Particular care is required when narrow path cells are used to ensure that the beam is accurately aligned. Unusually flat-topped peaks in chromatograms indicate that stray light or electronic effects are distorting the results. Random scatter, or gradual changes, in the absorbance measurements of column fractions may be due to differential evaporation of fractions, especially when small volumes have been exposed at room temperature for several hours in the fraction collector. On the other hand, if fractions have been stored in the cold they should be allowed to reach room temperature before measurement of the absorbances. If the background absorbance is high, short path-length cells may be used to advantage: the loss in absolute sensitivity may be more than offset by a decrease in background scatter.

Some examples of the use of low-wavelength absorbance for the detection of peptides are given in Figs. 4.4, 4.5 and 4.7.

5.1.2. Reaction with ninhydrin in solution

Ninhydrin reacts with primary amino groups, particularly the α-NH_2 groups of amino acids and peptides, to generate a blue colour:

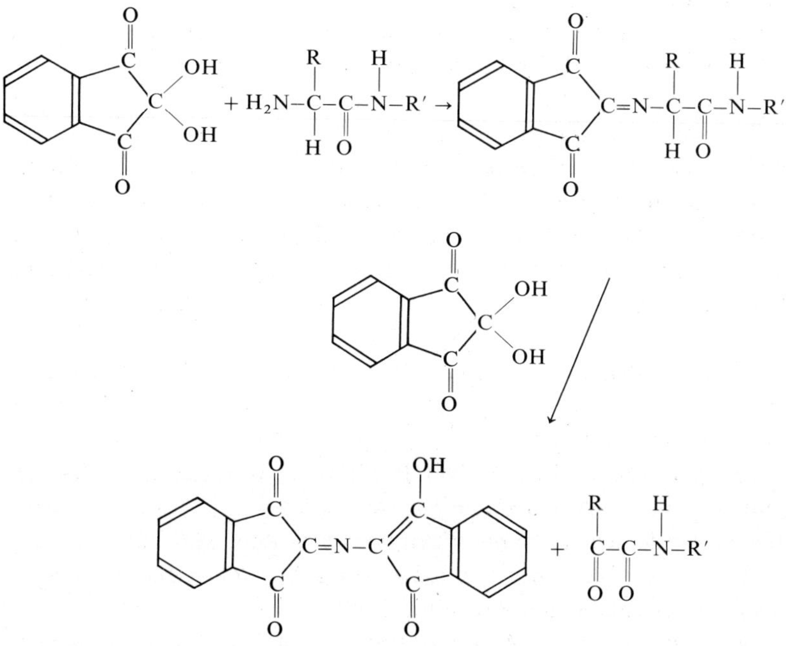

Ruhemann's purple (blue colour)

Since all peptides generated by cleavage of peptide bonds in a protein possess α-NH_2 groups, apart from any blocked N-terminal peptides or peptides with N-terminal proline residues, the method is capable of detecting almost all peptides. The ϵ-NH_2 groups of lysine residues also contribute to the colour development. However, peptides with N-terminal glutamine may become blocked during the isolation of peptides at acid pH, and unreactive towards ninhydrin. Peptides with N-terminal proline fail to give a blue product (measured by absorbance at 570 nm) but to react give absorbance at 420 nm. In order to detect all peptides and to increase the sensitivity, hydrolysis of the peptides to their constituent amino acids, or at least to di- and tri-peptides, prior to detection with ninhydrin, is preferable. Alkaline hydrolysis is used and contaminating NH_3 or volatile amines, which also react with ninhydrin, are driven off.

Details of the technique are given below (§ 5.1.2.1). The procedure may be automated for continuous monitoring of column eluates (§ 4.6.3).

The major limitation of the ninhydrin technique is that it can not be used for the detection of peptides in solutions containing ammonia, urea (which is hydrolyzed to ammonia under the conditions of the reaction) or primary or secondary amines. The sensitivity is lower than that of the fluorescent techniques described below (§§ 5.3, 5.4), which are therefore to be preferred.

5.1.2.1. Alkaline hydrolysis and reaction with ninhydrin Hydrolysis (Hirs, 1967b): Samples, containing maximally about 300 nmol amino acid residues, are placed in polypropylene tubes and dried in an oven at 110 °C. 13.5 M NaOH (0.15 ml) is added to each tube, and the tubes are placed in an autoclave at 121 °C for 20 min. If no autoclave is available, heating in a boiling-water bath for 2 h may be used. The tubes are cooled, and 0.25 ml glacial acetic acid is added to each to neutralize the NaOH.

Ninhydrin reagent (Moore and Stein, 1948): 0.8 g $SnCl_2 \cdot 2\ H_2O$ is dissolved in 500 ml of citrate buffer, pH 5 (prepared from 21.0 g citric acid monohydrate and 200 ml of 1 M NaOH, and made up to 500 ml with water). To this solution is added 20 g ninhydrin dissolved in 500 ml peroxide-free methyl cellosolve (2-methoxy-ethanol). The solution is filtered through glass fibre, saturated with N_2 and kept at 4 °C in a dark bottle.

Ninhydrin reaction: To each tube is added 0.50 ml ninhydrin reagent, and the tubes are vortexed to mix the solutions. The tubes are covered and heated for 20 min in a boiling-water bath. The tubes are cooled, and 2.5 ml 50% (v/v) propan-2-ol is added to each tube. The contents of each tube are thoroughly mixed by vortexing, and the absorbance at 570 nm is measured within 1 h, using 1 cm path-length cells.

Controls may also be prepared, using leucine (10–200 nmol). The approximate molar extinction coefficient, ϵ_{570}, is 2×10^4. Blank tubes should give little absorbance ($\leqslant 0.05$).

5.1.3. Reaction with fluorescamine

The fluorogenic reagent fluorescamine (4-phenylspiro-(furan-2(3H),1′-phthalan)-3,3′-dione), introduced by Udenfriend et al. (1972), is one or two orders of magnitude more sensitive for the detection of peptides than is ninhydrin. Primary amines, including amino acids and peptides, react rapidly with the reagent to form fluorescent products:

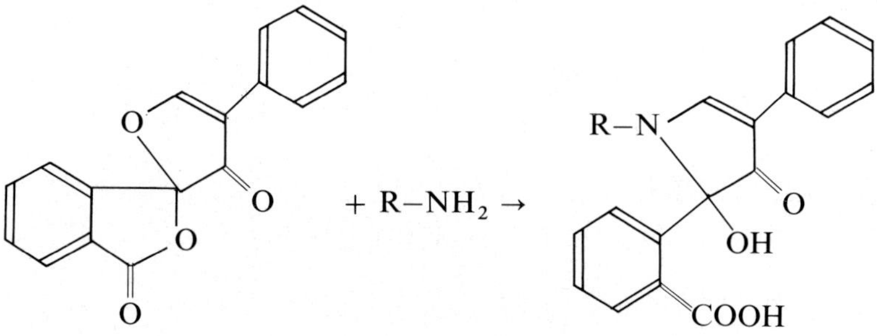

Ammonia also reacts with fluorescamine, but the product yields little fluorescence. The fluorescence intensity of the products is dependent upon the structures of the peptides; both the rate and the extent of reaction also depend upon the structure (De Bernardo et al., 1974). It is therefore preferable to hydrolyze the peptide mixture with alkali before the reaction with fluorescamine. The following method follows that of Lai (1977).

Samples containing 0.2–5 nmol of peptides are placed in polypropylene or borosilicate glass tubes, and dried in an oven at 110 °C for 2 h. 0.5 M NaOH solution (0.2 ml) is added to each tube, and the open tubes are autoclaved at 120 °C for 20 min. After cooling, 0.5 M HCl (0.2 ml) is added to each tube, followed by 1.0 ml of 0.5 M sodium borate buffer, pH 8.5. The contents of each tube are vigorously mixed by vortexing, while fluorescamine solution (20 mg/100 ml of dry acetone) (0.15 ml) is added. The mixing is continued for several seconds. The fluorescence emission at 475 nm is measured within 1 h, with excitation at 390 nm. Disposable glass tubes may be used if preferred, with direct measurement of the fluorescence in the tubes. Controls (0–10 nmol leucine) may be carried through the same procedure.

5.1.4. Reaction with o-*phthalaldehyde and 2-mercaptoethanol*

In alkaline solution, primary amines, including amino acids and peptides, react with a mixture of *o*-phthalaldehyde and 2-mercaptoethanol to give fluorescent products (Roth, 1971; Simons and Johnson, 1976):

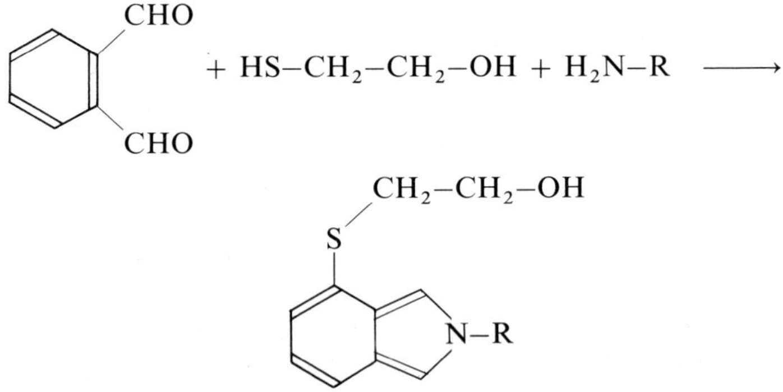

The use of the reagent has been reviewed (Lee and Drescher, 1978). In a comparison with fluorescamine for the detection of amino acids in the amino-acid analyzer, the phthalaldehyde reagent was found to be the more sensitive (Benson and Hare, 1975). This was partly because aqueous solutions of the reagent may be used, in contrast with fluorescamine, which is hydrolyzed upon contact with water. Phthalaldehyde also has an advantage in price over fluorescamine. However, a fluorescent product is formed with ammonia, which may lead to high background values if buffer solutions are contaminated with ammonia. The fluorescence intensity of the products formed with peptides may be low (Joys and Kim, 1979; Chen et al., 1979), and prior alkaline hydrolysis of the peptide samples is required for the detection of all peptides. The following method is taken from Mendez and Gavilanes (1976). The sample is dried in a 13 mm × 100 mm borosilicate glass tube at 110 °C. Hydrolysis and neutralization may be performed as described above for use with fluorescamine. To each tube is added 1.6 ml 0.25 M sodium borate buffer, pH 10.0, containing 0.05% (v/v) 2-mercaptoethanol, 0.4 ml of a solution of *o*-phthalaldehyde in water (0.3 mg/ml) is added, and the contents are mixed by vortexing for 2–3 s. After about 30 min at room temperature the fluorescence emission is measured at 455 nm, with excitation at 340 nm. Controls (0–10 nmol leucine) may be taken through the same procedure.

5.1.5. Hydrolysis followed by amino acid analysis

Quantitative analysis of peptides in column fractions is possible by amino-acid analysis after hydrolysis of samples with 6 M HCl for 24 h at 110 °C. In general this method is too time-consuming for regular use, but for certain applications, such as the search for peptides containing particular residues which are not easily detected by other methods (e.g. methionine), the method is valuable. If such peptides are being sought, the amino-acid analyses of samples from peaks detected by other methods suffice. To ensure that no peptides are discarded in fractions in which no peptides can be detected by other methods, the hydrolysis and amino-acid analysis may be performed on samples of these combined eluate fractions.

Samples containing 1–20 nmol of peptide (depending on the sensitivity of the analyzer) are dried in vacuo in borosilicate glass tubes. The peptides are hydrolyzed and analyzed by the usual method (§ 2.5).

As an alternative to quantitative analysis of the hydrolyzates, semi-quantitative analysis may be performed by thin-layer electrophoresis (1–2 nmol of each amino acid) or paper electrophoresis (10–20 nmol of each amino acid) at pH 1.9. The hydrolyzates of samples from column fractions are applied as small spots to the origin on thin-layer plates and subjected to electrophoresis at 300 V for 1 h, using 20 cm long plates (or 3 kV for 30 min on paper 50 cm in length). Standard amino acid mixtures are also applied beside the samples. The dried plates or papers are stained with the ninhydrin/Cd reagent (§ 5.2.1.3). Quantitative amino-acid analyses may be performed after this initial screening, if desired.

5.1.6. *Reaction with dansyl chloride*

The sensitive and rapid dansylation technique may be used for the detection of N-terminal residues of peptides in column fractions. Aliquots containing 0.2–2 nmol of individual peptides are dried in borosilicate glass tubes, 3 mm i.d. Dansylation and identification of the N-terminal residues are performed as described below (§ 6.2.1). A maximum of about 60 samples may be analyzed in one day: the method is too time-consuming for general use, and is more usefully reserved for the determination of N-terminal residues in mixtures of peptides after the combination of fractions comprising chromatographic peaks detected by other methods. Cataloguing the N-terminal residues present in peptide mixtures during purification schedules may be advantageous (Henschen-Edman, 1977).

5.1.7. *Thin-layer chromatography or electrophoresis*

Aliquots of fractions from chromatography columns may be transferred to thin-layer plates, and the peptides further separated by chromatography or electrophoresis, followed by detection with one or more of the reagents described in section 5.2. This is one of the most useful approaches to the detection of peptides in column eluates, since it provides information about the number of peptides present, the likelihood of their purification by thin-layer methods, and, through the use of a number of detection methods, information about the structures of the peptides. In practice it is limited to peptides of less than about 30 residues, since longer peptides rarely migrate satisfactorily on thin layers.

Aliquots containing 1–5 nmol of individual peptides are applied as small spots at the origin of the chromatogram. If the volume required is greater than 3 μl, the requisite aliquots may be transferred to microtitre trays and dried in vacuo. The dried samples are then taken up in 2 μl of 20% (v/v) pyridine for application to the thin layer. Electrophoresis or chromatography are performed as described above (§ 4.10), and peptides are detected by the methods described below (§ 5.2).

If few peptides are present in the eluate, spot tests on samples applied to thin-layer cellulose may be sufficient for the detection of peaks.

5.1.8. *Polyacrylamide gel electrophoresis*

Larger peptides (mol.wt. > about 3,000) may be detected and their purity estimated by polyacrylamide gel electrophoresis of samples from column fractions, followed by staining with Coomassie blue, or drying and radioautographing if radiolabel is incorporated. The main limitation to detection is the difficulty of fixing and staining small peptides, but the method is sensitive for peptides which are fixed and stained, allowing the detection of 1 μg of a peptide (0.1 nmol of a peptide with mol.wt. 10,000). For the estimation of purity at least 10-fold greater amounts are required. Two methods have been described above (§§ 4.3.1, 4.3.2). An alternative detection procedure involves labelling the peptide mixture with a fluorogenic or chromogenic reagent, such as dansyl chloride (Kato et al., 1975) or 2-methoxy-2,4-diphenyl-3(2H)furanone (Barger et al., 1976), before electrophoresis.

Isoelectric focusing in polyacrylamide gels is in general less useful for the detection of peptides (§ 4.3.3). A method for the detection by a modified Pauly reagent of histidine-containing peptides separated by isoelectric focusing has recently been described (Faupel and von Arx, 1978).

5.2. General methods for the detection of peptides adsorbed to paper or thin-layer supports

Most of the methods for the detection of peptides on paper or thin layers are destructive, and are therefore suitable only for analytical peptide maps or strips cut from the edges of one-dimensional chromatograms or electrophoretograms. The disadvantage of the latter procedure is that the peptides may not migrate evenly across the chromatogram; one or more strips should also be cut from the middle of the chromatogram for staining, so that the areas of the peptide zones are more accurately delineated. A non-destructive method, such as the dilute fluorescamine reagent (§ 5.2.2.2) or radioautography, may also be used on the whole sheet. For preparative peptide maps, only the non-destructive methods may be used; light spraying with ninhydrin may be used, but this is not recommended. Peptides which are not revealed on the preparative maps by dilute fluorescamine or radioautography may be located by comparison with another peptide map produced under identical conditions but stained with one or more of the destructive methods, such as ninhydrin and Cl_2/*o*-tolidine.

5.2.1. Reaction with ninhydrin

The most widely used reagent for the detection of peptides on thin-layer plates or on paper is ninhydrin, and this may be used in combination with other reagents which enhance or stabilize the colour produced, or give different colours depending upon the structures of the peptides. Either spraying or dipping may be used, the latter being generally preferable, since it allows a rapid, even application of the reagent, avoiding the production of noxious aerosols and not requiring a fume hood reserved for the purpose. Spraying is, however, preferable when only a light staining of the surface is required and when there is the possibility that some of the peptides may dissolve in the solvent used. A suitable amount of peptide for detection with ninhydrin is about 5 $nmol \cdot cm^{-2}$, equivalent to 1–2 nmol on thin-layer peptide maps or 10–20 nmol on peptide maps on paper, depending on the sizes of the spots. Three recipes are given below.

Ninhydrin gives colours with free amino-terminal groups of peptides; the ϵ-NH_2 groups of lysine residues produce a weak colour which appears slowly. N-terminal valine and isoleucine residues also give slow colour development. It is useful to note the rates of development of the colours of spots. N-terminal proline residues give only a weak yellow colour with the ninhydrin reagents. Most of the commonly used proteases do not cleave peptide bonds involving the secondary amine groups of proline residues, so few peptides obtained during sequence analysis will be expected to have N-terminal proline residues. However, acid-catalyzed cleavage of Asp-Pro bonds, and CNBr cleavage of Met-Pro bonds will yield peptides with N-terminal proline residues, and alternative detection methods (e.g. chlorination, § 5.2.4, isatin, § 5.4.8) should also be used. Peptides with blocked N-termini (e.g. N^{α}-acetyl or pyroglutamyl peptides) are not detected with ninhydrin (except for a slow reaction with any ϵ-NH_2 groups of lysine residues which may be present), and the chlorination method should be used for the detection of such peptides.

5.2.1.1. Ninhydrin/acetic acid reagent 0.2 g ninhydrin in 100 ml of 95% ethanol/acetic acid (20 : 1, v/v). The paper or thin layer is dipped into or sprayed with this solution and dried at 60 °C for 30 min. Alternatively, colour development takes place at room temperature in the dark in an ammonia-free area overnight.

5.2.1.2. Ninhydrin/collidine 0.1 g ninhydrin in 100 ml of 95% ethanol/2,4,6-collidine (95 : 5, v/v). The paper or thin layer is treated in similar fashion to that described above (§ 5.2.1.1). A variety of colours, ranging from yellowish grey to violet, depending on the amino-terminal residues of the peptides, are produced. Technical grade collidine usually gives better results than the pure reagent; presumably impurities contribute to the variety of colours produced. A 'neutral' ninhydrin reagent may be prepared with both collidine and acetic acid present. A disadvantage of the ninhydrin/collidine reagent is the unpleasant smell, which is retained during storage of the chromatograms.

5.2.1.3. Ninhydrin/cadmium reagent (Heilmann et al., 1957) (A) 0.25% (w/v) ninhydrin in acetone, (B) 1 g cadmium acetate · 2 H_2O in 50 ml glacial acetic acid + 100 ml H_2O. Before use, mix 100 ml of (A) and 15 ml (B). Dip or spray papers or thin layers. The colours develop overnight at room temperature in an ammonia-free atmosphere in the dark, or may be developed in 15 min at 70 °C. The colours and their rates of development and intensities should be noted. Most N-terminal residues yield pink spots, valine and isoleucine giving slow development and weak colour. N-terminal glycine and threonine residues give yellow colours, along with weak yellow colour from N-terminal proline residues. N-terminal asparagine gives yellow, turning pink. A similar effect is noted with peptides containing N-terminal glycine or threonine residues as well as lysine residues: the ϵ-NH_2 groups of the lysine residues give a slowly developing pink colour which may obscure the yellow initially produced. Peptides with N-terminal serine or carboxymethylcysteine give orange colours. The colours are stable for months in the dark away from ammonia. This method is recommended, but the toxicity of cadmium is a disadvantage.

5.2.2. Fluorescamine

Fluorescamine (§ 5.1.3) is extremely useful, particularly for the detection of peptides on preparative peptide maps, since very low concentrations of the reagent are required, and peptides may be eluted in high yield from the stained area with little destruction of the amino-terminal residues. For sensitive detection of peptides on analytical peptide maps, the reagent may be used at higher concentrations. The reaction is pH-dependent, and pre-treatment of papers or thin layers with a solution of triethylamine in acetone may be required, especially after electrophoresis or chromatography under acidic conditions. The fluorescence intensity is dependent upon the structures of the peptides: acidic peptides, especially those with N-terminal aspartic acid, carboxymethylcysteine, cysteic acid or glutamic acid residues are weakly detected. Strong fluorescence is usually observed with peptides with N-terminal hydrophobic residues, including valine and isoleucine which tend to give only weak colours with ninhydrin reagents. Fluorescamine is therefore particularly suitable for the detection of peptides erived by digestion with thermolysin. Free amino acids are poorly detected unless quite strongly alkaline conditions are used. Peptides with N-terminal proline residues, or with blocked N-termini, are not detected, except through reaction with the ϵ-NH_2 groups of lysine residues, under more alkaline conditions than 1% pyridine in acetone.

Two procedures, suitable for analytical and preparative peptide maps, respectively, are given here:.

5.2.2.1. Fluorescamine for analytical peptide maps (Lai, 1977) The dried paper or thin layer is washed with acetone and dried. It is then dipped in 1% (v/v) triethylamine in acetone, dried in air for 5 min, dipped in fluorescamine solution (10 mg in 100 ml of dry acetone) and allowed to dry. Peptides are revealed as fluorescent spots by illumination with UV light (336 nm). The limit of detection is about 0.1 $nmol \cdot cm^{-2}$ on Whatman No. 3MM paper, but is dependent upon the structure of the peptide. With silica gel G thin layers, pre-treatment with a more concentrated solution of triethylamine is recommended (Felix and Jimenez, 1974).

5.2.2.2. Fluorescamine for preparative peptide maps (Vandekerkhove and Van Montagu, 1974) A fluorescamine stock solution (10 mg/ 100 ml of dry acetone) is prepared; this may be kept dry for months at −20 °C. Before use, 1 ml of this solution is added to 100 ml of acetone and 1 ml of pyridine. The dried paper or thin layer is dipped in this solution and dried in air. Peptides are revealed as above. The sensitivity is lower, and comparable with that of ninhydrin reagents.

5.2.3. Hydrolysis of peptides followed by reaction with ninhydrin

As with the detection of peptides in column fractions, it is possible to hydrolyze peptides on thin layers to increase the sensitivity (e.g. Vandekerkhove and Weber, 1978a). The thin layer is placed in a desiccator containing a vessel with 6 M HCl. The desiccator is evacuated and incubated at 100 °C for 6–16 h. The acid is removed from the thin layer in vacuo over NaOH pellets, then dipped in a 5% solution of triethylamine in acetone and dried in warm air, to neutralize residual acid. The ninhydrin reagents are then used as above. Although all peptides may be detected, the method is not very convenient.

5.2.4. Chlorination

All compounds containing the NH functional group may be detected by conversion to *N*-chloro-compounds followed by reaction with *o*-tolidine or starch/KI. This method is particularly useful for the detection of blocked peptides. The chlorination of peptide bonds may be accomplished with chlorine gas, obtained either from a cylinder or from the reaction of $KMnO_4$ with conc. HCl, or with a solution of *t*-butyl hypochlorite. All traces of the chlorination reagent must be removed before the second stage of the reaction, or a high background colour is produced. The yield of colour is roughly proportional to the number of peptide bonds in the peptide. The method may be used after staining with ninhydrin, since the colour developed with ninhydrin is bleached by chlorine and does not interfere.

5.2.4.1. Reaction with chlorine (Reindel and Hoppe, 1954) The paper or thin layer is placed in a tank containing Cl_2 gas, which may be generated in situ by the addition of a few millilitres of 12 M HCl to a few grams of $KMnO_4$ in a beaker. The paper or thin layer should not be completely dry, but equilibration with atmospheric humidity (about 50%) is sufficient. No solvents capable of reaction with chlorine, such as acetone or pyridine, should be present. After exposure to Cl_2 for 10 min, the chromatogram is removed and thoroughly aerated in a fume hood (note that chlorine is highly toxic) until no trace of chlorine can be detected by smell or until there is a negative reaction of a corner of the chromatogram with the detection reagent (about 30 min). Exposure to reducing atmospheres, such as acetone vapour, should be avoided. The chlorinated peptides are then revealed with *o*-tolidine or starch/KI.

5.2.4.2. Reaction with t-*butyl hypochlorite* (Mazur et al., 1962) The dry paper or thin layer is dipped in a 1% (v/v) solution of *t*-butyl hypochlorite in cyclohexane and is aerated for 1 h in the fume hood. The chlorinated peptides are detected with *o*-tolidine or starch/KI.

5.2.4.3. Detection with o-*tolidine* (Reindel and Hoppe, 1954) The paper or thin layer is dipped in a freshly prepared 1% (w/v) solution of *o*-tolidine in 95% ethanol (**Care! carcinogen**). A blue colour (blue-green on silica gel) is produced within a few seconds by chlorinated peptides. The background steadily darkens also, and the positions of peptides should be recorded immediately. The materials are then disposed of safely, to avoid possible contact with the carcinogen. The sensitivity is similar to that of ninhydrin for small peptides, but considerably greater for large peptides.

5.2.4.4. Detection with starch/KI (Rydon and Smith, 1952) Because of the carcinogenicity of *o*-tolidine, spraying the chromatogram with a 1% solution of KI in 1% soluble starch may be preferred, although it is less convenient. The solution is prepared by the addition of boiling water to starch suspended in a little cold water, then KI is added. The paper or thin layer is lightly sprayed with the fresh reagent, while it is still warm. Heavy spraying must be avoided, since peptides may dissolve and spread. Chlorinated peptides are revealed as blue spots, with a sensitivity similar to that with *o*-tolidine.

5.3. Detection methods specific for certain amino acid residues in column fractions

5.3.1. UV absorbance

Measurement of the ultraviolet absorbance of chromatographic eluates at 280 nm is useful for the detection of peptides containing tyrosine or tryptophan residues. Cystine and phenylalanine residues also absorb, but weakly. Other chromophoric groups, either present in vivo or introduced by chemical modification, may also be detected at appropriate wavelengths (e.g. 250 nm for the maleylamino group, ϵ_{250} 3360 [Glazer et al., 1975]). If no other chromophores are present, the concentrations of tyrosine and tryptophan residues in column fractions may be determined approximately from the absorbances at 280 nm and 288 nm (§ 2.5.3; Edelhoch, 1967). Continuous monitoring of the eluates from columns at 280 nm is often convenient; simultaneous detection at two wavelengths is possible with two flow-cells and a dual-channel chart recorder.

5.3.2. Fluorescence

Tryptophan residues in proteins and peptides emit fluorescent light at 320–360 nm upon excitation at 280 nm. The quantum yield, which is usually low, and the wavelength of maximum emission are dependent upon the structures of the peptides, and UV absorbance is more suitable for the detection of tryptophan residues. More intensely fluorescent groups introduced by chemical modification, such as the dansyl group or tryptophan residues oxidized by performic acid, may however be best detected by fluorescence measurement. Peptides from glycogen phosphorylase containing ϵ-*N*-phosphopyridoxyl-lysine residues were detected in chromatographic eluates by their fluorescence emission at 395 nm upon excitation at 330 nm (Forrey et al., 1971).

5.3.3. *Liquid-scintillation counting or gamma-counting* (see Fox, this series, 1976, for full details)

A simple, quantitative detection method for radio-labelled peptides derived from proteins which have been intrinsically labelled in vivo or in vitro, or chemically modified with a radioactive reagent such as iodo[^{14}C]acetate, is liquid-scintillation counting on samples taken from each fraction. This is suitable for the commonly used isotopes, ^{3}H, ^{14}C, ^{35}S and ^{32}P. Gamma-counting, which is non-destructive of samples, may be used for peptides labelled with ^{125}I or ^{131}I. For work with single isotopic labels, a simple scintillation cocktail, such as 2,5-*bis*-(5-*t*-butyl-benzoxazol-2-yl)thiophen (BBOT)(4.9 g)/toluene (750 ml)/2-methoxyethanol (250 ml) suffices. In double-labelling experiments, a more efficient scintillation cocktail, such as Bray's fluid (Bray, 1960), is required, and the possibility of quenching by the chromatographic solvents should be checked. Prepared scintillation cocktails are commercially available. Normally small quantities of peptides in aliquots (10–50 μl) of aqueous solutions dissolve or disperse readily in the scintillation cocktail. Samples containing high concentrations of salts, such as guanidinium chloride, may give precipitates, possibly with occluded peptides, which could lead to a serious reduction in the counting efficiency. Dilution with water may prevent precipitation; otherwise scintillants capable of dissolving larger quantities of aqueous samples, such as those based on Triton X-100 or dioxan, or gel scintillants, may be more satisfactory.

5.3.4. *Detection of cystine residues*

For the identification of disulphide cross-links in proteins, the isolation of cystine-containing peptides is required (§ 7.7.1). These peptides may be detected in column fractions by reduction followed by reaction with 5,5′-dithio*bis*(2-nitrobenzoate) (DTNB) (Zahler and Cleland, 1968; Henschen, 1978). The procedure may be automated (Walsh et al., 1970). The following procedure is that of Henschen.

Portions of fractions (0.1–0.5 ml, containing up to 50 nmol cystine residues) are evaporated to dryness in 5 ml glass tubes in a stream of N_2. To each tube is added 0.1 ml of 2.5% (w/v) $NaBH_4$, and the tubes are incubated at 50 °C for 1 h. Excess borohydride is destroyed by the addition of 0.05 ml of 1 M HCl and incubation at 20 °C for 30 min. 1 ml of a freshly prepared 0.13% (w/v) DTNB solution in 0.04 M Na_2HPO_4 buffer (adjusted to pH 8.1 with NaOH) is added, and the contents of the tube are mixed. The absorbance at 410 nm is measured immediately. Standards of cystine or oxidized glutathione (0–50 nmol) may be prepared, and blank tubes must be tested. The blank absorbance increases with time, owing to hydrolysis of DTNB.

The molar absorption coefficient of reduced DTNB is about 14,000 (ϵ_{412} 13,600 [Ellman, 1959]; $\Delta\epsilon_{410}$ 14,140 [Collier, 1973]).

5.3.5. Detection of glycopeptides

Several methods are available for the detection of carbohydrate groups in solution. A simple method for the detection of hexoses is the phenol/sulphuric acid reaction (Dubois et al., 1956). The sample, containing 5–50 μg of hexose (typically about 10 nmol of peptide with an asparagine-linked complex oligosaccharide) in 10–50 μl aqueous solution, is placed in a clean glass tube (10 ml). Freshly prepared 5% (w/v) aqueous phenol solution (0.1 ml) is added, and the solutions are mixed. Concentrated H_2SO_4 (0.5 ml) is added, and the contents of the tube are mixed thoroughly by vortexing. After 20 min, the absorbance at 490 nm is measured in 1 cm path-length cells. A calibration curve may be constructed using 5–100 μg glucose. The blank absorbance should be low; fine particles of cellulose or Sephadex, or traces of sucrose in the tubes give high and erratic background absorbance, and thorough cleaning of the tubes, preferably by annealing at 500 °C, is essential.

5.3.6. Other specific detection methods

Other methods for the detection of particular residues in peptide mixtures are more conveniently performed after application of samples to thin-layer plates and further separation by chromatography (§ 5.1.7).

5.4. *Detection methods specific for certain amino-acid residues on paper or thin-layer supports*

5.4.1. *Fluorescence emission*

Inspection of papers or thin layers under UV light (254 nm or 280 nm) may reveal faint blue fluorescence due to tryptophan residues, but the method is insensitive. Oxidation of proteins with performic acid converts tryptophan residues to highly fluorescent products, and peptides containing these derivatives may be detected under UV light (366 nm) (Vandekerkhove and Weber, 1978a; Ambler and Scott, 1978). During the isolation of peptides, autoxidation of tryptophan residues may occur, and weakly fluorescent spots or bands on paper or thin layers may correspond to small amounts of these oxidized peptides.

Peptides containing fluorophores introduced by chemical modification may also be detected under UV light.

5.4.2. *Radioautography*

Radioactive peptides are conveniently located on paper or thin-layer chromatograms by radioautography. The dried chromatogram is marked with radioactive ink (prepared by the addition of waste ^{35}S or ^{14}C isotopes to ink, giving about 1000 cpm/μl) to aid re-alignment of the developed film, and is overlaid by an X-ray film (e.g. Kodak No-Screen film). The film and chromatogram are kept in close contact in a light-tight envelope evenly weighted with a lead sheet or wooden block, or in a commercially-available spring-loaded cassette, for the requisite time. Exposure for 16 h is sufficient for the detection of 10,000 cpm of ^{14}C per cm^2. The film is then developed and fixed with the reagents recommended by the manufacturer, with a final rinse in distilled water before drying.

Peptides labelled with ^{35}S or ^{32}P are detected as efficiently as those labelled with ^{14}C, but ^{3}H-labelled peptides are not readily detected by radioautography. The efficiency is greatly increased, however, by coating the paper or thin layer with a scintillator, such as PPO (diphenyloxazole). The paper is dipped in a solution of PPO in diethyl ether, and the solvent is allowed to evaporate. The paper is then placed in contact with a blue-sensitive film at −70 °C for the requisite period before the film is developed. Fluorography also decreases the time required for exposure with the other isotopes, but has the disadvantage of making the recovery of peptides difficult. Careful experimental technique is required for good reproducibility using fluorography; various factors influencing the results have been studied by Randerath (1970).

If several chromatograms are exposed simultaneously, care should be taken that each film is shielded from radiation emitted by other chromatograms; this is particularly important with ^{32}P; sheets of lead about 2 mm thick should be used to separate the cassettes, together with blocks of wood 2 cm thick. Such blocks also serve to apply pressure during the exposure to ensure close contact of film and chromatogram.

The darkening of the film in radioautography is not directly proportional to the number of disintegrations per cm^2 in the samples, there being a lag phase. For quantitative work, spots detected by radioautography may be excised, and the radioactivity determined by liquid-scintillation counting. If the papers or thin layers are not thoroughly dried after chromatography or electrophoresis, or after a staining reaction, chemical reaction of volatile components remaining on the chromatograms, such as acids, oxidizing reagents or reducing reagents, will affect the film, possibly inhibiting the radiation-sensitive process or leading to general film darkening.

As an alternative to radioautography, side strips cut from one-dimensional chromatograms may be divided into short lengths, and the radioactivity in these detected by liquid-scintillation counting. For highly active samples, the paper may be scanned manually using a Geiger counter, or an automatic scanning device may be used. This method is particularly effective for locating ^{32}P-labelled peptides.

5.4.3. Phenanthrenequinone reagent for arginine

Guanidine and several derivatives, including arginine residues in peptides, react with phenanthrenequinone under alkaline conditions to give strongly fluorescent products. The following sensitive test for arginine residues is that of Yamada and Itano (1966).

Two stock solutions are prepared: (A) 0.02% (w/v) phenanthrene quinone in absolute ethanol; may be stored at 4 °C in the dark for months. (B) 10% (w/v) NaOH in 60% (v/v) aqueous ethanol (may be stored for up to 1 month).

Immediately before use, equal parts of (A) and (B) are mixed. The paper or thin layer is dipped rapidly through the solution, and placed on a flat glass plate to dry at room temperature for about 20 min. The chromatogram is examined under the UV lamp (254 nm or 366 nm). Arginine-containing peptides give a strong greenish-white fluorescence against a dark background. The limit of detection is about 0.1 nmol · cm^{-2} on cellulose thin-layer plates, but the sensitivity is considerably lower on silica gel G plates.

5.4.4. Ehrlich test for tryptophan residues

Tryptophan is not determined by amino-acid analysis after hydrolysis in 6 M HCl. It is therefore useful to screen all peptides, or peptide fractions, for this amino acid by other methods, if there is a possibility of its presence. In solution in UV-transparent solvents tryptophan residues may be determined spectrophotometrically (§ 5.3.1). On paper or thin layers the Ehrlich test may be used.

A 2% (w/v) solution of *p*-dimethylaminobenzaldehyde in acetone is prepared. Immediately before use, this solution (9 vols) is mixed with conc. HCl (1 vol.). The paper or thin layer is dipped through this mixture. After a few minutes at room temperature, tryptophan-containing peptides yield purple spots, which gradually fade. The sensitivity is about 5 nmol · cm^{-2} on paper, cellulose or silica gel thin layers. This test may be performed after staining with ninhydrin; the ninhydrin colours fade after treatment with the acidic reagent.

Specific detection methods for other residues are less important for the determination of peptide sequences, since amino acid analyses of isolated peptides provide direct, quantitative information. However, several methods which may be useful in extending the information gained from peptide maps, or in the selective isolation of peptides containing particular residues, are given below. A large number of alternative methods is given by Smith and Seakins (1976). Easley (1965) has summarized useful combinations of specific colour reactions in peptide mapping.

5.4.5. Pauly test for histidine residues (Block et al., 1955)

The following solutions are required (prepared freshly or stored at 4 °C): (A) 1 g sulphanilamide and 10 ml of 12 M HCl in 90 ml H_2O; (B) 5 g $NaNO_2$ in 100 ml H_2O; (C) 50 ml saturated Na_2CO_3 plus 50 ml H_2O.

5 ml of (A) and 5 ml of (B) are mixed at room temperature. After 1 min, 40 ml *n*-butanol is added, and the mixture is shaken thoroughly for 1 min, then allowed to settle for 4 min. The butanol layer is decanted into a shallow trough or tray. The chromatogram is dipped in this solution and allowed to dry in air for 5 min. A cherry red colour is developed with histidine and other imidazoles upon spraying the treated chromatogram with solution C. The sensitivity is about 5 nmol · cm^{-2}. Tyrosine residues give a dull brown colour.

5.4.6. Nitrosonaphthol test for tyrosine residues (Easley, 1965)

Two solutions are prepared: (A) 0.1% (w/v) α-nitroso-β-naphthol in acetone, (B) acetone/conc. HNO_3 (9 : 1, v/v), freshly prepared.

The paper or thin layer is dipped through solution (A) and dried; it is then dipped through solution (B). After drying at room temperature for 5–10 min the paper or thin layer is heated carefully over a hot plate or with hot air. The background clears from dark yellow to light yellow and tyrosine-containing peptides give a rose colour, which gradually fades. The results are traced immediately and the chromatogram is discarded. About 10 nmol · cm^{-2} is required.

This test may be performed after staining with ninhydrin.

5.4.7. *Iodoplatinate test for methionine residues* (Easley, 1965)

This test should be on freshly prepared chromatograms, with no previous staining, and no sulphur compounds (e.g. thiodiglycol or 2-mercaptoethanol) should be used in the electrophoresis buffers or chromatography solvents.

Three solutions are prepared: (A) 2 mM chloroplatinic acid ($H_2PtCl_6 \cdot 6\ H_2O$, 1 mg/ml H_2O); (B) 1.0 M KI (freshly prepared); (C) 2 M HCl. Just before use, the solutions are mixed in the following order: (A) (4 ml), (B) 0.25 ml, (C) 0.4 ml, and acetone (76 ml) is added. The paper or thin layer is dipped and dried. Reducing sulphur compounds appear as white areas on a pink background. Methionine reacts immediately, while *S*-alkylcysteine derivatives react more slowly. Cysteine gives a yellow spot.

5.4.8. *Isatin test for N-terminal proline residues* (Acher et al., 1950)

The paper or thin layer is dipped in 0.2% isatin in *n*-butanol/acetic acid (25 : 1, v/v), and dried at 70 °C, then kept for 24 h in the dark. Heating at 100 °C for 10 min gives quicker development of proline as a blue spot on a pale yellow background. Other amino acids and peptides give a weak colour. The test is not satisfactory after ninhydrin staining.

5.4.9. *The technique of dipping chromatograms*

An even distribution of staining reagents is more easily achieved by dipping than by spraying. Spraying must, however, be used for aqueous reagents, in which peptides are soluble. Papers are drawn through a shallow trough containing about 1–2 cm depth of reagent solution, as shown in Fig. 5.1a. The trough should be constructed from a material such as polyethylene, which is not corroded or dissolved in the reagent solutions. A steady, continuous movement of the paper from one end to the other is used. The paper is then suspended from a stainless steel or glass rod with non-corroding clips (e.g. clothes-pegs) to dry.

Plastic-backed thin-layer plates may be coated evenly with a reagent solution by dipping in a flat-bottomed glass tray containing about 20 ml of the solution, as shown in Fig. 5.1b. The plates are dipped with the layer downwards, so that only that side of the plates are wetted. The layers may then be placed flat, with the thin layer upwards, to dry.

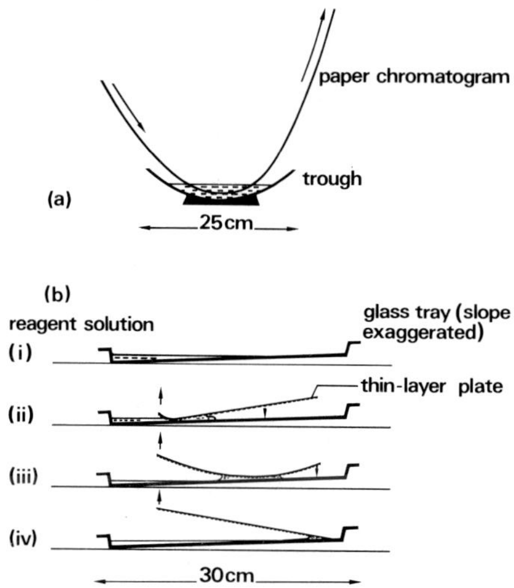

Fig. 5.1. Dipping chromatograms into detection reagent solutions. (a) Paper (cross section). (b) (i)–(iv), stages in dipping plastic-backed thin-layer chromatograms. The solution is drawn across the thin layers in a smooth movement in about 1 s.

Determination of peptide sequences

The choice of method for peptide sequence analysis will govern to a large extent the whole approach to cleavage of the protein and separation of peptides. If extensive use is to be made of an automated spinning-cup sequencer with the usual methods for identification of phenylthiohydantoin derivatives, quantities of several micromoles of protein will be required for the preparation of large fragments. On the other hand, if the sensitive 'dansyl-Edman' or Dabitc methods are to be used, a larger number of smaller peptides will be isolated after cleavage with enzymes such as trypsin, or after subfragmentation of large peptides, and less starting material will, in general, be required.

Pure peptides are required for the application of all methods for structural analysis, with the exception of the mass spectrometric method, which, in the hands of an experienced investigator, may be applied to the determination of the structures of a small number of peptides in a mixture (Morris and Dell, 1975). Methods used to assess the purity of peptides are first discussed, followed by methods for determining amino- and carboxy-terminal residues and amino-acid sequences.

6.1. Methods for the estimation of the purity of peptides

For the purposes of amino-acid sequence analysis, a peptide is sufficiently pure if the impurities do not interfere with the correct interpretation of the structural data. Non-peptide impurities (such as residues of solvents or small amounts of salts) which do not interfere with analytical procedures may be tolerated. For very small peptides, peptide impurities totalling up to 30% of the concentration of the major component may be tolerated, although a maximum limit of impurities of 10% should be used as a general guide. For larger peptides, with more than 20 residues, more reliance must be placed upon the quantitative amino-acid analysis, and the concentrations of impurities must be lower; however, impurities present in a molar ratio of less than 5% will be unlikely to interfere with the interpretation of the results. Thus the methods used to test the purity of peptides need be sensitive enough only to detect impurities at the level of about 5% of that of the major peptide component.

There is no single method for the estimation of the purity of peptides, but rather a series of observations made during the isolation of peptides and subsequent characterization. The final test of purity is the observation of a single sequence; for example, two peptides which are identical but for the order of amino-acid residues will probably satisfy other criteria of purity, such as chromatographic properties, amino-terminal analysis and amino-acid analysis. However, peptides which are so similar will be encountered very rarely, and satisfactory fulfilment of the following criteria (§§ 6.1.1.–6.1.4) is good evidence for the purity of a peptide.

6.1.1. Elution profiles from chromatographic columns

Inspection of the shape of a peak eluted from a chromatographic column may indicate whether a single peptide comprises the peak. Unless the chromatographic conditions are abruptly changed during the elution of gel filtration or ion-exchange columns, a pure peptide will be eluted as a symmetrical peak of width characteristic of the column. Tailing of the peak may, however, occur with some peptides. The ratio of the values of such parameters as absorbance at different wavelengths and radioactivity for a series of fractions across the width of the peak is constant if the peptide is pure.

6.1.2. Analytical chromatography or electrophoresis

Many of the chromatographic and electrophoretic methods described in Chapter 4 may be employed for the investigation of the purity of peptides. The major component of the peptide mixture must be present in at least a 10-fold greater amount than the limit of detection of the method. Thus, for thin-layer chromatography followed by staining with ninhydrin, where the practical detection limit is about 0.3 nmol, at least 3 nmol of the peptide is required.

Clearly, different conditions must be used for the estimation of purity than have been used in the isolation procedures. A useful analytical procedure is thin-layer electrophoresis on cellulose MN300 at pH 6.5, since the values of the electrophoretic mobilities of the peptides determined by this method give useful information on the structures of the peptides (§ 6.1.2.1); however, in this system, neutral peptides are not separated.

The staining procedure may also provide evidence of purity. For example, two peptides which give overlapping spots on thin-layer chromatography may sometimes be distinguished by differences in the colour, or the rate of development of colour, with ninhydrin reagents, and if one is radioactive the radioautographic spot may not coincide precisely with the ninhydrin-positive spot. A method capable of detecting blocked peptides should also be used.

An already purified peptide may give rise to more than one spot upon chromatography or electrophoresis, particularly after prolonged storage. The most common reason for this is partial autoxidation of *S*-alkylcysteine and methionine residues (peptides containing the sulphoxide derivatives have lower R_F values in most chromatographic solvents on cellulose or silica gel thin layers). Partial deamidation of asparagine or glutamine residues, partial cyclization of N-terminal glutamine or *S*-carboxymethylcysteine residues and interconversion of homoserine and homoserine lactone residues at the C-termini of CNBr-derived peptides may also occur during storage. Thus these analytical methods do not give unambiguous evidence for the purity of peptides.

Polyacrylamide gel electrophoresis is a useful analytical method for larger peptides, but staining with Coomassie blue is extremely variable, and impurities may not be detected. Pre-labelling of the peptides with a fluorophore (§ 5.1.8) may be useful in this respect.

6.1.2.1. The use of electrophoretic mobilities of peptides Offord (1966) described the use of logarithmic plots of electrophoretic mobilities of peptides on paper at pH 6.5 versus their molecular weights for determining the net charge on the molecules, primarily for the assignment of acid or amide groups in small peptides. The method has been used widely, mainly in conjunction with the 'dansyl-Edman' method of sequence determination (§ 6.4.4), and Offord (1977) has recently given a thorough description of the technique, with many examples and cautionary notes.

Most small peptides have mobilities relative to that of aspartic acid which lie close to one of the lines in Fig. 6.1, depending on the

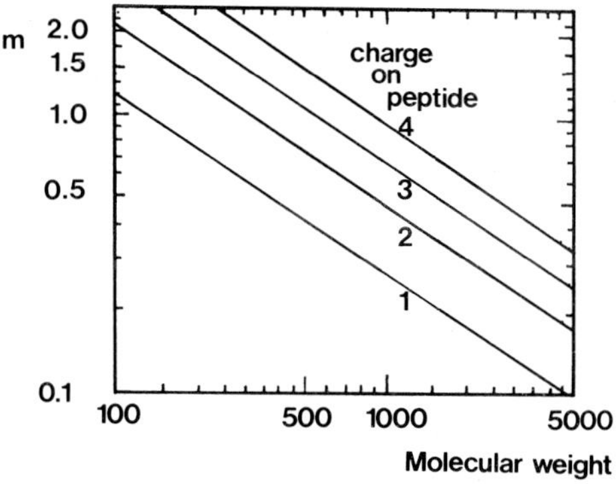

Fig. 6.1. Mobilities of peptides at pH 6.5 relative to aspartic acid (from Offord, 1966). See text for details.

net charge on the peptide molecules. This figure is also applicable for thin-layer electrophoresis on cellulose MN300 plates (Bates et al., 1975). Peptides containing histidine residues do not have mobilities lying on one of the lines, since the pK_a of the imidazoyl group is close to 6.5, and the net charge is not integral. Some other peptides behave anomalously, particularly those with N-terminal cysteic acid, methionine sulphoxide, arginine, lysine, asparagine, serine and threonine, and have a slightly more negative (or less positive) charge than expected, probably because the amino-terminal groups have low pK_a values. In most peptides, the α-amino group is essentially fully protonated at pH 6.5. Other peptides containing cysteic acid residues may also give anomalous results. The results may be difficult to interpret for peptides with C-terminal homoserine, derived by CNBr cleavage of a protein, because of the interconversion with the lactone form.

The electrophoretic mobilities are measured as the ratios of the displacement of the peptides from the neutral amino-acid marker (itself displaced from the origin by electroendosmosis) to the displacement of aspartic acid from the neutral marker. The electrophoresis must be terminated before either aspartic acid or the peptides approach within 5 cm of the electrode buffers (or within 2 cm of the wicks with thin-layer electrophoresis), since migration is affected by the increased wetness at the ends of the paper. Care should be taken that migration is even across the width of the paper; the aspartic acid and neutral markers should be applied close to the samples. Salt-free samples must be used. Larger peptides, particularly acidic peptides, often trail badly; the interpretation of the results in such cases is unreliable, although the leading edge of the spot may correspond closely to the expected position.

Examples of the use of this method for the determination of acid or amide groups in peptides are given in Fig. 6.10.

6.1.3. Determination of the N-terminal residue

The observation of a single N-terminal residue by the dansyl chloride or Dabitc methods (§ 6.2) is one of the most useful indicators of purity. About 1 nmol of peptide is required, since the practical limit of detection in both cases is about 50 pmol.

6.1.4. Amino-acid analysis

Amino-acid analysis is one of the few quantitative procedures used during sequence determinations, and is the most important test for the purity of peptides. Modern analyzers capable of accurate determination of 1 nmol of each amino acid can detect about 20 pmol; less sensitive instruments which require about 20 nmol for accurate determination will detect about 1 nmol. The major problem associated with the use of the analyzers of the highest sensitivity is contamination of the peptide with impurities derived not from the protein but from such sources as dust, fingerprints or chromatographic materials. Unless very stringent precautions are taken it is probably best to consider 1 nmol as a practical lower limit for amino-acid analysis.

The technique of amino-acid analysis of peptides is the same as that of proteins (§ 2.5), except that for small peptides rather lower accuracy may be tolerated, and it is not usually necessary to perform hydrolyses for different periods of time. Low values for valine and isoleucine will be expected for peptides hydrolyzed under standard conditions (6 M HCl/0.1% (w/v) phenol at 110 °C for 24 h in vacuo) if the peptide contains Val-Val, Val-Ile, Ile-Val or Ile-Ile sequences; only about 50% hydrolysis of the Ile-Ile bond occurs under these conditions. If valine and isoleucine residues are present in the peptide, careful inspection of the results of sequence analysis will be necessary to ensure that Val-Val or Ile-Ile sequences are correctly identified; the dansyl method (§ 6.4.4) has the advantage that such dipeptide sequences are clearly identified. If there is any doubt concerning the correct amino-acid composition of a peptide, hydrolysis of a second sample for 72 h should be performed before analysis. Methionine and carboxymethylcysteine (or other *S*-alkylcysteine derivatives) are often partially oxidized during the isolation of small quantities of peptides, even if thiodiglycol is included in solvents. The addition of 2-mercaptoethanol (0.05%, v/v) to the HCl used for hydrolysis results in higher yields of these amino acids (Keutmann and Potts, 1969). It should be noted that the presence of thiodiglycol during the hydrolysis results in lower yields of some amino acids (Doscher and Hirs, 1967). Alternative procedures are required for the determination of tryptophan (§ 2.5).

A qualitative determination of the amino-acid composition of small peptides may be made by dansylation of the hydrolysate of about 0.5 nmol of the peptide and submitting the dansylated amino acids to two-dimensional chromatography (§ 6.2.1). This procedure is rapid and may be used to check large numbers of peptide hydrolysates before application to an amino-acid analyzer, to ensure that instrument time is not wasted on samples containing too little material. The hydrolyzate is dansylated in $NaHCO_3$ solution and the solution is dried in vacuo as described below (§ 6.2.1). The reaction product is then acidified with 5 µl of 6 M HCl, and the acid is removed over NaOH pellets in vacuo. The residue is dissolved in 50% pyridine and chromatographed on polyamide layers (§ 6.2.1).

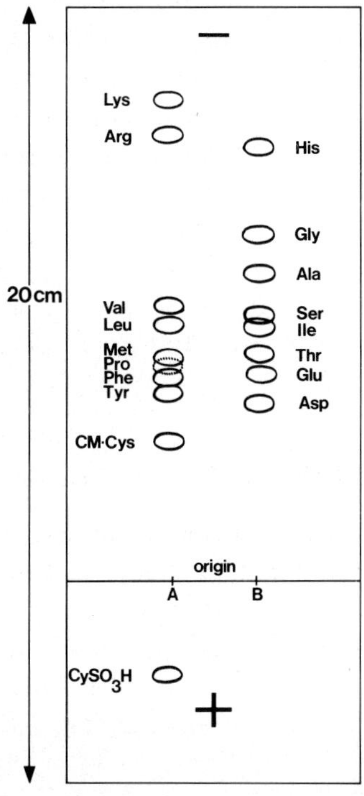

Fig. 6.2. Thin-layer electrophoresis of amino acids at pH 2.0. Two amino-acid mixtures (A and B), containing 1 nmol of each amino acid, prepared with the compositions shown, were separated on thin-layer cellulose MN300 in acetic acid/formic acid/water (8:2:90, by vol.) by electrophoresis at 20 V/cm for 1 h. Abbreviations: CM-Cys, *S*-carboxymethylcysteine; $CySO_3H$, cysteic acid. A similar pattern is obtained by paper electrophoresis (Stevenson, 1971).

Histidine is converted to the *bis*-dansyl derivative under these conditions, which gives a characteristic yellow-orange fluorescing spot (Fig. 6.4).

An alternative procedure for the qualitative analysis of peptide hydrolyzates is thin-layer electrophoresis at pH 2 followed by staining with 1% (w/v) ninhydrin-cadmium reagent (§ 5.2.1.3). Mixtures of 1 nmol of each standard amino acid (made up in two solutions, A and B) are applied alongside the samples for the comparison of the positions and intensities of the developed spots. The separation of the amino acids is shown in Fig. 6.2. Complete resolution of all amino acids is not obtained, but as with the dansyl method thin-layer electrophoresis is useful for the rapid screening of peptide hydrolyzates to ensure that suitable quantities are applied to the amino-acid analyzer. 1–3 nmol of peptide hydrolyzate is required for the thin-layer electrophoresis.

With experience, both the dansyl method and the ninhydrin-staining method can give reliable results for small peptides; levels of impurities may be estimated and approximate ratios of amino acids determined.

6.2. Determination of the N-terminal residues of peptides

As described above (§ 2.7), many methods are available for the determination of N-terminal residues in peptides and proteins. Experimental techniques suitable for the study of large polypeptides (mol.wt. > 10,000) and proteins (§ 2.7) are different from those suitable for smaller peptides; two of the latter methods described here have been chosen because of their high sensitivity (§§ 6.2.1–6.2.2).

6.2.1. Dansyl chloride method

The use of dansyl chloride as an N-terminal reagent has been described in Chapter 2 (§ 2.7). An authoritative description of the technique applied to peptides has been given by Gray (1972). The method described here is essentially the high-sensitivity modification of Bruton and Hartley (1970).

The following solutions are prepared: (A) dansyl chloride, 2.5 mg/ml in dry acetone; stable for months in the dark at −20 °C in sealed vials. The reagent is completely soluble, but commercial samples may contain some white, insoluble material (the hydrolysis product, dansic acid), and this product may also form slowly during storage of the reagent. If the extent of hydrolysis is small there is no significant interference with the method. (B) 0.2 M $NaHCO_3$. (C) 6 M HCl.

A sample of the peptide, 0.2–1 nmol, is dried in the bottom of a medium wall borosilicate glass tube (3 mm i.d. × 40 mm). The peptide must be free from compounds that react with dansyl chloride, such as amines, ammonia or phenol, and from acids. Small quantities of neutral salts may be tolerated, but the peptide should preferably be completely salt-free. 0.2 M $NaHCO_3$ solution (3 μl) is placed in the tube and transferred to the bottom by brief centrifugation. The peptide is dissolved by vortexing, and the contents of the tube are dried in vacuo over P_2O_5 and NaOH pellets (10 min). H_2O (3 μl) is added and the peptide and $NaHCO_3$ dissolved. Dansyl chloride solution (3 μl) is added, and the tube is centrifuged briefly. After mixing the contents, the tube is covered with parafilm and aluminium foil and incubated at 20 °C for 45 min in the dark. The reaction may be performed in 20 min at 37 °C if preferred. The reaction mixture, which should be almost colourless, is dried in vacuo over P_2O_5 and NaOH (10 min). 6 M HCl (10 μl) is added to the tube, and transferred to the bottom by centrifugation. The top of the tube is sealed in an oxygen flame, and the tube is placed in an oven or heating block at 105 °C for 16 h. Alternatively, hydrolysis at 140 °C for 1 h may be employed. It is important that the tube be heated evenly: a cooler top may lead to complete evaporation of the acid from the sample and condensation at the top, so that hydrolysis is incomplete. The cooled tube is scratched with a diamond pencil and broken open, and the acid is removed in vacuo over NaOH pellets (1 h). Complete removal of acid is essential.

The residue is dissolved in 1 μl of 50% aqueous pyridine, and half of the solution is applied as a spot about 3 mm in diameter to the origin on each side of a double-sided 5 cm × 5 cm polyamide layer plate (Schleicher and Schuell), about 1 cm from one corner. The spots are dried in warm air. Fine capillaries for the application of the sample may be drawn from melting-point tubes. The polyamide layers are conveniently handled with a sprung clothes peg, at the corner opposite the origin, and holes punched or drilled about 7 mm from the corner allow easy handling of up to ten layers simultan-

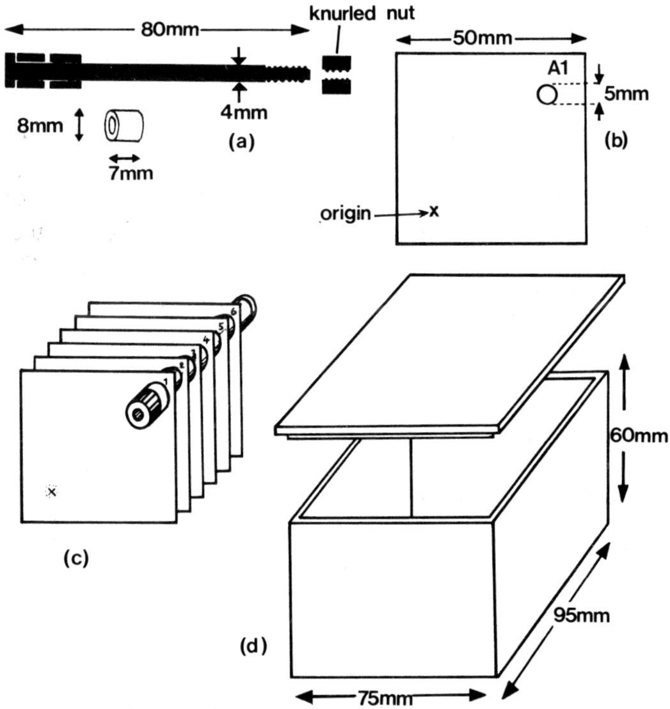

Fig. 6.3. Chromatography of dansyl-amino acids. (a) Stainless steel support rod and spacers. (b) Polyamide layer with hole punched or drilled for support rod. (c) Several double-sided polyamide layers held on the support rod for simultaneous chromatography and for drying in a stream of warm air. (d) Chromatography tank, internal dimensions. A flat-bottomed histological staining tank is suitable. About 15 ml of solvent is placed in each tank.

eously during chromatography and drying, through the use of a stainless steel support rod and spacers (Fig. 6.3). A standard mixture of seven dansyl amino acids (Dns-Pro, -Ile, -Phe, -Gly, -Glu, -Ser, and -Arg) in 50% (v/v) aqueous ethanol (0.1 μl, containing 0.1 nmol of each) is applied to the origin on one side only.

Chromatography is performed (following Woods and Wang, 1967) in solvent I (1.5% v/v aqueous formic acid) until the solvent front just reaches the top of the plate (about 5 min). The plate is dried in a stream of warm air (below 70 °C), and chromatography is performed in the second dimension in solvent II (toluene/acetic acid, 10 : 1, v/v), until the solvent front just reaches the top of the plate (about 5 min). The plate is dried in warm air, and inspected under the UV lamp (254 nm). The following dansyl amino acids may be identified: Dns-Pro, -Ile, -Leu, -Val, -Phe, -Gly, -Met, -Ala (overlapping the Dns-NH_2 spot), *bis*-Dns-Tyr and *bis*-Dns-Lys. In addition, the following sets of dansyl amino acids may be detected: Dns-Asp + Dns-Glu, Dns-Thr + Dns-Ser, and Dns-Arg, ϵ-Dns-Lys, α-Dns-Lys and α-Dns-His. N-terminal ϵ-*N*-(3-carboxypropionyl)-lysine residues (from succinylated proteins) yield two dansyl derivatives, one of which (the imide, succinyl-lysine, derivative) migrates close to Dns-Ala, while the other is α-Dns-Lys.

Particular note should be made of spots due to dansyl dipeptides, often observed when Val and Ile are the N-terminal residues. The distribution of dansyl amino acids and some dansyl dipeptides is shown in Fig. 6.4a. Clear distinction is made between Dns-Leu and Dns-Ile by reference to the standard spot of Dns-Ile on one side of the plate; if a single spot is observed in that position on both sides of the plate, Dns-Ile is present, while if two spots are present on the side with the standard markers, the sample contains Dns-Leu. Similar inspection allows a distinction to be made between Dns-Phe and Dns-Ile-Val, and other pairs of derivatives which migrate close together.

When all the spots (including the positions of unusual spots) have been recorded, chromatography is performed in solvent III (ethyl acetate/methanol/acetic acid, 20 : 1 : 1, by vol.) in the second

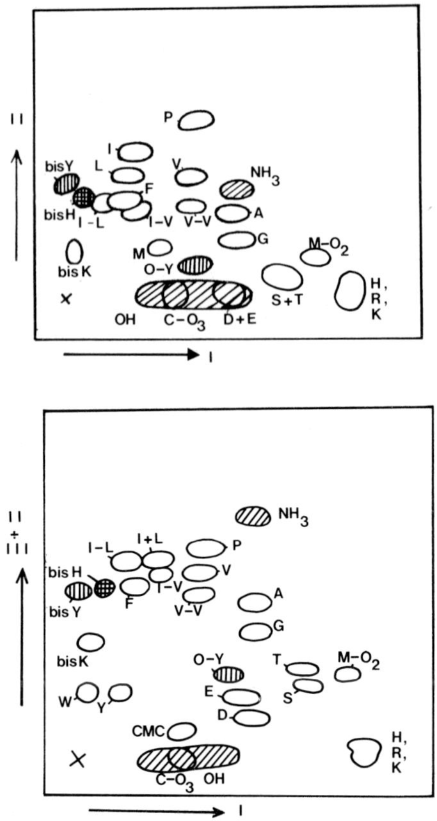

(a)

(a)

Fig. 6.4. Chromatography of dansyl-amino acids on polyamide layers. (a) After solvents I and II; (b) after solvents I, II and III (see text for details). Abbreviations: A, Dns-Ala; CMC, Dns-S-carboxymethyl-cysteine; C-O_3, Dns-cysteic acid; D, Dns-Asp; E, Dns-Glu; F, Dns-Phe; G, Dns-Gly; H, α-Dns-His; bisH, *bis*-Dns-His (not seen after acid hydrolysis); I, Dns-Ile; I-L, Dns-Ile-Leu; I-V, Dns-Ile-Val; K, α-Dns-Lys and ϵ-Dns-Lys; bisK, *bis*-Dns-Lys; L, Dns-Leu; M, Dns-Met; M-O_2, Dns-methionine sulphone; NH_3, Dns-NH_2; OH, dansic acid; O-Y, O-Dns-Tyr; P, Dns-Pro; R, Dns-Arg; S, Dns-Ser; T, Dns-Thr; V, Dns-Val; V-V, Dns-Val-Val; W, Dns-Trp; Y, α-Dns-Tyr; bisY, bis-Dns-Tyr. Dns-Ile-Ile chromatographs with Dns-Phe after solvents I + II, and with Dns-Leu after solvents I + II + III. The relative positions of some of the spots vary with different batches of polyamide layers. In particular, O-Dns-Tyr may overlap Dns-Glu, and Dns-methionine sulphone may lie between Dns-Ser and Dns-Thr. All spots fluoresce green (the shade depending on the dryness of the layer), except for Dns-NH_2 (blue-green), Dns-OH (blue), O-Dns-Tyr and *bis*-Dns-Tyr (yellow) and *bis*-Dns-His (orange). Impurities in solvents yield by-products which mostly have high R_F values in solvents II and III, and which do not interfere with the identifications.

dimension until the solvent front just reaches the top of the plate (about 4 min). The plate is dried and again inspected under the UV lamp. The following amino acid derivatives may be identified (Fig. 6.4b): Dns-Phe, -Ala, -Gly, -Thr, -Ser, -Met, -$MeSO_2$, -Glu, -Asp, -CMCys (usually weak), *bis*-Dns-Tyr, *bis*-Dns-Lys and O-Dns-Tyr. The imide derivative, α-Dns-ε-succinyllysine, migrates slightly further than Dns-Ala in solvent III. α-Dns-Tyr, which may be formed in small amounts from N-terminal tyrosine residues, is also identified. Dns-Leu and Dns-Ile are close, as are Dns-Val and Dns-Pro, and the monosubstituted derivatives of the basic amino acids are still not separated. Commonly encountered dipeptide derivatives of Val and Ile migrate further than Dns-Phe in solvent III, and may be clearly distinguished from the latter derivative even if this was not possible after chromatography in solvent II. Dns-Cysteic acid is obscured by the strong dansic acid spot. For the identification of this derivative and those of the basic amino acids, additional solvents for chromatography in the second dimension are required. Solvent IV (pyridine/acetic acid/water/ethanol, 1 : 2 : 100 : 34 by vol.) resolves α-Dns-His from ε-Dns-Lys and Dns-Arg. Another solvent, 50 mM Na_3PO_4/ethanol (3 : 1, v/v), resolves α-Dns-His, ε-Dns-Lys and Dns-Arg (Hartley, 1970), but often poorly. Other solvents have been described by Metrione (1978): H_2O/ethanol/ 35% NH_3 (17 : 2 : 1, by vol.) resolves these three basic derivatives. Dns-Trp is destroyed during the hydrolysis, yielding traces of Dns-Ser, -Ala and -Gly, but mainly Dns-NH_2 and Dns-OH. Other products are formed after performic acid oxidation of tryptophan residues, and Inglis et al. have recently described the identification of Dns-kynurenine, which is stable to acid hydrolysis (Inglis et al., 1979). The identification of 18 dansyl dipeptides has been described (Sutton and Bradshaw, 1978).

6.2.2. *Dabitc method*

The method for the use of the coloured derivative of phenylisothiocyanate, 4-*N*,*N*-dimethylaminoazobenzene-4′-isothiocyanate (Dabitc), for the determination of amino-terminal residues in peptides is essentially the same as that for the determination of amino-acid sequences (§ 6.5.1), except that the addition of phenylisothiocyanate is omitted. About 1 nmol of peptide is required, and the method has the advantage over the dansyl method that asparagine and glutamine are distinguished from aspartic and glutamic acids, respectively, and that tryptophan may be determined. The determination of N-terminal serine, threonine and lysine is less satisfactory, however, and additional, less sensitive chromatography on silica gel is required for differentiating leucine and isoleucine derivatives.

6.3. Determination of peptide sequences by the Edman degradation

By far the most important technique used for the determination of amino-acid sequences in peptides is the phenylisothiocyanate degradation procedure introduced and developed by Edman (1950, 1953, 1956; Ilse and Edman, 1963; Edman and Begg, 1967) and which bears his name. Because of the importance of the method, the chemistry of the degradation will be discussed in some detail, and several alternative procedures for sequence determination will then be described.

6.3.1. Chemistry of the Edman degradation

The chemistry has been discussed in great detail by Edman and Henschen (1975) and by Tarr (1977) among others. The degradation is a cyclic procedure, by which amino-acid residues are cleaved one at a time from the N-terminus of the peptide and identified as the phenylthiohydantoin derivatives. There are three steps in each cycle: *coupling* of phenylisothiocyanate (Pitc) with the amino-terminal residue, *cleavage* of the amino-terminal residue via cyclization in acidic media, and *conversion* of the thiazolinone derivative formed to the more stable thiohydantoin (Pth) derivative, which may be identified chromatographically.

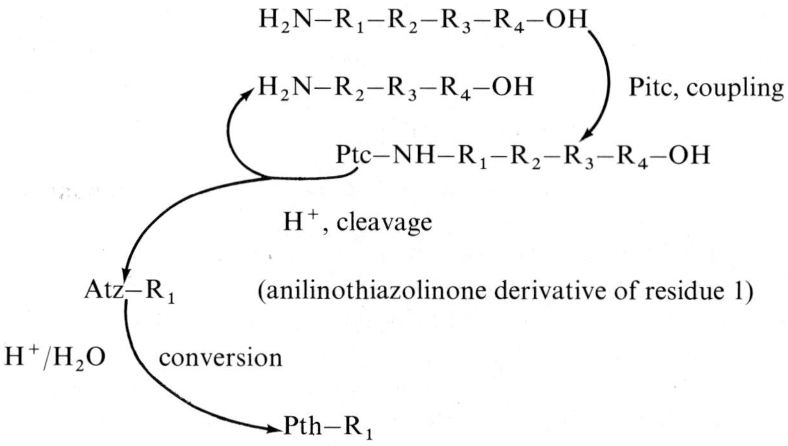

6.3.1.1. Coupling

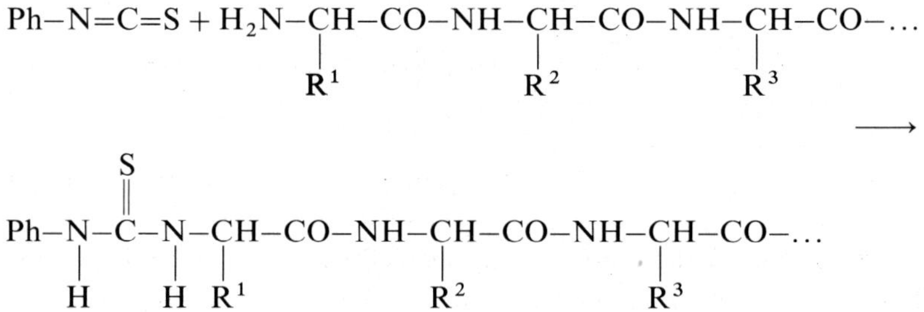

phenylthiocarbamyl- (Ptc-)peptide

The coupling reaction takes place in alkaline solution: the free amino group is required. The average pK_a of the terminal amino groups of peptides is 7.8 (Steinhardt and Beychok, 1964), and the pH of the solution should thus be above 8. Very alkaline conditions ($pH > 10$) should be avoided, since competing side reactions, such as hydrolysis of phenylisothiocyanate, are base-catalyzed. Hydrolysis of the reagent leads to the formation of aniline, which can react with a second molecule of phenylisothiocyanate:

$$Ph\text{–}NH_2 + Ph\text{–}N{=}C{=}S \longrightarrow Ph\text{–}NH\text{–}CS\text{–}NH\text{–}Ph \quad \text{(diphenylthiourea).}$$

Traces of ammonia (and primary and secondary amines) contaminating the peptide also react, e.g.:

$$NH_3 + Ph\text{–}N{=}C{=}S \longrightarrow Ph\text{–}NH\text{–}CS\text{–}NH_2 \quad \text{(phenylthiourea).}$$

Apart from consuming the reagent, these side reactions give products which may be difficult to remove from the coupled peptide, and which may interfere with the identification of the phenylthiohydantoin derivative.

The ϵ-NH_2 groups of lysine residues in the peptide also react with phenylisothiocyanate under alkaline conditions:

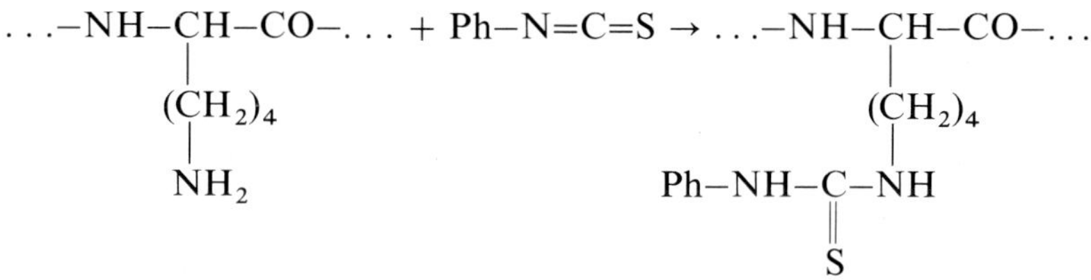

(N^{ϵ}-phenylthiocarbamyl-lysine residue)

Since the pK_a of the ϵ-NH_2 group of lysine residues is considerably higher than that of the α-NH_2 group of the peptide, coupling under slightly alkaline conditions (pH 8) leads to incomplete reaction at the ϵ-NH_2 groups, which are mainly protonated at this pH.

The solubility of phenylisothiocyanate in water is low, but a high concentration of the reagent is required to drive the reaction with the α-NH_2 group of the peptide to completion. Mixed organic-aqueous solvents are therefore used. Suitable media for the coupling reaction are pyridine/water (1 : 1, v/v) or 3 M *N*,*N*-dimethylallylamine solution, adjusted to pH 9.5 with trifluoroacetic acid. These solutions are good solvents both for phenylisothiocyanate and for many peptides. Many alternative solvents have been used.

Edman has studied the rates of reaction in some detail; different

N-terminal residues react at different rates, but essentially complete reaction of all N-terminal residues is obtained after 30 min at 50 °C with 2.5% (v/v) phenylisothiocyanate in the solvents mentioned above. Prolonged reaction times and higher temperatures lead to greater extents of undesirable side reactions.

Several factors may reduce the yield of the desired phenylthiocarbamyl peptide. Aldehyde impurities in the buffer solution cause partial blocking of the amino groups through the formation of Schiff bases and related products. Oxidizing reagents, including atmospheric O_2, may lead to desulphurization (Ilse and Edman, 1963). Acidic compounds contaminating the peptide, which may accumulate during many cycles of degradation, lower the pH of the solution and reduce the rate of reaction. Incomplete mixing of the peptide and reagent, perhaps due to poor solubility of the peptide, may also lead to low yields.

After the coupling step, excess reagent, solvents and by-products must be removed from the phenylthiocarbamyl peptide. If direct identification of the phenylthiohydantoin derivatives is to be made after the cleavage and conversion reactions, this removal of by-products must be efficient. (In the indirect 'dansyl-Edman' method, the new N-terminal amino acids are determined separately on small aliquots of the peptide after each cycle; the anilino thiazolinones are discarded and complete extraction of the by-products is not required. This is one of the major advantages of the indirect method.) The excess phenylisothiocyanate and the volatile solvents used may be removed in vacuo at 50 °C over P_2O_5 and NaOH pellets in 1 h, providing a good vacuum (<0.1 mm Hg) is maintained. The by-products, mainly phenylthiourea and diphenylthiourea, are not removed in this way, and solvent extraction is required. Several procedures for this extraction have been described. The dry reaction product may be extracted with an organic solvent, or a two-phase aqueous/organic extraction may be performed. Hydrophobic peptides may be lost into the organic solvent by either method. After the extraction, the coupled peptide is dried over P_2O_5 and NaOH pellets before the cleavage step.

6.3.1.2. Cleavage

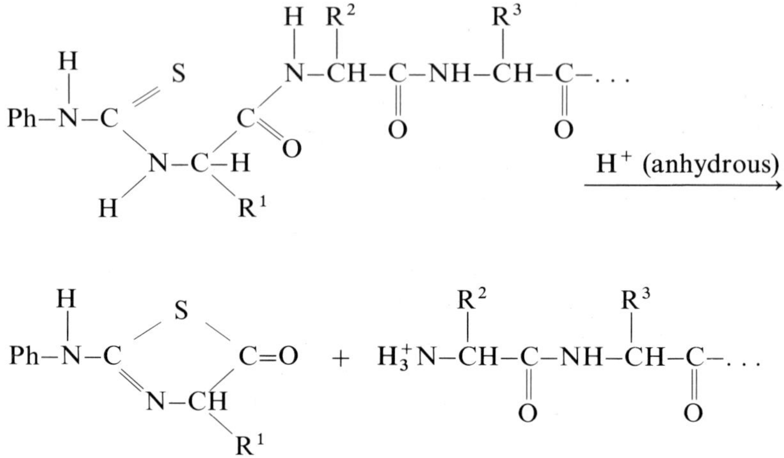

(anilinothiazolinone)

(salt of released peptide, with new N-terminal residue)

Most procedures in current use for the cleavage reaction employ anhydrous perfluorinated carboxylic acids, usually trifluoroacetic acid for the manual methods. Other reagents have included HCl in nitromethane (Edman, 1950), HCl in glacial acetic acid (Edman, 1953) and concentrated aqueous HCl (Tarr; 1977). The presence of water in the cleavage medium is generally avoided, because of the danger of acid-catalyzed hydrolysis of sensitive peptide bonds. It has, however, been observed that up to 4% water in heptafluorobutyric acid was not detrimental in the automated sequencer, while attempts to use absolutely anhydrous acid (by the addition of the anhydride) led to poorer results (Henschen-Edman, 1980). Conditions which may lead to the esterification of the hydroxyl groups of serine and threonine residues in the peptide, such as HCl in glacial acetic acid, should be avoided, for two reasons. Firstly, any O-acetyl groups bound to serine or threonine residues are rapidly transferred to the α-NH_2 group in the alkaline media used for coupling when these residues become N-terminal during the degradation:

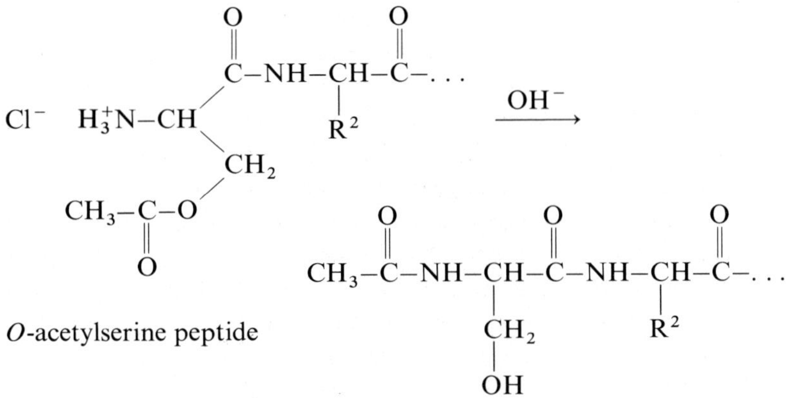

O-acetylserine peptide

N-acetylserine peptide.

The *N*-acetylserine peptide is blocked to further degradation. Secondly, the acetylated anilinothiazolinone or phenylthiohydantoin derivatives of serine and threonine are more susceptible to β-elimination during the cleavage and conversion reactions:

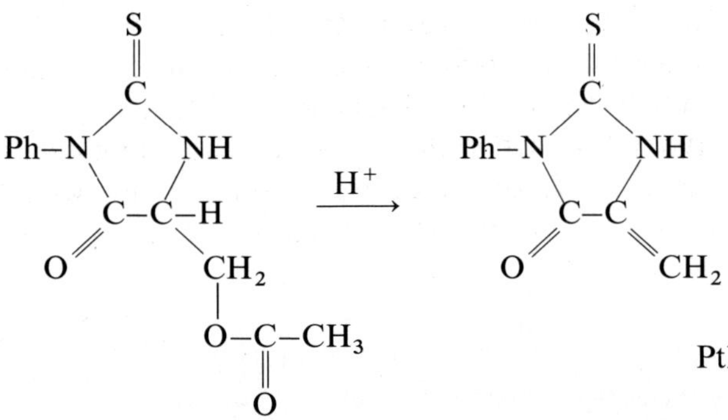

Pth-dehydroserine

further reactions

O-acetyl-Pth-serine

polymeric products

This β-elimination reaction of derivatives of serine and threonine residues is one of the major problems in the direct method. Tarr (1977) suggests that any acid capable of esterifying hydroxyl groups leads to this problem, and has therefore pursued an alternative reaction pathway, using concentrated aqueous HCl. Traces of water in the trifluoroacetic acid commonly used for cleavage may improve the yields of the phenylthiohydantoin derivatives of serine and threonine.

After the cleavage reaction, the volatile acid is removed in vacuo and the anilinothiazolinone derivative of the released amino acid is extracted from the salt of the residual peptide. Extraction of the dry residue, or a two-phase extraction, may be performed. The latter method is technically easier, especially when extremely small quantities of peptide are being degraded. Under acidic conditions, little extraction of peptides into butyl acetate is usually observed, although the presence of residual trifluoroacetic acid may enhance the solubility of peptides in some organic solvents. The organic phase, containing the anilinothiazolinone derivative, is dried in a stream of N_2 and converted to the phenylthiohydantoin.

6.3.1.3. Conversion The anilinothiazolinone derivatives of amino acids are not sufficiently stable to allow of reliable chromatographic identification. They are therefore converted to the more stable phenylthiohydantoin (Pth) derivatives. A commonly used method is heating at 80 °C for 10 min in 1.0 M HCl under N_2. Edman (1956; Ilse and Edman, 1963) showed that the thiazolinone ring first opens to give the intermediate phenylthiocarbamyl amino acid, which cyclises in a second step to the phenylthiohydantoin:

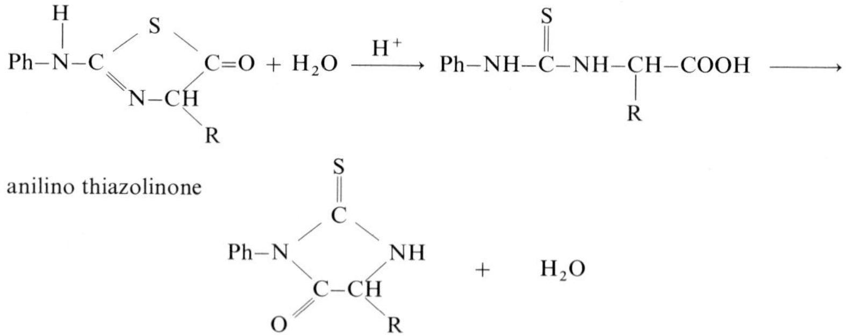

The rate-limiting step is the cyclization. Under the above conditions this is essentially complete for all derivatives except for that of glycine, which is unusually slow to cyclise (Ilse and Edman, 1963).

The side-chain amide groups of asparagine and glutamine derivatives are partially hydrolyzed under these conditions. Tarr (1975) has observed that the cyclization reaction is faster if methanol is used as the solvent, but conversion of the derivatives of aspartic acid and glutamic acid to their methyl esters also occurs.

An alternative method for the direct identification of released amino acids, which avoids some of the problems associated with the conversion to and determination of the phenylthiohydantoins, is reaction of the anilinothiazolinones with methylamine. The methyl amides of the phenylthiocarbamyl amino acids which are formed may be identified chromatographically (Jörnvall et al., 1978):

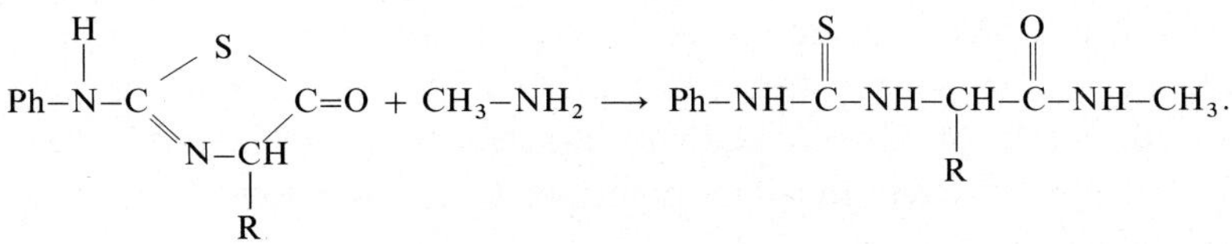

6.3.2. Some problems in the use of the Edman degradation

6.3.2.1. Incomplete coupling and cleavage reactions The rates of the reactions described above are affected by the structures of the side chains of the amino-acid residues at the N-terminus and, for the cleavage reaction, at the second position. The rates of reactions have been discussed by Tarr (1977), and the general conclusion is that cleavage of only Ptc-Gly-His- and Ptc-Pro-His- will be significantly incomplete under the conditions usually employed. Yields of 98% are common under optimum conditions for the whole cycle.

A gradually increasing proportion of the preceding residue is observed as the degradation proceeds through many cycles. This ‘overlap’ is due to incomplete coupling and cleavage reactions at each cycle. If quantitative determinations of the phenylthiohydantoins are made, the sequence may be determined correctly even when the newly released phenylthiohydantoin is present in a yield barely greater than that of the overlapping residue. Together with a gradually increasing 'background' resulting from small amounts of non-specific peptide bond cleavage during the degradation, the incompleteness of the reactions is the major factor limiting the application of the Edman degradation to at the most about 70 residues, using the automated liquid-phase sequencer, or about 30 residues using manual methods.

6.3.2.2. Histidine residues Phenylthiocarbamyl-histidine peptides are very rapidly cleaved, and this cleavage may occur in the coupling reaction mixture, allowing partial coupling of the following residue to take place. Two residues are therefore detected at this step and at each subsequent step. The 'preview' of the following residue is in contrast to the normally encountered overlap by the preceding residue. This behaviour of histidine residues, together with the abnormally slow cleavage when histidine is present at the second position, and the difficulty of identification of Pth-His by some methods, lead to some problems in the determination of the sequences of histidine-containing peptides.

The abnormal 'pre-cleavage' of histidine residues has often been observed (e.g. Blombäck et al., 1967; Tarr, 1977), and the conditions giving rise to this were investigated by Thomsen et al. (1972). During removal of volatile buffers in vacuo after the coupling reaction, the effective pH may decrease to about 6, which is sufficiently acidic to allow of the cleavage of some Ptc-His-peptides, particularly those with proline or glycine at the second position. Although the kinetics and mechanism of the reaction were not thoroughly examined, the side-chain imidazole group presumably catalyzes the reaction, possibly by the mechanism:

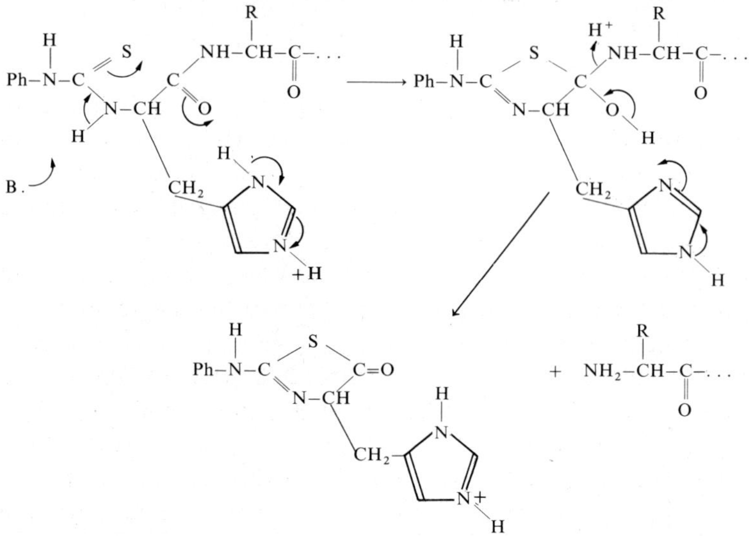

If non-volatile buffers, such as quadrol, are used, this abnormal cleavage of Ptc-histidine residues is not observed (Edman and Henschen, 1975).

It is interesting that a similar cleavage in the coupling medium of phenylthiocarbamyl-Nα-methyl-peptides has been observed (Chang, 1978).

6.3.2.3. Asparaginyl-glycine peptide bonds Under acidic conditions, the Asn-Gly sequence in peptides is converted to an imide structure; hydrolysis of the imide under alkaline conditions leads preferentially to the formation of the β-aspartyl peptide bond:

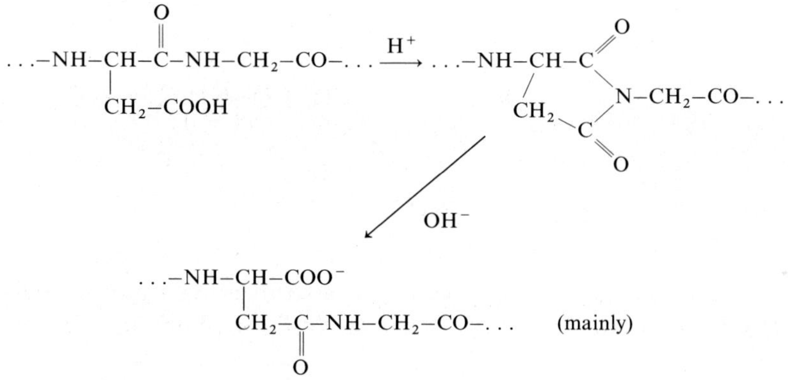

The side chains of residues other than glycine sterically inhibit the formation of the imide in other sequences containing asparagine residues. Treatment of peptides containing aspartic acid with carbodiimides or other reagents activating the carboxyl group may also lead to the formation of imide structures.

Both the imide and the β-aspartyl linkages are resistant to the Edman degradation, and it is often observed that degradation of a peptide or protein through Asn-Gly sequences is associated with a drastic reduction in yield of the liberated phenylthiohydantoins. The imide may, however, be cleaved with hydroxylamine (§ 3.2.2.1; Bornstein, 1970), and cleavage of the protein by this reagent, followed by the determination of the sequences of the resulting peptides, is the best way of circumventing this problem, especially when automated sequencing methods are used.

6.3.2.4. Glutamine and S-*carboxymethylcysteine residues* Partial blockage to further degradation may be observed when glutamine residues are N-terminal, through cyclization under acidic conditions to pyrrolidonecarboxylyl (pyroglutamyl) residues (Smyth et al., 1963):

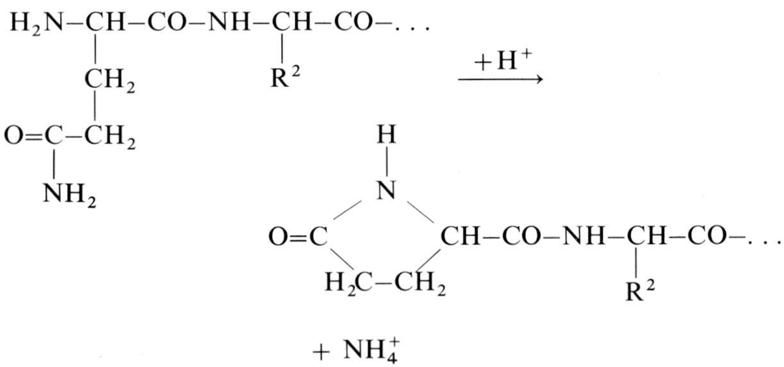

A similar cyclization may occur with N-terminal *S*-carboxymethylcysteine residues (Smyth and Utsumi, 1967; Bradbury and Smyth,

1973). In practice this cyclization is not a serious problem if the peptide is not allowed to remain under acidic conditions for more than about 15 min at each cycle of the degradation. These reactions give greater problems during the purification of peptides, however (Sanger and Thompson, 1953; Smyth et al., 1962; § 6.1.2), and make the determination of the sequences of peptides with N-terminal glutamine difficult (§ 7.5.2).

6.3.2.5. Tryptophan residues Tryptophan residues are unstable under acidic conditions, especially in the presence of oxidizing reagents, including atmospheric O_2. Some of the reaction products may block the Edman degradation, and the destruction of tryptophan also leads to problems in the identification of this residue. Another problem may be oxidative cleavage of tryptophanyl peptide bonds under acidic conditions. The destruction of tryptophan residues in trifluoroacetic acid was studied by Uphaus et al. (1959); one of the products from N^{α}-acetyl tryptophan had the structure:

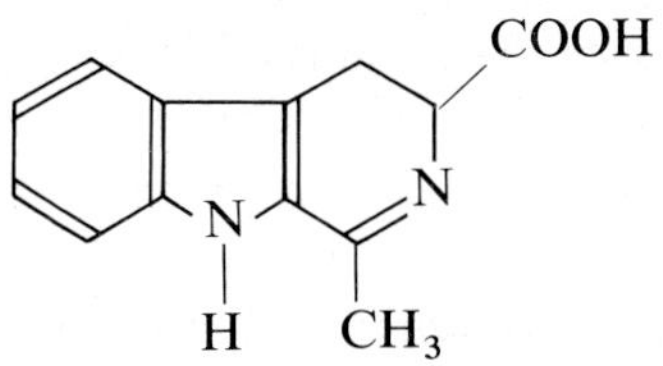

In practice, satisfactory degradation through tryptophan residues is usually achieved, especially using the automated spinning-cup sequencer, where a complete barrier against O_2 is maintained, but the identification of this residue becomes difficult after many cycles of degradation.

Inglis et al. (1979) have recently described the identification of the derivatives of tryptophan which arise after periodate oxidation of the protein.

6.3.2.6. Glycosylated residues The determination of the sequences of peptides bearing O-glycosyl groups on serine or threonine residues often presents difficulties, especially when several such glycosyl groups are present. Examples include erythrocyte glycophorin (Tomita et al., 1978) and antifreeze glycoproteins from antarctic fish (Komatsu et al., 1970; Morris et al., 1978). Apart from the difficulty of identification of the glycosylated residues (the hydrophilic glycosyl groups prevent extraction of the anilinothiazolinone derivatives), the ready β-elimination of such residues, and possibly the aldehyde groups of released carbohydrate, may contribute to these problems. Degradation through N-glycosyl asparagine residues proceeds normally, but the anilinothiazolinone is not extracted, and the glycosylated residue is not normally identified directly. Methods for the investigation of glycosylation sites are described later (§ 7.4).

6.3.2.7. Other blocking reactions The blockage to further degradation caused by O→N migration of acyl groups on serine and threonine residues has been mentioned above (§ 6.3.1.2). Similar blockage occurs if O-acyl groups are present in the isolated protein, or introduced by chemical modification such as maleylation (Jörnvall, 1970) or succinylation (Allen and Harris, 1976). *S*-Acylcysteine residues would give similar problems, but cysteine residues are usually converted to more stable derivatives for protein structure determination.

The possibility of interference with the Edman degradation by other groups introduced by post-translational modification should also be considered, and careful inspection of the results for losses in yield or unusual chromatographic spots during sequential degradation may reveal unsuspected modifications.

6.4. Experimental procedures for the Edman degradation

Many procedures for the direct Edman degradation, with identification of the phenylthiohydantoins, have been described. There are three major techniques: manual liquid phase, automated liquid phase (using a spinning cup), and automated solid phase. Each of these techniques yields phenylthiohydantoin derivatives which may be identified in a variety of ways. The manual methods will be discussed first, together with methods for the identification of phenylthiohydantoins which are applicable to all of the direct procedures. Then the automated liquid phase and solid phase methods will be described without going into the detailed technology of the machines themselves. Finally the indirect methods ('dansyl-Edman' and 'subtractive Edman' degradations, §§ 6.4.4, 6.4.5) will be described. Alternative degradative methods are discussed in sections 6.5–6.8.

6.4.1. Manual methods and identification of phenylthiohydantoins

A large number of methods, differing in the composition of the coupling buffer solution, methods of extraction of by-products and anilinothiazolinones, and in other details, for the direct manual phenylisothiocyanate degradation have been used. These methods generally require 50–250 nmol of peptide, although with identification of the phenylthiohydantoins by the recently developed high-performance liquid chromatographic (hplc) methods the sensitivity is higher. One procedure, essentially that of Peterson et al. (1972), has been described by Glazer et al. (1975) in a volume in this series, and the details are not repeated here. Other procedures which have given satisfactory results have been described by Tarr (1977), Sauer et al. (1974) and Edman and Henschen (1975).

For all of these procedures, scrupulous experimental technique is required, and the identification of the phenylthiohydantoins at high sensitivity is difficult. Other methods, using 4-*N*,*N*-dimethylamino-azobenzene-4′-isothiocyanate (§ 6.5) or dansyl chloride (§ 6.4.4), are therefore recommended.

Each of the direct Edman procedures yields anilinothiazolinone derivatives, which are converted to the phenylthiohydantoins by incubation in 1 M HCl at 80 °C for 10 min. The phenylthiohydantoins are extracted into ethyl acetate, which is then evaporated in a stream of nitrogen. The phenylthiohydantoin derivatives of arginine, histidine and cysteic acid remain in the aqueous phase, and separate identification of these is required after freeze-drying.

No single method for the identification of the phenylthiohydantoins is completely satisfactory (although hplc is almost so), and the results of at least two methods are usually considered together. Quantitation is important when long sequences are being degraded; three quantitative methods frequently used are high performance liquid chromatography (hplc), gas-liquid chromatography (glc) and 'back hydrolysis' to regenerate the amino acid followed by amino-acid analysis. Several systems for the thin-layer chromatography (tlc) of phenylthiohydantoins have been described. Mass spectrometric identification of phenylthiohydantoins has also been used. Radioactive labelling of some residues, especially of cysteine derivatives such as *S*-carboxymethylcysteine, greatly aids their identification.

The quantitative methods for the determination of phenylthiohydantoins have been developed almost exclusively for use with the automated sequencers, and while they may be used in conjunction with manual methods they do not greatly improve the quality of the results, which are limited much more by the inefficiency of the manual degradation procedures. For this reason, only brief outlines of the quantitative methods are given here, and the reader is referred to more specialized texts (Niall, 1977; Bridgen et al., 1975; Zimmerman et al., 1977) for experimental details.

6.4.1.1. High-performance liquid chromatography This recently developed technique has replaced gas-liquid chromatography as the primary method for the quantitative determination of phenylthiohydantoins in many laboratories. The reverse-phase mode, in which the stationary phase is hydrophobic, consisting of a coating of long-chain alkyl groups bonded to silica microspheres, and the moving phase is aqueous, is most suitable. Many systems have been described, and examples are given in Table 6.1. A gradient of increasing concentration of a polar organic solvent, either methanol or acetonitrile (Zimmerman et al., 1977), is usually employed, but isocratic elution has certain advantages. Most of the phenylthiohydantoin derivatives of amino acids may be identified in a single chromatographic run, but the resolution of some pairs of derivatives in a particular system

TABLE 6.1
Examples of recently reported systems for high-performance liquid chromatography of phenylthiohydantoin derivatives of amino acids

Zimmerman et al. (1977)	Zorbax ODS column, gradient of CH_3CN in 0.01 M sodium acetate buffer, pH 4.5.
Bhown et al. (1978b)	Gradient of MeOH, Pth-Val and -Met coelute;
(1978a)	Application to radio-labelled derivatives.
Abrahamson et al. (1978)	LiChrosorb column, isocratic elution with 35% CH_3CN, Pth-Met and Pro coelute.
Devillers-Thiery and Blobel (1978)	Modification of published procedures for use with radio-labelled derivatives.
Elion et al. (1978)	MicroBondapak C_{18} column, gradient of methanol in acetate buffer, Pth-Val and Pth-Met coelute.
McKean and Maurer (1978)	Zorbax ODS column, gradient of CH_3CN in 0.01 M sodium acetate, pH 3.5, applied to radio-labelled derivatives, Pth-Met and -Val coelute.
Margolies and Brauer (1978)	MicroBondapak C_{18}, gradient of CH_3CN, use of A_{313} to determine Pth-dehydrothreonine.
Van Beeumen et al. (1978)	Micro-hplc column of ODS-18 (0.5 mm diameter column).
Zeeuws and Strosberg (1978)	Similar system to that of Elion et al.
Annan (1979)	Use of diethylenetriamine–trichloroacetate buffer.
Gates et al. (1979)	Zorbax ODS column, modification of the method of Zimmerman et al., with isocratic elution.
Moser and Rickli (1979)	Spherisorb S5-ODS column, complex gradient of CH_3CN in lithium acetate buffer.
Hunkapiller and Hood (1980)	Zorbax CN column; sensitivity 5 pmol, using computer baseline correction.

may be incomplete (e.g. those of valine and methionine in the system of McKean and Maurer [1978]). Slight adjustments of temperature, pH, acetonitrile concentration, or salt concentration may be required for optimal resolution on different columns (Gates et al., 1979). An example of the separation of phenylthiohydantoin amino acids by hplc is given in Fig. 6.5.

The hplc equipment is rather expensive, and will often only be available for routine use in laboratories using automated sequencing instruments. A typical chromatographic run requires 20 min, so hplc

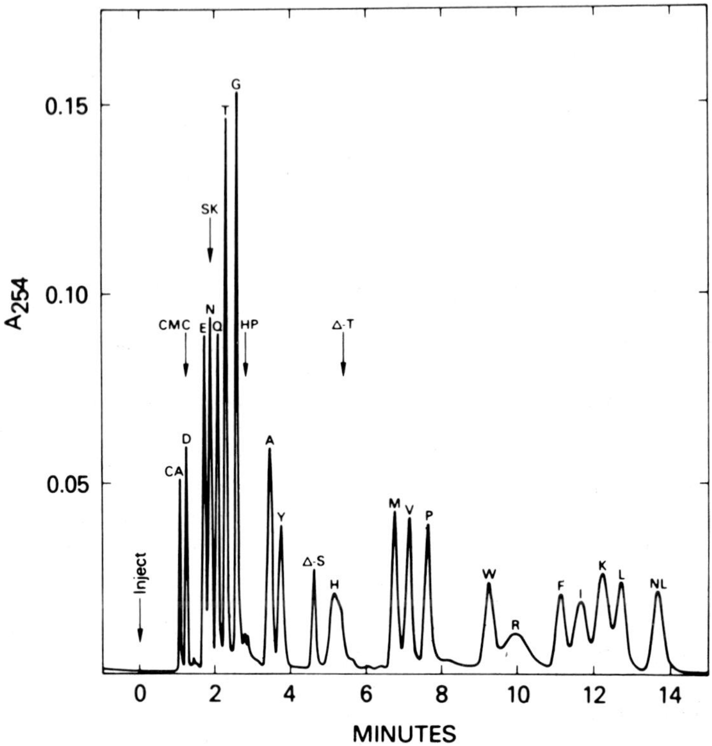

Fig. 6.5. High-performance liquid chromatography of phenylthiohydantoin-amino acids. A standard containing 1 mg/ml Pth-amino acids in 3 μl of ethyl acetate/methanol (1 : 1) was analyzed on a Waters Associates Model 204 liquid chromatograph equipped with a DuPont Zorbax ODS (25 cm × 4 mm) column with a 55°C water jacket. The column was eluted isocratically by using 32% acetonitrile and 68% 0.02 M sodium acetate, pH 4.52, containing 1% acetonitrile, at 2 ml/min. Arrows show the location of additional Pth amino acid derivatives. CA, Pth-cysteic acid; CMC, Pth-(carboxymethyl)cysteine; SK, Pth-succinyllysine; HP, Pth-hydroxyproline; Δ-S, Pth-dehydroserine; Δ-T, Pth-dehydrothreonine; NL, Pth-norleucine. Pth-serine elutes slightly behind Pth-glutamine. Pth-Cys elutes in two peaks: slightly behind Pth-Gly and slightly before Pth-Δ-Ser. The standard one-letter code is used for other derivatives. Reprinted with permission from Gates, F.T., III, Coligan, J.E. and Kindt, T.J. (1979) Biochemistry, *18*, 2267–2272. Copyright 1979 American Chemical Society.

is compatible with these machines, in which the degradative cycle time is about 1 h. For routine use, 1–10 nmol of each phenylthiohydantoin may be quantitated readily; high-sensitivity applications are described below (§ 6.4.2.2).

6.4.1.2. Gas-liquid chromatography Until recently, this was perhaps the most important method for the quantitative determination of phenylthiohydantoins. However, the more polar phenylthiohydantoins cannot be identified directly by this method, requiring conversion to more volatile derivatives (e.g. by trimethylsilylation), and phenylthiohydantoin derivatives of arginine and cysteic acid are not identified at all. One of the more widely used gas chromatographic procedures is that of Pisano and Bronzert (1969). Experimental details have been given by Pisano (1972), Niall (1973) and Bridgen et al. (1975).

6.4.1.3. Back hydrolysis and amino-acid analysis Phenylthiohydantoin derivatives of amino acids may be hydrolytically converted back to the amino acids. The conditions required are severe, and lead to the destruction of some derivatives and poor recoveries of several amino acids. Either acidic or alkaline conditions may be used; Table 6.2 lists some of the procedures.

Hydrolysis with HCl in the presence of $SnCl_2$ (Mendez and Lai, 1975) is a significant improvement on the widely used method with HI (Smithies et al., 1971). Both these procedures convert serine and

TABLE 6.2

Methods for the hydrolysis of phenylthiohydantoin derivatives to amino acids.

All hydrolyses are performed with highly purified reagents in sealed glass tubes in vacuo
6 M HCl, 150 °C, 24 h (Van Orden and Carpenter, 1964)
0.1 M NaOH, 120 °C, 12 h (Africa and Carpenter, 1966)
55% HI, 130 °C, 20 h (Smithies et al., 1971)
0.2 M NaOH, 0.1 M $Na_2S_2O_4$, 127 °C, 3.5 h (Smithies et al., 1971)
5.7 M HCl, 0.1% (w/v) $SnCl_2$, 150 °C, 4 h (Mendez and Lai, 1975)

S-carboxymethylcysteine phenylthiohydantoins to alanine, which must therefore be separately determined by a non-hydrolytic method, threonine phenylthiohydantoin to α-aminobutyric acid, which chromatographs between alanine and valine on many analyzers (Bridgen et al., 1975), and isoleucine phenylthiohydantoin to a mixture of isoleucine and *allo*-isoleucine. Tryptophan phenylthiohydantoin is destroyed by both methods, and methionine phenylthiohydantoin by the HI method. The derivatives of asparagine and glutamine are converted to aspartic acid and glutamic acid, respectively, by all the hydrolytic methods. Alkaline hydrolysis with dithionite allows the determination of tryptophan and methionine, but the phenylthiohydantoins of serine, threonine, arginine and derivatives of cysteine are destroyed (Smithies et al., 1971).

A major disadvantage of amino-acid analysis after hydrolysis for the identification of phenylthiohydantoins is the time required for each analysis: if most samples from an automated sequencer are to be analyzed by this method, a modern, high-speed analyzer totally dedicated to this purpose is required. In addition, a residue builds up on the resin, which necessitates frequent re-packing of the top of the chromatography column, if the HI method is used (Bridgen et al., 1975). In practice, therefore, this method is often reserved for the analysis of samples which have not yielded definitive results using other methods of identification, and it is particularly useful for the determination of the phenylthiohydantoins of histidine, arginine and cysteic acid, which are difficult to identify by thin-layer chromatography or gas chromatography.

6.4.1.4. Thin-layer chromatography The method for the identification of phenylthiohydantoins most commonly employed in conjunction with the direct manual Edman degradation is thin-layer chromatography. Many chromatographic systems have been described, with either silica gel or polyamide as the stationary phase. A combination of at least two solvent systems is required for the identification of most of the phenylthiohydantoins. One set of solvents used widely is that of Edman and Henschen (1975), and a large number of systems has been reviewed by Rosmus and Deyl (1972). One of the simpler systems, which is nevertheless capable of routine separation of all the commonly encountered phenylthiohydantoins except those of histidine and arginine, is that described by Bridgen et al. (1975).

Silica gel plates containing fluorescent indicator (Merck, aluminium foil backed, No. 5554), 20 cm in height, are used. Solvent I is chloroform/ethanol (98 : 2, v/v) (Analar grade chloroform from BDH, containing 2% ethanol as a preservative). Solvent II is chloroform/ethanol/methanol (88.2 : 1.8 : 10, by vol.), prepared by mixing solvent I with methanol in the ratio 9 : 1. The solvent tanks are equilibrated for at least 1 h before use, and fresh solvents are used daily.

The samples are dissolved in ethyl acetate, to give a phenylthiohydantoin concentration of about 1 nmol/μl, and 5–10 μl of each sample is applied to the thin layer. Up to 20 samples may be applied on a single 20 cm × 20 cm thin-layer plate. Several spots of marker solution, containing 10 nmol each of the phenylthiohydantoins of all amino acids except histidine, cysteine and arginine (the lysine derivative is present as the ε-*N*-phenylthiocarbamyl compound) are also placed at the origin, among the sample spots. Appropriate derivatives of cysteine and other chemically modified residues, such as N^{ϵ}-3-carboxypropionyl-lysine, may be included in the marker solution.

The development in solvent I takes about 2 h. When the solvent front reaches the top of the plate, the plate is dried in a stream of air and inspected under the UV lamp (254 nm). Phenylthiohydantoins are detected as dark spots against the fluorescent background, and are identified by comparison with the marker spots. A permanent record is made by photography or tracing. The plate is then developed in the same direction with solvent II, dried, and similarly inspected.

The separation of the phenylthiohydantoins is shown in Fig. 6.6. Very careful comparison with the standards is required to discriminate between the phenylthiohydantoin derivatives of leucine and isoleucine, valine and phenylalanine, glycine and tryptophan, and glutamine and asparagine. Small amounts of the phenylthiohydantoins of aspartic acid and glutamic acid, derived by partial hydrolysis of the derivatives of asparagine and glutamine, respectively, during the conversion reaction, aid in the identification of the latter derivatives. Partial destruction via β-elimination of phenylthiohydantoins of threonine and serine leads to low recoveries of these derivatives.

The dehydrothreonine derivative may be identified chromatograpically, but the dehydroserine derivative is unstable and is itself converted to unidentified products (cf. the dimethylaminoazobenzene analogue, § 6.5). Methionine residues in proteins may be partially oxidized to the sulphoxide during the isolation of peptides, or converted to methionine sulphone by performic acid oxidation. Homoserine residues, derived from methionine residues by treatment of the protein with CNBr, yield homoserine phenylthiohydantoin, which may also be identified by tlc. Derivatives of cysteine are best identified by liquid-scintillation counting after alkylation of the protein with a radioactive reagent. The phenylthiohydantoins of all these derivatives should be subjected to chromatography in the same systems so that they may be recognized if encountered during sequence determination, rather than being possibly mis-identified.

By-products of the reactions of the Edman degradation, principally phenylthiourea and diphenylthiourea, are also detected on fluorescent plates under UV light, and care is required to discriminate between these and phenylthiohydantoins of hydrophobic amino acids.

The identification of the phenylthiohydantoins may be aided by spraying with reagents which yield different colours with different derivatives. Ninhydrin/collidine (Roseau and Pantel, 1969; Inglis and Nicholls, 1973) and iodine/azide (Cherbuliez et al., 1964) each produce a variety of colours, as described by Bridgen et al. (1975).

The phenylthiohydantoin derivatives of histidine and arginine, which remain in the aqueous phase after the conversion reaction, may be identified by tlc on silica gel (Eastman Chromogram 6060) in xylene/95% ethanol/acetic acid (50:50:0.5 by vol.) (Inagami,

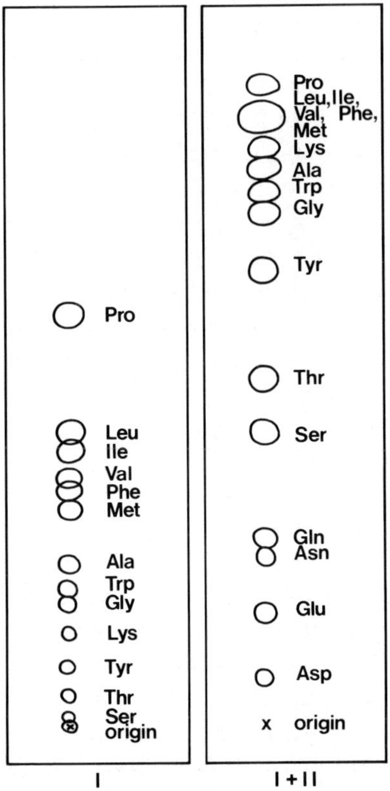

Fig. 6.6. Chromatography of phenylthiohydantoin-amino acids on silica gel plates. A standard mixture of Pth-amino acids (10 nmol each) (except Pth-His, -Arg, and -Cys), separated in chloroform/ethanol (98 : 2, v/v) (I), followed by chloroform/ethanol/methanol (88.2 : 1.8 : 10, by vol.) (II), as described by Bridgen et al. (1975).

1973). Confirmation of the identifications may be made with a modified Pauly reagent for histidine and a phenanthrenequinone reagent for arginine (Easley et al., 1969; Reed and McKay, 1978). Alternatively, thin-layer electrophoresis (Edman and Begg, 1967) followed by detection with these reagents or with the ninhydrin or iodine/azide reagents may be used. A more reliable method is hydrolysis and amino-acid analysis (§ 6.4.1.3).

For the clear identification of the phenylthiohydantoins by these tlc methods, allowing for poor yields of some derivatives, particularly that of serine, about 50 nmol of the phenylthiohydantoins produced during sequencing is required. Many more sensitive procedures have been described. Chromatography on polyamide layers, described below, allows detection of less than 1 nmol of each phenylthiohydantoin, but discrimination between the derivatives of leucine and isoleucine and of histidine and arginine is difficult to achieve. Other techniques for detection at high sensitivities by tlc are listed in Table 6.3. In practice, these methods (except those using intrinsic radioactive labelling) cannot be used to full advantage, since the presence of by-products of the Edman degradation limits the amount of sample which can be applied to the thin layers without distortion of the development or obscuration of the phenylthiohydantoin spots.

Kulbe (197, 1974) and Summers et al. (1973) have described the separation of phenylthiohydantoins on polyamide layers coated with a fluorescent indicator. The following procedure is that of Kulbe (1974).

Double-sided micropolyamide sheets (Schleicher and Schuell) are cut to 5 cm × 5 cm. Three chromatography tanks are required, with the solvents I, toluene/*n*-pentane/acetic acid (60 : 30 : 16, by vol.); II, 25% aqueous acetic acid; III, 40% aqueous pyridine/acetic acid (9 : 1, v/v). The ratio of the volume of the solvent to that of the tank should be about 1 : 50. Solvent I contains in addition 250 mg butyl-PBD (2-(4′-*t*-butylphenyl)-5-(4″-biphenylyl)-1,3,4-oxadiazole) (from Koch-Light Laboratories) per litre. The tanks are pre-equilibrated for 10 min before chromatography, and each solvent may be used at least ten times before being renewed.

The sample is applied as a small spot (2 mm diameter) on one side of the polyamide sheet. On the other side, exactly opposite, the sample is applied together with a marker mixture consisting of the phenylthiohydantoin derivatives of methionine, aspartic acid, asparagine, glycine and leucine, and the by-products of the coupling reaction, *N*-phenylthiourea and *N*,*N*′-diphenylthiourea. The marker solution is made to a concentration of 2 mM of each component in methanol, and about 0.2 μl (0.4 nmol) is applied to the thin layer.

TABLE 6.3

Examples of methods for high-sensitivity detection of phenylthiohydantoin amino acids by thin-layer chromatography.

(1) Use of [^{35}S]phenylisothiocyanate in the coupling reaction, followed by two-dimensional chromatography of the [^{35}S]phenylthiohydantoins on polyamide layers, with detection by radioautography (Bridgen, 1977). This method is suitable for use with the solid-phase sequencer. High backgrounds were observed if [^{35}S]phenylisothiocyanate of very high specific activity was used (Waterfield and Bridgen, 1975; Bridgen, 1977).

(2) Use of [^{35}S]phenylisothiocyanate and chromatography on silica gel followed by extraction of spots (at positions detected using carrier phenylthiohydantoins) and liquid-scintillation counting (Jacobs and Niall, 1975).

(3) High-performance thin-layer chromatography (hptlc) on Merck HPTLC silica gel 60 F_{254} (Bucher, 1977). Three solvents, I, chloroform/ethanol/ethyl acetate (88.65 : 1.35 : 10), II, chloroform/ethanol/acetic acid (88.65 : 11.35 : 2), III, 1,2-dichloroethane/ethyl acetate (4 : 1) (all by vol.), were used for chromatography in a single dimension, with inspection under a UV lamp (254 nm) after each development. All amino-acid phenylthiohydantoins normally encountered during sequencing, except for those of arginine and histidine, as well as *N*-phenylthiourea and *N*,*N*′-diphenylthiourea, could be identified in less than 1 h with a sensitivity better than 1 nmol.

(4) Use of two-dimensional chromatography on polyamide layers containing a mixture of fluorescent additives (polyamide FM plates, Wako, Osaka, Japan) (Nakamura et al., 1979). The identification of derivatives was aided by different colours seen under a UV lamp, and the limit of detection was 0.08 nmol for each phenylthiohydantoin.

(5) Use of calcein and Pd (II) for detection of phenylthiohydantoins (Inglis and Nicholls, 1974).

(6) Treatment of thin layers with iodine before inspection under the UV lamp (Inglis et al., 1974).

(7) Use of intrinsically labelled (^{3}H) proteins (Coligan et al., 1978). Total cell protein was labelled with ^{3}H-labelled amino acids in tissue culture. The radio-labelled protein of interest was purified and degraded in the automated sequencer. The [^{3}H]phenylthiohydantoin derivatives were separated on polyamide layers by the method of Summers et al. (1973) in the presence of unlabelled carrier phenylthiohydantoins. The spots detected under the UV lamp were cut out and the radioactivity was determined by liquid-scintillation counting (Coligan et al., 1978).

The thin layer is placed vertically in solvent I. When the solvent front just reaches the top of the plate (about 8 min), the plate is removed and completely dried in warm air. Chromatography is then performed in the second dimension in solvent II.

The dried plate is inspected under the UV lamp (254 nm). The phenylthiohydantoins are detected as dark spots against a luminous background. Identification of the phenylthiohydantoins of histidine, tryptophan and tyrosine is facilitated by their yellow colour. The sensitivity is claimed to be in the range 0.05–0.2 nmol (Summers et al., 1973), but 0.5–1 nmol is more reliably detected. A permanent record may be made by photographing the plates under UV light through a Wratten No. 45 filter.

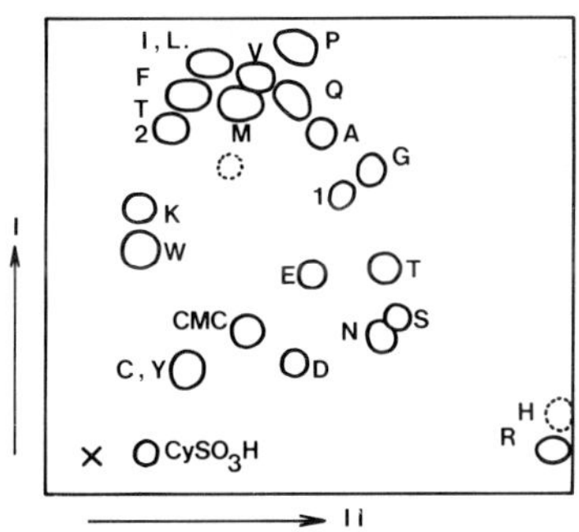

Fig. 6.7. Separation of phenylthiohydantoin-amino acids on polyamide layers (Kulbe, 1974). Solvent I, toluene/*n*-pentane/acetic acid (60 : 30 : 16, by vol.); solvent II, 25% aqueous acetic acid. The single-letter abbreviations for amino acids are used; CMC, Pth-(carboxymethyl)cysteine; $CySO_3H$, Pth-cysteic acid; 1, *N*-phenylthiourea: 2, *N,N*′-diphenylthiourea.

The distribution of the phenylthiohydantoins is shown in Fig. 6.7. The pairs of derivatives of histidine and arginine, leucine and isoleucine, and cysteine and tyrosine are not resolved; the last of these is not serious, since cysteine residues are normally modified before sequence analysis. In addition, the dehydrothreonine and phenylalanine derivatives are not resolved. If one of these pairs is identified, the derivatives are resolved by chromatography of a second portion of the sample in solvent III on plates precoated with butyl-PBD. Appropriate markers (phenylthiohydantoins of leucine, isoleucine, threonine, dehydrothreonine and phenylalanine) are applied beside the sample for this one-dimensional separation.

The phenylthiohydantoins of histidine and arginine, which remain in the aqueous phase after extraction of the conversion mixture with ethyl acetate, are extracted from the dried aqueous layer after the addition of a little ammonia solution, and chromatographed in solvent II. The separation is poor, and the less sensitive colour reactions described above, or amino-acid analysis after hydrolysis, are more reliable.

Kulbe (1974) has also described a series of one-dimensional separations of phenylthiohydantoins on polyamide layers.

6.4.2. Automated liquid-phase sequence determination

Commercially produced automated sequencing instruments modelled on that described by Edman and Begg (1967), in which the coupling and cleavage reactions are performed in a spinning cup, are widely used. As noted in the Preface, detailed descriptions of automated methods are not given here; excellent accounts have been given by Niall (1973, 1977), Waterfield and Bridgen (1975) and Edman and Henschen (1975). A brief outline of the technique is presented here, together with a discussion of some of the recently introduced improvements.

There are several technical problems in the direct manual method of degradation, including the difficulties of maintaining an effective barrier against atmospheric O_2 and performing effective extractions. Only about four cycles per day may be performed. The automated instrument solves these problems, and allows about 24 cycles to be performed per day. The speed of the automated approach for peptide degradation should not be overemphasized, however, since many peptides may be degraded in parallel by manual methods. Degradation of proteins and large polypeptides (of more than about 50 residues) is not satisfactorily performed by manual liquid-phase methods, owing to the difficulties of extractions, insolubility of the protein in the coupling buffer and thorough drying of the bulky material. In the spinning cup, the protein forms a thin film, through which the reagents may diffuse even if the protein is not soluble. In favourable cases, up to 70 amino acid residues may be identified at the N-termini of proteins, but reliable results are more often obtained for only the N-terminal 20–50 residues (Niall, 1977).

In outline, the sequence of operations is as follows. The protein is dissolved in heptafluorobutyric acid and introduced into the spinning cup. The vacuum is applied so that the protein dries as a thin film around the sides of the cup. A solution of phenylisothiocyanate is then introduced, followed by the coupling buffer (1 M Quadrol* [*N,N,N′,N′*-tetrakist(2-hydroxypropyl)ethylenediamine] in 1-propanol/water [3 : 2, v/v], pH 9.0). After the coupling reaction (about 30 min at 50 °C), the excess reagent and solvents are removed in vacuo, and the Quadrol and by-products are extracted with benzene and ethyl acetate. The film of phenylthiocarbamyl protein is dried thoroughly, and heptafluorobutyric acid is introduced. The cleavage reaction takes place in this acid in about 3 min. Most of the acid is removed in vacuo. The cleaved anilinothiazolinone is extracted with 1-chlorobutane and collected in a fraction-collector tube. The dried protein film is then ready for the second cycle of degradation. The anilinothiazolinone is converted to the phenylthiohydantoin and identified as described above.

A number of technical improvements to the instrument have been made in recent years, including the redesign of the vacuum system, with the incorporation of a cold trap and a continuous record of the vacuum in the cup (Wittmann-Liebold, 1973; Wittmann-Liebold et al., 1975), and the addition of an automatic conversion device (Wittmann-Liebold et al., 1976). Similar modifications, together with the use of very highly purified reagents and with identification of phenylthiohydantoins by a sensitive hplc method have allowed

* Quadrol is a registered trade mark of the Wyandotte Chemicals Corporation, Wyandotte, Michigan.

quite long sequences to be obtained at the sub-nanomolar level (Hunkapiller and Hood, 1978; 1980). Frank (1979) has described a cheap and simple method for the purification of sequencer reagents.

For the efficient operation of the sequencer, regular cleaning and careful maintenance are required, with particular attention payed to vacuum seals, frequent renewal of the pump oil, correct operation of valves and freedom from corrosion of all components. Determination of the sequence at the N-terminus of a standard protein, such as myoglobin, is routinely performed as a check on the functioning of the instrument. In the hands of experienced operators, spinning-cup sequencers have proved extremely valuable, and for many applications are almost indispensible.

Application of the technique to small peptides and small quantities (less than about 2 mg) of proteins requires modifications to the standard procedures.

6.4.2.1. Application to peptides In order to extract completely the Quadrol after the coupling step, a relatively polar solvent, ethyl acetate, is required, and coupled peptides may be washed from the cup in this solvent. Substitution of the Quadrol by volatile tertiary amines, such as dimethylallylamine (Niall et al., 1969), or tertiary amines soluble in benzene, such as dimethylbenzylamine (Hermodson et al., 1972) allows the ethyl acetate extraction to be omitted. More thorough removal of the heptafluorobutyric acid after the cleavage reaction is also needed, to reduce the loss of peptides from the cup in the 1-chlorobutane (Niall, 1973).

A second method of aiding the retention of peptides in the cup is chemical modification of the peptides to render them more hydrophilic; Braunitzer and co-workers (1970, 1971) reacted the ϵ-NH_2 groups of lysine residues with sulphonated isothiocyanates, such as 1,5-disulphonylnaphthalene-3-isothiocyanate. The N-terminal residue also reacts, but is cleaved in acid before the degradation begins. Peptides containing arginine residues tend to be retained in the cup (Waterfield and Bridgen, 1975). Other methods for converting peptides to more hydrophilic derivatives have been proposed by Foster et al. (1973), (carbodiimide coupling of carboxyl groups with hydrophilic amines) and Braunitzer and Pfletschinger (1978) (for treatment of CNBr peptides with C-terminal homoserine lactone).

Thirdly, the use of a film-stabilizing substance, such as Polybrene*, reduces wash-out of peptides as well as small quantities of proteins, as described below. This appears to be the method of choice, but on the whole the difficulties of sequencing small peptides (less than 20 residues) in the spinning-cup sequencer are better avoided through the use of manual techniques.

6.4.2.2. High-sensitivity applications When small quantities (less than 1–2 mg) of proteins are subjected to the standard sequencing procedures, the repetitive yield is low because the protein is extracted from the cup. The use of a dilute (0.1 M) Quadrol buffer with less vigorous extractions reduces the loss of the protein (Brauer et al., 1975), but the most successful approach has been the addition of a film-stabilizing carrier, such as a copolymer of norleucine and arginine (Niall et al., 1974), a partially sulphonylated derivative of polyornithine (Waterfield and Bridgen, 1975), or a synthetic poly-quaternary amine, Polybrene (Tarr et al., 1978; Klapper et al., 1978). The last of these is particularly satisfactory, but the addition of the Polybrene (5 mg) together with glycylglycine (50–100 nmol) to the cup, and preliminary performance of three cycles of degradation are recommended before the addition of the protein or peptide (Hunkapiller and Hood, 1978). This serves to block any reactive groups in the polymer which might interfere with the degradation or with the identification of the phenylthiohydantoins.

In addition to the requirement for a carrier, high-sensitivity application of the spinning-cup sequencer requires a sensitive method for the determination of phenylthiohydantoins. High-performance liquid chromatography, thin-layer chromatography on polyamide

* Polybrene is a registered trade mark for 1,5-dimethyl-1,5-diazaundecamethylene polymethobromide (Aldrich Chemical Co.).

layers or hptlc plates, and high-sensitivity amino-acid analysis after hydrolysis are suitable for analysis at the 1 nmol level. After allowing for the difficulty in identifying some residues, especially serine, for the decreasing yield of the phenylthiohydantoins as the degradation proceeds through a large number of cycles, and for the need to use two methods for reliable determinations, a reasonable minimum quantity of protein required for extended sequence determination using these methods is about 25 nmol. Hunkapiller and Hood (1978) were, however, able to identify the N-terminal 77 residues in an immunoglobulin light chain from only 5 nmol of the protein. Using an improved instrument and highly purified reagents for the degradation and liquid chromatography of the phenylthiohydantoins (Hunkapiller and Hood, 1980), they were able to determine the amino-terminal 20 residues of human lymphoblastoid interferon from about 500 pmol of protein (Zoon et al., 1980).

As the quantity of protein is reduced, interference by impurities in the reagents and by by-products with the detection of the phenylthiohydantoins becomes a limiting factor. Radiochemical methods have been developed in order to overcome this problem.

Jacobs and Niall (1975) used [^{35}S]phenylisothiocyanate of high specific activity, followed by thin-layer chromatography in the presence of unlabelled carrier phenylthiohydantoins, extraction of the spots and liquid-scintillation counting. This method is applicable at the 5 nmol level (Jacobs et al., 1973). The use of [^{35}S]phenylisothiocyanate of very high specific activity, which would be required for a further increase in sensitivity, results in unacceptably high backgrounds (Waterfield and Bridgen, 1975), and is also expensive.

An effective way of increasing the sensitivity to the sub-nanomolar range is through the use of radio-labelled protein in the presence of unlabelled carrier proteins and phenylthiohydantoins. The radioactivity may be introduced into the protein in three ways, in vivo, in vitro using mRNA-dependent cell-free translation, or by chemical modification. The latter two methods are applicable only where limited sequence information, such as N-terminal sequences of nascent proteins (§ 10.5) and identification of positions in the sequence of residues (e.g. cysteine) which may be selectively modified, is required. The use of in vivo (tissue culture) radio-labelling to high specific activity of proteins which may be obtained in only small amounts (of the order of 1 nmol) has been described by McKean et al. (1974), Ballou et al. (1976) and Coligan et al. (1978). The method requires very large amounts (several hundred millicuries) of tritium-labelled amino acids for complete sequence determinations, and difficulties are encountered with some amino acids due to metabolic conversions. Quantitative determination of the phenylthiohydantoins is required, which necessitates the use of high-performance liquid chromatographs or amino-acid analyzers coupled with fraction collectors so that each phenylthiohydantoin may be separated and subjected to liquid-scintillation counting. Careful correction for relative specific activities (after metabolic interconversions) is required. It should be noted that some hydrogen atoms in amino-acid derivatives are exchangeable, particularly under the conditions of hydrolysis of phenylthiohydantoins, so that low specific activity may be retained in, for example, [^{3}H]glycine derivatives (both α-^{3}H atoms are exchangeable in glycine phenylthiohydantoin).

For the determination of the amino-acid sequences of the murine major histocompatibility antigens (H-2K and H-2D gene products), Nathenson and co-workers (Coligan et al., 1978) have used labelling with groups of amino acids selected to minimize problems arising from metabolic conversions and to aid in the identification of some amino-acid phenylthiohydantoins which are not unambiguously identified chromatographically. Double-isotope labelling is also used to advantage. In addition to the expenditure on equipment and radiochemicals, considerably more work is required for the fractionation of peptides derived from the protein, although the ready detection of the radioactive peptides is some compensation.

The alternative approach to the determination of amino-acid sequences of proteins which are present in only small amounts in cells is to use very large numbers of cells; the technical difficulties are then transferred from the field of sequence analysis to that of protein purification and possibly that of tissue culture. The very high degrees of purification which may be obtained using modern techniques of affinity chromatography (Lowe, 1979; Turková, 1978) may make this the alternative of choice. For example, the sequence of a human major histocompatibility antigen, from a high-producer cell line, has been determined by chemical, rather than radiochemical, techniques (Orr et al., 1979).

6.4.3. *Solid-phase sequencing*

An alternative approach to the solution of problems incurred during the extraction of excess reagents, by-products and anilino thiazolinones from the peptide is to use covalent attachment of the peptide to an insoluble support (Laursen, 1966). Efficient extractions may then be performed without the risk of loss of the peptide, and the degradation may be automated relatively simply. The insoluble support is placed in a column fitted with lines for the sequential delivery of reagents and collection of the anilinothiazolinone derivatives (Laursen, 1971, 1972; Laursen et al., 1975). Automated solid-phase sequencing machines are commercially available. Satisfactory operation of these requires a good understanding of the chemistry involved, and particularly of the means of attachment of peptides to the support. Two useful compendia (Laursen, 1975; Previero and Colletti-Previero, 1977) contain descriptions of many techniques for attachment and degradation of peptides and proteins by this approach.

Detailed descriptions of the construction and operation of automated solid-phase sequencers have been given by Laursen (1971; Laursen et al., 1975); only the chemistry is discussed here.

6.4.3.1. Types of solid support In the original work of Laursen (1966), polystyrene resin, modified by chloromethylation or nitration and reduction, as introduced by Merrifield (1964) for solid-phase peptide synthesis, was used. Although the polystyrene resins are often used for peptide sequence determination, they have pronounced disadvantages, including deformability, differential swelling in different solvents, and a small pore size, resulting in very low attachment efficiency for larger peptides and proteins. The physical disadvantages may be circumvented by admixture with a large excess of glass beads for packing into the column (Laursen et al., 1975), but this leads to larger reaction volumes and higher consumption of reagents than would otherwise be needed. An alternative resin, designed for sequence work, has been described (Atherton et al., 1976).

1. Aminopolystyrene

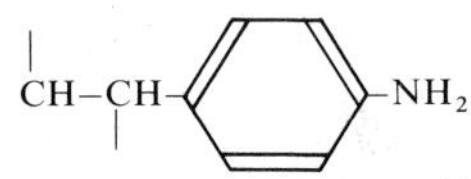

2. Triethylenetetramine-polystyrene

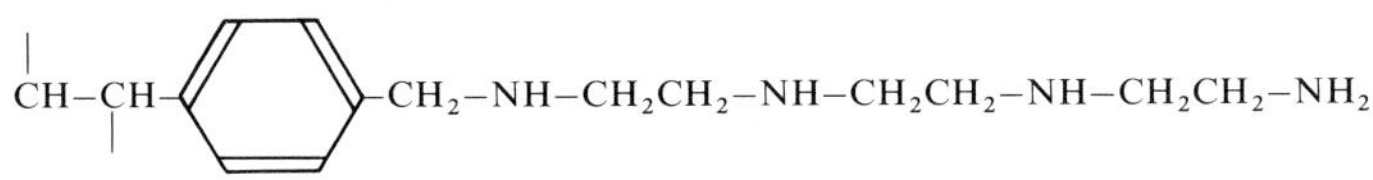

3. 3-amino-propyl-glass

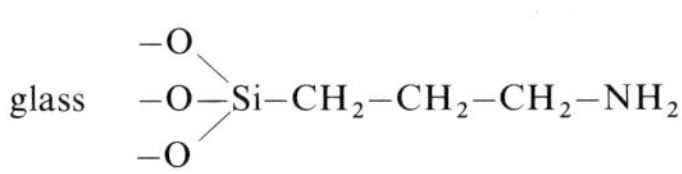

4. Isothiocyanato-glass

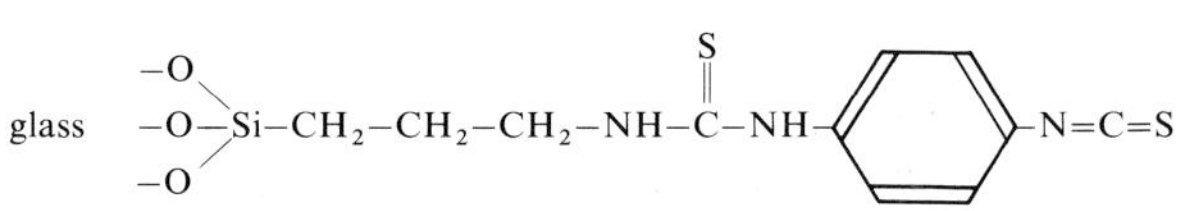

5. Iodoacetyl glass

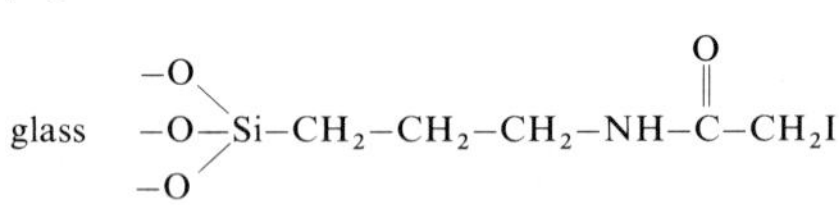

Fig. 6.8. Some types of support for solid-phase sequencing.

Porous glass beads have proved to be extremely valuable for the attachment of larger peptides and proteins (Wachter et al., 1973; Machleidt et al., 1973); their use has been reviewed in detail (Wachter et al., 1975; Machleidt et al., 1975; Machleidt and Wachter, 1977). The glass beads have excellent physical properties. A slight chemical instability under alkaline conditions does not seriously detract from their usefulness. The surface of the glass may be converted to amino-glass (Robinson et al., 1971) which may then be further treated to give covalently bound reactive groups, such as isothiocyanato (Bridgen, 1975, 1976) or iodoacetyl (Chang et al., 1977a), suitable for direct attachment of proteins and peptides. The structures of various solid supports are given in Fig. 6.8.

Triethylenetetramine polystyrene beads may be prepared via chloromethylation of Bio-Beads S-X1, minus 400 mesh (Bio-Rad Laboratories, Ltd.) (Laursen, 1971) or from commercially available chloromethyl polystyrene beads containing about 1.5 mEq chloromethyl groups per gram of resin. The dry resin (1 g) is stirred with triethylenetetramine (12.5 ml) at room temperature for 30 min, then at 100 °C for 1.5 h. The beads are washed thoroughly with methanol, stirred with triethylamine, washed with methanol, water, and methanol again and dried at 60 °C overnight (Horn and Laursen, 1973). The triethylenetetramine resin is unstable, and is stored for up to 1 month at −20 °C.

3-Aminopropyl-glass (from Robinson et al., 1971) Controlled-pore glass beads, 200–400 mesh, with a mean pore size of 7.5 nm (e.g. Corning CPG 10–75Å, CPG-75 from Serva, Heidelberg, or from Pierce Chemical Co.) are immersed in a 4% solution of 3-aminopropyltriethoxysilane (Pierce) in acetone. The mixture is incubated at 45 °C for 24 h. The beads are washed thoroughly with acetone, with methanol, and dried in vacuo (care is required to prevent the beads from blowing away in the stream of vapour: a glass sinter or filter-paper cover to the vial is needed).

N-(2-Aminoethyl)-3-aminopropyl glass may be prepared similarly, using *N*-(2-aminoethyl)-3-aminopropyltrimethoxysilane (Bridgen, 1976).

Isothiocyanato-glass (Bridgen, 1976) Amino-propyl glass beads are added in small portions to a 5% solution of *p*-phenylene diisothiocyanate (5 mmol, 960 mg/g of glass) in pure dimethyl formamide over 30 min. The mixture is stirred gently for a further 30 min at 20 °C. The beads are then filtered, washed with dimethyl formamide and with methanol and dried in vacuo (care: see above). The isothiocyanato-glass may be stored for at least 6 months at −20 °C under N_2.

Iodoacetyl glass (Chang et al., 1977a) 3-Aminopropyl glass (600 mg) is suspended in ethyl acetate (4 ml) containing iodoacetic acid (90 mg), and dicyclohexylcarbodiimide (100 mg) is added. The mixture is incubated at 20 °C, with gentle swirling, in the dark for 2 h. The iodoacetyl glass beads are collected on a sintered glass funnel and extensively washed with water, with acetone, and dried in vacuo (care: see above). The iodoacetyl glass is stored at −20 °C in the dark. The iodoacetyl groups may be determined by reaction with excess glutathione, followed by thorough washing, acid hydrolysis and amino-acid analysis.

6.4.3.2. Attachment of peptides to the solid support The ideal point of attachment of the peptide to the support is solely through the C-terminal residue. Such attachment would allow the complete sequence to be determined, providing that other factors, such as declining yields, a high background or extensive overlap did not first call a halt to the degradation. Unfortunately there is no satisfactory general procedure for the attachment of the C-terminal residue alone. The use of carbonyldiimidazole (Laursen, 1971) or water-soluble carbodiimides (Previero et al., 1973) for activation of the terminal carboxyl group for attachment to aminopolystyrene has not been found reliable, although satisfactory attachment is often obtained (Laursen, 1975). Apart from the activation of the side-chain carboxyl groups of glutamic and aspartic acid residues, leading to attachment through these residues in addition to the C-terminus, a serious problem is the formation of a cyclic imide structure with activated aspartyl peptides, preventing the Edman degradation from proceeding beyond the aspartyl residue:

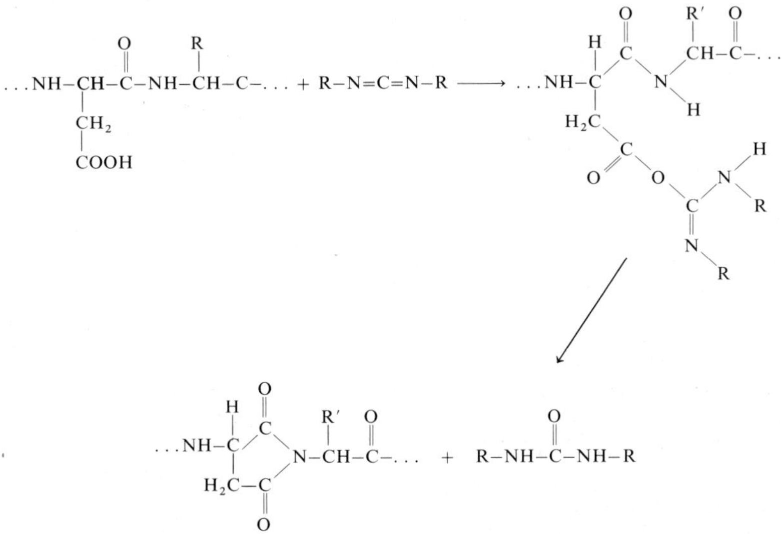

The imide opens preferentially in aqueous alkali to give the β-aspartyl peptide bond, also incapable of cleavage in the Edman degradation (§ 6.3.2.3).

Peptides containing carboxyl-terminal homoserine residues, derived by cleavage of the protein at methionine residues with CNBr, may be attached, generally in high yields, to amino-substituted resins or glass beads by aminolysis of the lactone form of the peptide which is produced by prior treatment with anhydrous trifluoroacetic acid (Horn and Laursen, 1973):

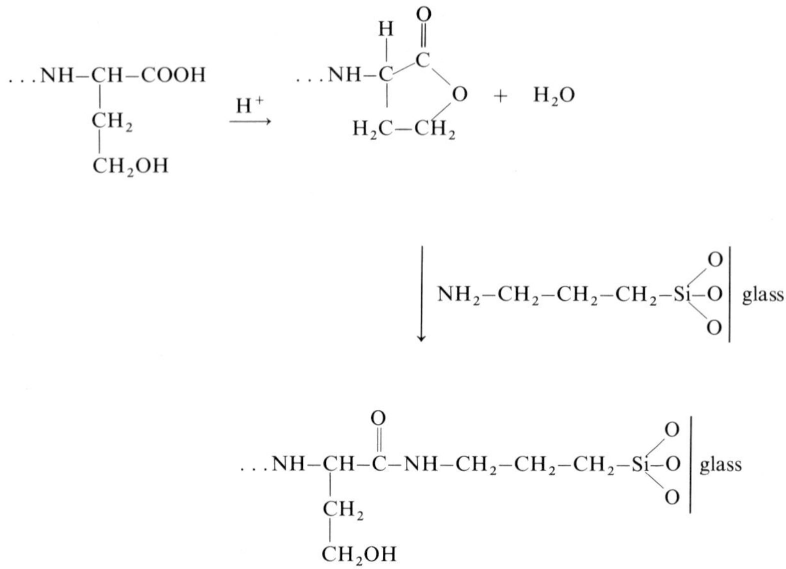

The side reactions, hydrolysis of the lactone and reaction of the amino groups in the peptide with the lactone, are minimized through the use of a dipolar aprotic solvent where possible, and the presence of a large excess of the amino-groups on the solid support.

Peptides with C-terminal lysine residues (e.g. many tryptic peptides) may be attached through the ϵ-NH_2 groups of these residues to isothiocyanato-glass or -resins. The terminal α-NH_2 group also reacts, but during treatment with trifluoroacetic acid the N-terminal amino acid is cleaved off, while attachment through the ϵ-NH_2 group is stable. The derivative of the amino-terminal residue remains attached to the support and is not identified; sequential degradation starts at the second residue and may continue up to the penultimate residue, the C-terminal lysine residue not being released from the support. The attachment reactions are:

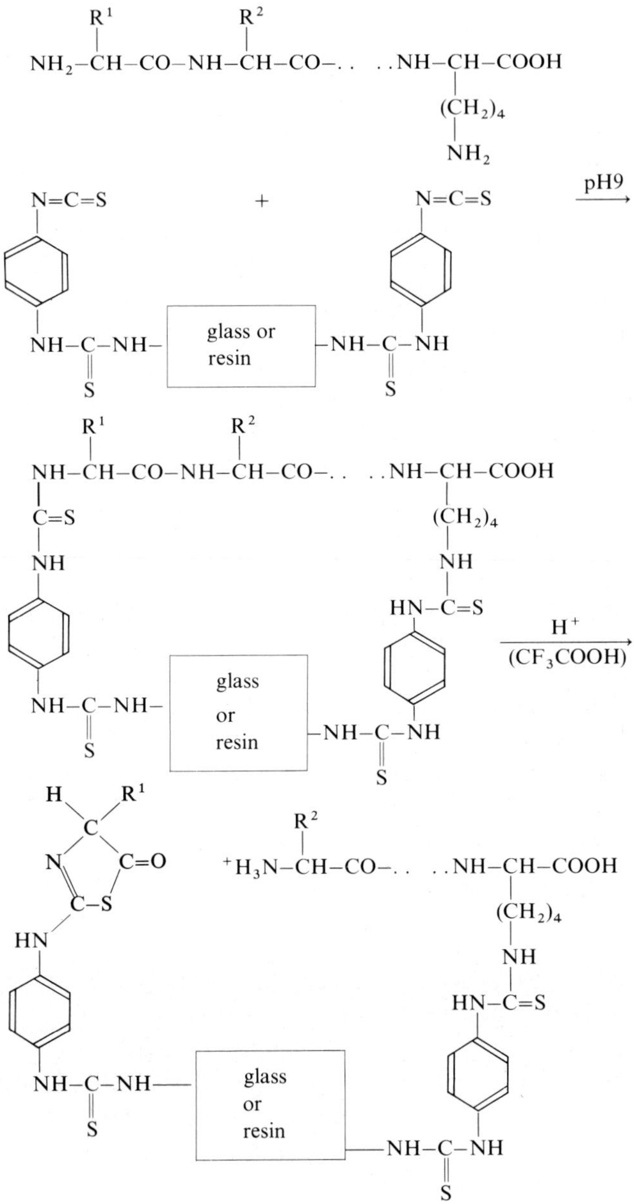
R1
R2
NH2–CH–CO–NH–CH–CO–. . . .NH–CH–COOH
(CH2)4
NH2
N=C=S
+
N=C=S
pH9
NH–C–NH–
glass or resin
–NH–C–NH
S
S
R1
R2
NH–CH–CO–NH–CH–CO–. . . .NH–CH–COOH
C=S
NH
(CH2)4
NH
HN–C=S
H+
(CF3COOH)
NH–C–NH–
glass or resin
–NH–C–NH
S
S
H
R1
C
N
C=O
C–S
HN
R2
+H3N–CH–CO–. . . .NH–CH–COOH
(CH2)4
NH
HN–C=S
NH–C–NH–
glass or resin
–NH–C–NH
S
S

An alternative method for attachment of peptides through amino-groups is to treat the peptides with *p*-phenylene diisothiocyanate, followed by reaction with amino-resin. The sequences of *E. coli* ribosomal proteins L7 and L12 were determined almost exclusively by this method (Terhorst et al., 1973).

This method has been extended for use with tryptic peptides with C-terminal arginine residues, via conversion of the arginine residues to ornithine residues with hydrazine (Laursen et al., 1972), but the yields are low and there are deleterious side reactions.

With longer peptides, or proteins, where it is not anticipated that complete sequences would be determined directly, attachment at several points along the peptide chain is generally used, since this is more efficient than attempting attachment at a single group at the C-terminus. Attachment through the side chains of lysine residues to isothiocyanato-glass is generally applicable. When lysine residues are encountered during the degradation, gaps in the sequence, or low yields of the lysine derivative, will be observed.

An alternative procedure for proteins containing cysteine residues is attachment to iodoacetyl glass (Chang et al., 1977a). This reaction is rapid and specific, and since cysteine residues are relatively rare in proteins little information is lost through the inability to identify these attached residues.

Laursen (1977) has described several attachment methods.

Attachment of peptides with C-terminal homoserine to 3-aminopropyl-glass (modified from Horn and Laursen, 1973) The peptide (100–300 nmol), which must be free from ammonia or amines, is treated with dry trifluoroacetic acid (0.2 ml) for 1 h at 40 °C. The acid is removed in a stream of N_2. The peptide is dissolved in dry dimethyl-formamide (0.5 ml) and aminopropyl-glass (100 mg) is added. It has been observed that the solubility of peptides in dimethyl-formamide is often enhanced by treatment with trifluoroacetic acid (Wachter et al., 1975), but if necessary a small amount of water may be added to help to dissolve the peptide. Triethylamine (0.1 ml) is added, and the mixture is swirled gently for 2 h at 40 °C. Application of water-pump vacuum followed by a return to atmospheric pressure is used to displace air trapped within the beads, which would otherwise prevent efficient attachment.

Methyl isothiocyanate (0.1 ml) is then added to block excess amino groups on the glass, and gentle swirling is continued for 1 h at 40 °C under N_2. The beads are collected by filtration and washed thoroughly with dimethylformamide (1 ml) followed by methanol (5 ml) and dried in vacuo. Care is needed to prevent the glass beads from being carried off in the stream of vapour during drying: a filter-paper cover is fitted to the flask. The beads are then treated with trifluoroacetic acid, to cleave off the N-terminal residue as the methylthiazolinone, which is not identified. The peptide, minus the N-terminal residue, is then subjected to the Edman degradation.

Attachment of peptides and proteins to isothiocyanato-glass Isothiocyanato-glass (100 mg) (§ 6.4.3.1) is suspended in 50% (v/v) aqueous pyridine (0.2 ml), and a solution of the peptide (with C-terminal lysine) (100 nmol) in the same solvent (0.2 ml) is added. The peptide must be free from NH_3 or amines. The mixture is de-gassed under water-pump vacuum, then gently swirled under oxygen-free N_2 at 40 °C for 1 h. To ensure complete phenylthiocarbamylation of the α-NH_2 group, phenylisothiocyanate (0.01 ml) is added, and the incubation continued under N_2 for a further 30 min at 40 °C. The beads, with the attached peptide, are washed on a sintered glass funnel with 50% pyridine (1 ml), and resuspended in this solvent (0.2 ml). Remaining isothiocyanate groups on the glass beads, which could interfere with the subsequent degradation, are blocked by the addition of ethanolamine (0.05 ml) and incubation under N_2 for 30 min at 40 °C. The beads are then collected on a filter funnel, washed with 50% pyridine (1 ml) and with methanol (2 ml), and dried in vacuo. Trifluoroacetic acid (0.2 ml) is added, and the mixture is incubated under N_2 at 40 °C for 20 min. The glass beads are again collected by filtration, washed with methanol, and dried. The Edman degradation is then performed from the second residue.

Proteins which are not soluble in 50% aqueous pyridine or

equivalent solvents such as N-methylmorpholine/water (1 : 1, v/v) may be attached to isothiocyanato-glass in 6 M guanidinium chloride, or in sodium dodecyl sulphate solution. The latter solution is particularly suitable for the attachment of proteins extracted from polyacrylamide gels after electrophoresis in the presence of this detergent (Bridgen, 1976, 1977).

Attachment of proteins containing cysteine residues to iodoacetyl glass (Chang et al., 1977a) The protein (250 nmol) is dissolved in 1.5 ml of 0.4 M Tris/HCl, pH 8.25, containing 8 M urea and 2 mM EDTA. If cystine residues, but no free thiol groups, are present, a 2-fold molar excess over cystine residues of dithiothreitol is added, and reduction is allowed to proceed for 1 h at 37 °C under N_2. To the solution is added 200 mg iodoacetyl-glass beads (§ 6.4.3.1) and the mixture is stirred gently at 20 °C in the dark under N_2 for 30 min. 2-Mercaptoethanol (20 μl) is added, and stirring is continued for 30 min. The beads are washed with the Tris/urea buffer, with 50% aqueous pyridine and with H_2O. Coupling yields of about 70% were obtained.

The method may be scaled down, and 6 M guanidinium chloride may be used instead of urea.

6.4.3.3. Edman degradation procedures The same cycle of chemical reactions is performed in solid-phase sequencing as in liquid-phase techniques, but there is more freedom of choice of solvents for the coupling reaction and for extractions. Buffers need not be volatile or soluble in ethyl acetate, for example, and anhydrous conditions may be used throughout. Laursen et al. (1975) have noted that the purity of reagents is not so critical with the solid-phase as with the liquid-phase technique, possibly because reactive impurities are scavenged by groups on the solid support; it is, however, recommended that reagents of the highest purity available are used. A suitable coupling buffer (Laursen et al., 1975) is prepared from 3 parts of pyridine and 2 parts of *N*-methylmorpholine buffer (pH 8.1) by volume. The buffer is prepared from 28 ml *N*-methylmorpholine and 3.75 ml of trifluoroacetic acid, made up to 200 ml with water.

Since the peptide is covalently fixed to the support, a relatively polar solvent, such as methanol, may be used to wash through excess reagents and by-products. If methanol is still present when the trifluoroacetic acid used for the cleavage reaction is introduced, some conversion of the aspartic acid and glutamic acid anilinothiazolinones to the methyl ester derivatives may occur. The presence of methanol leads to higher yields of the phenylthiohydantoins of serine and threonine than are normally obtained, and it may be advantageous to retain methanol up to the addition of the trifluoroacetic acid.

The anilinothiazolinone derivative is released rapidly following the introduction of the trifluoroacetic acid, and is collected as a solution in this acid in a fraction-collector tube. The prior addition of water to the tubes helps to reduce the extent of β-elimination of the serine and threonine derivatives (Wittmann-Liebold et al., 1977).

Conversion to the phenylthiohydantoins may be performed in 20% aqueous trifluoroacetic acid (80 °C for 10 min under O_2-free N_2 in sealed tubes), or the trifluoroacetic acid may be evaporated in a stream of N_2 and the conversion performed as usual in 1 M HCl at 80 °C for 10 min. The identification of the phenylthiohydantoin derivatives is by the methods described above (§§ 6.4.1.1–6.4.1.4).

After the cleavage reaction, the column is washed with methanol, and buffer, followed by buffer containing phenylisothiocyanate, is introduced to begin the next cycle.

The identification of the phenylthiohydantoins is often made difficult by the presence of by-products, especially if phenylisothiocyanate is used to block excess amino groups on the support after the attachment of the peptide. Laursen et al. (1975) purified the phenylthiohydantoins by passage through small columns of Dowex 50-X2 (H^+ form). Because of the high backgrounds, and because low attachment yields are often observed with very small amounts of peptides, about 100–200 nmol of peptide is required for sequencing on the solid-phase sequencer. A much more sensitive solid-phase technique, using a coloured derivative of phenylisothiocyanate, is described below (§ 6.5.2).

6.4.4. The 'dansyl-Edman' method

Because of the difficulties, described above, of the direct Edman degradation, an indirect method, with identification of the newly released N-terminal residues at each cycle with dansyl chloride, has been used widely. A small aliquot of the peptide is removed at each cycle for reaction with dansyl chloride, but as little as 0.2 nmol is required and the repetitive yield is only marginally reduced by this. In ideal cases, up to 30 residues may be determined from about 50 nmol, or 10 residues from about 2 nmol of peptide.

The following procedure, adapted from that of Bruton and Hartley (1970), may be applied to up to 24 peptides simultaneously, with one or two cycles of degradation per day.

The salt-free peptide (1–50 nmol) is placed in a medium-walled borosilicate glass tube (5 mm o.d. × 40 mm) and dried in vacuo. (1) 50% aqueous pyridine (20 μl) is added, the peptide is dissolved by vortexing, and an aliquot, containing 0.2–0.5 nmol peptide is transferred with a calibrated microcapillary to a small borosilicate glass tube for N-terminal determination (§ 6.2.1). (2) Phenylisothiocyanate (2 μl) is added, and transferred to the bottom of the tube by brief centrifugation. Nitrogen is introduced into the tube for 5 s, to displace air, and the tube is sealed with parafilm and aluminium foil. (Alternatively, tubes fitted with ground-glass stoppers or Teflon-lined screw caps may be manufactured). The contents of the tube are mixed thoroughly by vortexing, and the two-phase mixture is incubated at 55 °C for 30 min. (3) The parafilm is removed, and the solvent and excess phenylisothiocyanate are removed in a desiccator, preheated to 55 °C, containing vessels with fresh P_2O_5 and NaOH pellets. A good vacuum (<0.1 mmHg) must be maintained, and the residue after 45 min should be completely dry, with a white, flaky appearance. Any oily material remaining would create problems later; the addition of ethanol (20 μl) followed by further evaporation may help to dry the residue. (4) To the residue is added anhydrous trifluoroacetic acid (10 μl). Care should be taken to avoid skin contact or inhalation of the vapours of this volatile, corrosive acid. Nitrogen may be blown briefly into the tube to displace oxygen, but the presence of air at this stage does not significantly affect the degradation. The tube is sealed with parafilm and foil, and incubated at 55 °C for 10 min. The trifluoroacetic acid is removed by evaporation over NaOH pellets, using a water-pump to avoid undue contamination of pump oil with the acid; residual acid in the tube is removed using the oil pump for 10 min. (5) To the residue are added water (25 μl) and *n*-butyl acetate (100 μl). The contents are mixed thoroughly by vortexing. If particulate matter adheres to the glass, this is suspended using a small glass rod, and vortexing is repeated, to ensure efficient extraction of the anilinothiazolinone derivative and by-products. The peptide, as the trifluoroacetate salt, remains in the aqueous phase, unless it is very hydrophobic. The two phases are separated by centrifugation, and the upper layer, containing the anilinothiazolinone, is removed, using a narrow Pasteur pipette with a bent tip, as shown in Fig. 6.9. The butyl acetate extract may be saved for further investigation (e.g. scintillation counting for the determination of [^{14}C]carboxymethylcysteine residues, or hydrolysis and amino-acid analysis for the determination of histidine and arginine residues), or may be

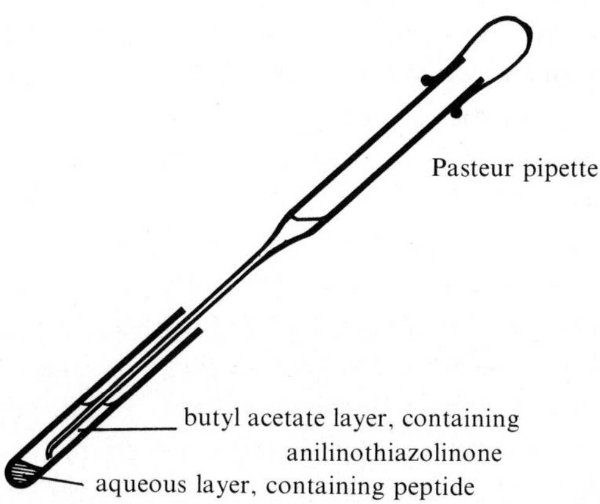

Fig. 6.9. Removal of the upper layer during butyl-acetate extraction of anilinothiazolinones.

discarded. The extraction is repeated with a further 100 μl of butyl acetate, and the upper layer is discarded. The aqueous layer is dried in vacuo over NaOH (careful application of the vacuum is required: samples may freeze and be ejected from the tubes if too low a pressure is used), and is then ready for the second cycle, at step (1).

Most amino-acid residues are clearly identified by chromatography on polyamide layers, as described previously (§ 6.2.1). However, as noted there, tryptophan is not identified (unless different hydrolysis conditions are employed), asparagine and glutamine are identified as aspartic acid and glutamic acid, respectively, carboxymethylcysteine is poorly identified, as are several other derivatives of cysteine, and differentiation of histidine and arginine requires a fourth solvent. Lysine residues are converted to ε-phenylthiocarbamyl-lysine residues during the degradation, and the phenylthiourea group is only partially hydrolyzed in 6 M HCl at 105 °C overnight, so two dansyl derivatives are observed: α-*N*-dansyl-ε-*N*-phenylthiocarbamyl-lysine, which comigrates with dansyl-leucine in solvents I and II, but with dansylphenylalanine after solvent III, and α-*N*-dansyl-lysine. Tryptophan residues are relatively rare in proteins. The observation of a blank at one position during the sequence determination of a small tryptophan-containing peptide, while all other positions are unambiguously determined and consistent with the amino-acid analysis, is good evidence for the location of a tryptophan residue; particular care is required when there may be more than one tryptophan residue, or some other unusual residue (introduced by post-translational modification) which is not easily identified. The Ehrlich test (§ 5.4.4) may be performed on samples of the peptide removed at each cycle of degradation: after the cleavage of the tryptophan residue the test becomes negative. In addition, the Ehrlich test may be performed after spotting the dried butyl acetate extracts onto silica gel thin layers. Neither method is sensitive. The identification of tryptophan residues should be confirmed by subfragmentation of the peptide, for example with chymotrypsin, which usually cleaves at the C-terminal side of tryptophan residues, and isolation and characterization of the fragments. Clearly, the unambiguous identification of tryptophan residues requires considerably more starting material than usual. The identification of tryptophan residues as oxidized derivatives has recently been described (Inglis et al., 1979).

The differentiation of acid and amide groups is probably the major problem with the dansyl-Edman method not encountered with direct methods. Identification of the extracted anilinothiazolinone after conversion to the phenylthiohydantoin is possible, but the method is insensitive and unreliable when the amounts of by-products present are large. Thin-layer electrophoresis at pH 6.5 of a sample of the peptide after each cycle of degradation usually gives a clear result with small peptides. About 1 nmol of the peptide is required at each cycle, so the method is not suitable for the highest sensitivities, and peptides of more than about 10 residues may not migrate to the positions expected according to the Offord plot (§ 6.1.2.1). Changes in relative mobility of the peptide after removal of each residue are more informative than the absolute mobilities: removal of a neutral residue gives little change in the mobility, while removal of an acidic or basic residue results in a significant change. The colour developed with the ninhydrin/Cd reagent (§ (§ 5.2.1.3) is also helpful in confirming assignments. Two examples of the use of this technique are given in Fig. 6.10.

The interpretation may be difficult if lysine residues are present, owing to their partial conversion to the ε-*N*-phenylthiocarbamyl derivatives, with the loss of a positive charge. Small peptides containing only a single acidic or amide residue present few problems, since electrophoresis of the whole peptide is sufficient for the determination of the charge on that residue. Caution is required, however, since a residue, such as γ-carboxyglutamic acid, which is destroyed upon acid hydrolysis, may be present and could lead to misinterpretation of the electrophoretic mobility (§ 7.1). With peptides longer than about 10 residues and containing several acid or amide residues, subfragmentation and further analysis of the fragments is required.

Probably the best way of identifying acid or amide groups is to use an alternative degradation method (§ 6.5.1) in addition to the dansyl-Edman method.

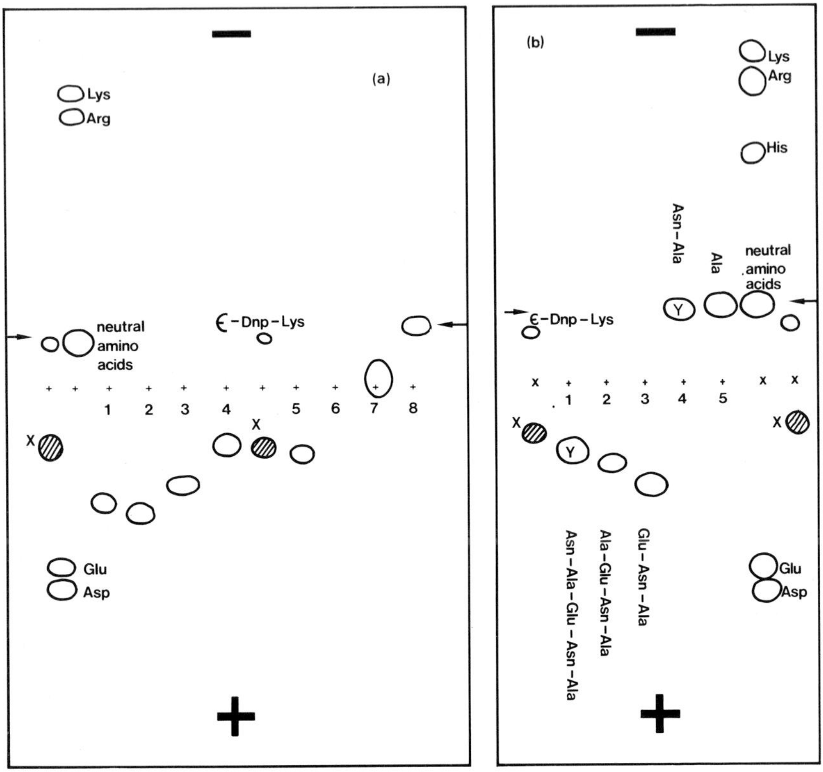

Fig. 6.10. Thin-layer electrophoresis at pH 6.5. Peptides isolated from succinylated Ca^{2+}-ATPase from rabbit sarcoplasmic reticulum were subjected to thin-layer electrophoresis on cellulose MN300 at pH 6.5 (for the determination of acid and amide groups) after each cycle of degradation using the dansyl-Edman technique (Allen, 1980). (a) Leu-Lys(Suc)-Glu-Tyr-Glu-Pro-Glu-Met-Gly-Lys(Suc)-Val-Tyr-Arg. Spot 1 is the whole peptide, spot 2 after the first cycle of degradation, etc. (b) Asn-Ala-Glu-Asn-Ala. The arrows mark the position of neutral compounds (displaced from the origin by electroendosmosis). ε-Dnp-Lys, ε-dinitrophenyl-lysine; X, xylene cyanol FF (blue dye marker); Y, spots giving an initial yellow colour with the ninhydrin-cadmium reagent. In peptide (a) the sixth residue was proline, and the peptide after five cycles of degradation was not detected with the ninhydrin-cadmium reagent.

6.4.5. *Subtractive Edman degradation* (Hirs et al., 1960; Konigsberg, 1972)

Another method for circumventing the difficulties of the direct manual Edman degradation involves amino-acid analysis after acid hydrolysis of a portion of the peptide after each cycle of degradation. The amino acid removed at a particular cycle is identified from the difference in the composition before and after that cycle. The method is time-consuming and insensitive, and is mainly of historical interest. It is, however, generally reliable for small peptides, although caution is required when Ile-Ile and related sequences are encountered, since the standard hydrolysis conditions do not cleave these bonds quantitatively. As with the dansyl-Edman method, tryptophan, asparagine and glutamine must be determined by additional methods.

The method may be useful for the determination of sequences containing modified residues, such as methylated histidine or lysine residues, of which the phenylthiohydantoin or dansyl derivatives are not readily identified.

6.5. *Alternative isothiocyanate reagents*

Phenylisothiocyanate is by far the most important reagent for sequential degradation, but several other isothiocyanates have been considered. Examples are given in Table 6.4. On the whole the claimed advantages, mainly the ease of detection of the thiohydantoin derivatives, are outweighed by disadvantages such as low volatility and low solubility in the coupling buffer. 4-*N*,*N*-Dimethylamino-azobenzene-4′-isothiocyanate (Dabitc) has, however, particular advantages, in that the by-products are mostly coloured differently from the thiohydantoins after chromatography on polyamide layers, so that detection is more reliable. Details of the reagent, its reactions with peptides and the properties of its thiohydantoin derivatives (Dabth-amino acids) are given in a series of papers by Chang and co-workers (Chang et al., 1976, 1977b, 1978; Chang, 1979b). The double-coupling methods for liquid-phase (Chang et al.,

TABLE 6.4
Some alternative isothiocyanate reagents proposed for sequential degradation of peptides.

Methyl isothiocyanate (Vance and Feingold, 1970). Identification by mass spectrometry (Richards and Lovins, 1972).

Pentafluorophenyl isothiocyanate (Lequin and Niall, 1972). Identification by gas-liquid chromatography, with electron-capture detection.

p-Phenylazophenyl isothiocyanate (Deyl, 1970). Coloured products.

4-*N*-Dimethylamino-1-naphthyl isothiocyanate (Ichikawa et al., 1970). Fluorescent thiocarbamyl derivatives.

p-Sulphophenyl isothiocyanate (Birr et al., 1970). Water-soluble reagent.

Diphenylindenonyl isothiocyanate (Ivanov and Mancheva, 1973, 1976). Coloured reagent; sensitive detection of products by colour and fluorescence on polyamide layers (Mancheva and Vladovska-Yukhnovska, 1978).

4-*N,N*-Dimethylaminoazobenzene-4′-isothiocyanate (Chang et al., 1976, 1978). Coloured reagent and products.

Fluorescein isothiocyanate (Muramoto et al., 1978). Fluorescent derivatives.

3-Isothiocyanato-4-methoxy-4′-nitrostilbene (Nowak et al., 1977). Fluorescent reagent.

1978; § 6.5.1) and solid-phase (Chang, 1979a; § 6.5.2) sequencing are effective at the 2–10 nmol level. As with the direct Edman degradation, discrimination between leucine and isoleucine is difficult to achieve by thin-layer chromatography, arginine and histidine residues are not reliably determined at high sensitivity, and threonine and (especially) serine are difficult to identify because of the β-elimination reactions. The studies of Chang (1979b) on the products of this reaction help in the identification. Chang et al. (1978) were able to identify the N-terminal 28 residues of the oxidized B-chain of insulin, starting with 3.6 nmol, apart from the cysteic acid residues.

A disadvantage of the liquid-phase method, as with the direct Edman degradation, is that short hydrophobic peptides are extracted into the organic phase after the coupling reaction; thus it is difficult to identify the C-terminal 2–3 residues of tryptic peptides with C-terminal lysine residues (converted to the hydrophobic thiocarbamyl derivatives) or of chymotryptic peptides.

The major advantage over the dansyl -Edman method is the direct determination of tryptophan, asparagine and glutamine residues. Since most of the disadvantages of the Dabitc method may be circumvented through the use of the dansyl-Edman method, the following strategy may be recommended for high-sensitivity (less than 10 nmol) manual sequence analysis: peptides containing aspartic acid, glutamic acid, their amides, or tryptophan are degraded by the Dabitc method. If serine or threonine residues are present, the degradation is performed by the dansyl-Edman method. With longer peptides, where subfragmentation is required, the Dabitc method may be performed on the total peptide, and the results may be confirmed and extended by the determination of the sequences of the fragments by the dansyl-Edman method. Further confirmation of the acid and amide residues is made using thin-layer electrophoresis of the fragments. Although double determination of the sequence may seem inefficient, the time taken for the determination of the sequences of peptides is only a small portion of the total time required for the determination of the structure of the whole protein, and this strategy has the pronounced advantage that the probability of error is greatly reduced.

6.5.1. Liquid-phase sequencing with Dabitc

The following method is that of Chang et al. (1978), but with reduced reaction volumes and slight modifications. (1) The peptide (1–10 nmol) is placed in a borosilicate glass tube (4 mm i.d. × 40 mm), and dissolved in 20 μl of 50% (v/v) aqueous pyridine. (2) A freshly prepared solution of Dabitc in pyridine (10 μl, 2.82 mg/ml) is added, the tube is flushed with oxygen-free N_2, sealed with parafilm and Al foil, and incubated at 55 °C for 20 min. (3) The tube is opened, phenylisothiocyanate (2 μl) is added, the tube is flushed with N_2, sealed, vortexed and incubated for a further 20 min at 55 °C. (4) Heptane/ethyl acetate (2 : 1, v/v) (150 μl) is added, and the phases are thoroughly mixed by vortexing for 10 s. The phases are separated by brief centrifugation, and the organic phase is removed with a narrow pipette (Fig. 6.9) and discarded. Care is required to avoid removal of any of the lower phase or of any solid material at the interface. The extraction is repeated with a further 150 μl of heptane/

ethyl acetate. A third extraction may be performed if there remains significant contamination by by-products, evident on the thin layers used for identification of the thiohydantoin derivatives. (5) The aqueous phase is dried thoroughly in vacuo over P_2O_5 and NaOH pellets, care being taken to avoid ejection of the sample from the tube through a sudden application of a high vacuum. (6) The dry residue is treated with anhydrous trifluoroacetic acid (about 10 μl). The tube is flushed briefly with N_2, sealed with parafilm and Al foil, and incubated at 55 °C for 10 min. The trifluoroacetic acid is removed in vacuo over NaOH pellets. (7) The residue is dissolved in water (25 μl), and *n*-butyl acetate (100 μl) is added. The cleaved thiazolinone derivative is extracted into the upper phase after vortexing for 10 s and separation of the phases by centrifugation. The butyl acetate extract is placed in a tube (4 mm i.d. × 40 mm) and the solvent is removed in vacuo. (8) After removal of the butyl acetate layer, the aqueous phase is dried in vacuo over P_2O_5 and NaOH, and is subjected to the next cycle of degradation, starting at step (1).

To the residue from the butyl acetate layer is added 50% (v/v) aqueous trifluoroacetic acid (20 μl), and the thiazolinone derivative is converted to the thiohydantoin derivative by incubation at 80 °C for 10 min under N_2. The sample is dried and redissolved in a suitable volume of ethanol (about 5 μl). A portion (1/40–1/5) of the ethanol solution, containing about 0.1 nmol thiohydantoin derivative, is used for thin-layer chromatography.

The sample is placed as a very small spot (2 mm diameter) about 8 mm from one corner of a 25 mm × 25 mm polyamide sheet (Schleicher and Schuell). A marker, Dabtc-diethylamine (4-*N*,*N*-dimethylaminoazobenzene-4′-thiocarbamyl diethylamine; 10–20 pmol; prepared by reaction of Dabitc with diethylamine), is applied to the same spot. The identification of the Dabth derivatives is aided by co-chromatography of the sample and a suitable mixture, consisting of 10–20 pmol of the Dabth derivatives of proline, phenylalanine, glutamine, aspartic acid and arginine on the reverse side of the thin layer, by analogy with the methods for the dansyl amino acids (§ 6.2.1) and phenylthiohydantoin amino acids (§ 6.4.1.4). With practice, however, the identification may be achieved simply by reference to the positions of the marker and by-product spots alone.

The solvent for the first dimension is acetic acid/water (1 : 2, v/v). The thin layer is removed from the tank and dried in warm air immediately the solvent front reaches the top of the layer. The thin layer is rotated through 90°, and placed in the second tank, containing toluene/*n*-hexane/acetic acid (2 : 1 : 1, by vol.). As soon as the solvent front reaches the top, the thin layer is removed and dried in warm air. The Dabth derivatives appear as faint yellow spots; strong red colours are produced when the thin layer is exposed to the vapour of concentrated HCl, while the marker Dabtc-diethylamine is blue. The distribution of Dabth amino acids is shown in Fig. 6.11.

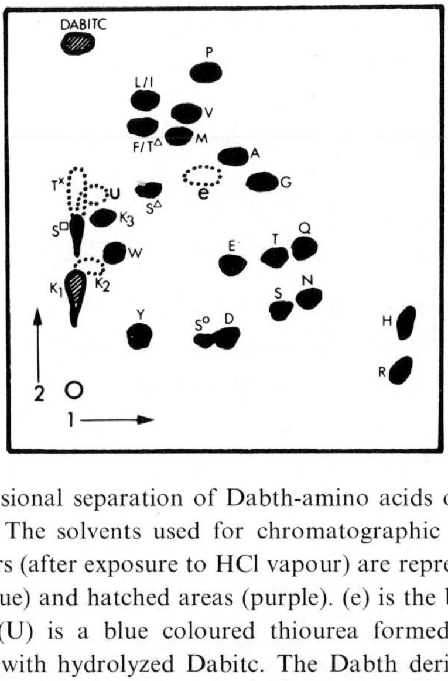

Fig. 6.11. Two-dimensional separation of Dabth-amino acids on a polyamide sheet (Chang et al., 1978). The solvents used for chromatographic separation are given in the text. The colours (after exposure to HCl vapour) are represented by solid areas (red), dotted areas (blue) and hatched areas (purple). (e) is the blue synthetic marker Dabtc-diethylamine. (U) is a blue coloured thiourea formed by the coupling of phenylisothiocyanate with hydrolyzed Dabitc. The Dabth derivatives of the amino acids are denoted by the single-letter code. T^{Δ}, Dabth-dehydrothreonine; S^{Δ}, S°, $S^{\square}$, products formed after β-elimination of Dabth-serine; $T^{\times}$, product formed after β-elimination of Dabth-threonine; K_1, α-Dabth-ϵ-Dabtc-Lys; K_2, α-Pth-ϵ-Dabtc-Lys; K_3, α-Dabth-ϵ-Ptc-Lys. Other blue spots are sometimes visible, due to incomplete conversion to Dabth derivatives, particularly with the glycine derivative. For further details see Chang et al. (1978) and Chang (1979b).

Methionine sulphone gives a spot close to that of threonine. Histidine and arginine derivatives are not always well resolved, and a by-product is sometimes observed in the position of Dabth-Arg.

For discriminating between Dabth-leucine and Dabth-isoleucine a second portion (at least 50 pmol) of the sample is applied to a 10 cm-high silica gel thin-layer plate (Merck, G60, without fluorescent indicator). Standards of Dabth-leucine and -isoleucine are applied beside the sample, and chromatography is performed in chloroform/ethanol (100 : 3, v/v). The dried plate is then exposed to HCl vapour to reveal the Dabth derivatives as red spots. A slight difference in R_F between the derivatives allows the identification to be made (Chang et al., 1977b).

Factors similar to those which affect identification of phenylthiohydantoin derivatives during sequencing affect the identification of the Dabth derivatives, namely the partial deamidation of asparagine and glutamine derivatives, low yields of serine and threonine derivatives together with their β-elimination products, and identification of lysine as ϵ-*N*-thiocarbamylated derivatives.

6.5.2. Manual solid-phase sequencing with Dabitc (Chang, 1979a)

The polypeptide or protein (5–10 nmol) is attached to isothiocyanato-glass (20–25 mg) as described above (§ 6.4.3.2). Other attachment methods may be used when appropriate. The glass beads with attached peptide are placed in a tube (10 mm i.d. × 50 mm) fitted with a C-10 glass stopper and a 2 mm × 6 mm magnetic stirring bar. Reactions are performed by placing the tube in a heating block (52 °C) over a magnetic stirrer. A C-10 stopper with a D2 sinter is used at every stage of vacuum drying to prevent the loss of glass beads.

To the tube are added 50% (v/v) aqueous pyridine (0.4 ml) and 0.2 ml of a solution of Dabitc in pyridine (15 mM; 4.23 mg/ml). The tube is flushed with N_2, stoppered, and placed in the heating block for 45 min. Phenylisothiocyanate (0.05 ml) is added, and incubation under N_2 is continued for 30 min. The supernatant is removed after gentle centrifugation, and the glass beads are washed with pyridine

(2 ml) followed by methanol (twice 2 ml), and dried in vacuo. Trifluoroacetic acid (0.2 ml) is added. The tube is flushed with N_2, stoppered, and placed in the heating block for 15 min. The acid is removed in vacuo over NaOH pellets. The thiazolinone derivative is extracted with methanol (0.3 ml). The beads are washed with methanol (2 ml), dried in vacuo and subjected to the next cycle of degradation. The extracted thiazolinone is dried in vacuo and converted to the Dabth derivative as described above. The Dabth derivative is identified by thin-layer chromatography. Up to 30 residues of a protein may be identified by this technique.

The method may also be adapted for use in columns. An important advantage of the manual technique is that several peptides may be degraded in parallel.

6.6. *Other chemical degradation methods for peptide sequence determination*

Several reagents other than isothiocyanates have been proposed for the sequential degradation of peptides. Some of these are listed in Table 6.5. None of the reagents for degradation from the N-terminus is as effective as isothiocyanates, although thioacetylthioglycolic acid has been used with some success (Takagi and Doolittle, 1975; Watt et al., 1979).

TABLE 6.5

Some examples of reagents other than isothiocyanates proposed for the sequential degradation of peptides

(a) *From the amino-terminus*
Thioacetylthioglycolic acid (Takagi and Doolittle, 1975; Previero, 1977; Largman and Mross, 1975).
Methyl dithioacetate (Previero, 1977).
N-Thiobenzoyl succinimide (Cavadore, 1978).
(b) *From the carboxyl-terminus*
Acetic anhydride/thiocyanate (Stark, 1968; 1972b; Cromwell and Stark, 1969) followed by strong acid or acetohydroxamic acid.

A method for degradation from the C-terminus would be extremely useful. The thiocyanate method of Stark (1968, 1972b; Cromwell and Stark, 1969), although inefficient, may be used for the identification of up to 6 residues in the sequence in very favourable cases. The solid-phase modification (Darbre and Rangarajan, 1975; Darbre, 1977) allows easier separation of the degradation products from the remaining peptide. However, the method is not generally applicable, has not been developed for high-sensitivity work, and cannot be recommended for general use.

6.7. Enzymic methods for peptide sequence determination

Limited sequence information may be derived through the use of exopeptidases. Carboxypeptidases have been described in Chapter 2; as with the degradation of proteins, a time course of release of amino acids from the peptide is required (determined with an amino acid analyzer). Contamination of the carboxypeptidases with endopeptidases is a less serious problem than with proteins, but the differences in the rates of release of different residues still make unambiguous deduction of the sequence difficult to achieve. Methods applicable to peptides are discussed by Ambler (1972b).

Aminopeptidases may be used analogously for degradation from the N-terminus of the peptide. The isothiocyanate degradation is far more efficient for most peptides, but for the identification of modified residues which are unstable under the alkaline coupling conditions or in the acid cleavage or conversion media, enzymic digestion is advantageous. As with the carboxypeptidases, aminopeptidases release amino acids at widely differing rates, and unambiguous results may not be obtainable. Leucine aminopeptidase releases preferentially hydrophobic amino acids, while acidic residues are released very slowly. Peptide bonds involving the secondary amino groups of proline residues are hydrolyzed at a negligible rate. Aminopeptidase M has a broader specificity. The use of aminopeptidases is described by Light (1972).

Dipeptidyl aminopeptidases, which release dipeptides sequentially from the amino-termini of peptides, may be used for sequence analysis (Callahan et al., 1972; Krutzsch and Pisano, 1977), but the method is of little practical significance.

6.7.1. *Carboxypeptidase A* (Ambler, 1972b)

The peptide (5–500 nmol, depending on the sensitivity of the analyzer) is dissolved in 0.2–0.5 ml of 0.2 M *N*-ethylmorpholine acetate buffer, pH 8.5. Carboxypeptidase A (typically 1–50 μg; the optimum amount is determined in preliminary experiments) is added and the mixture is incubated at 37 °C. If lysine or arginine are at, or close to, the C-terminus, carboxypeptidase B should be added in addition. Samples are removed after times from 10 min to 4 h, acidified to pH 2 and stored frozen until applied to the analyzer. A time course of the release of amino acids is plotted.

An example of results obtained with carboxypeptidase A is shown in Fig. 6.12.

Further information may be obtained if the mixture of residual peptides is resolved and each peptide analyzed, but this approach is time-consuming.

Carboxypeptidases C and Y, of broader specificity, are used in similar ways; suitable conditions for digestion are given in Chapter 2 (§ 2.8).

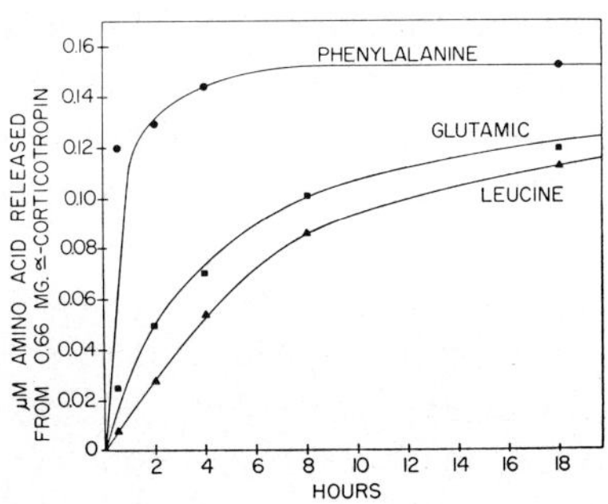

Fig. 6.12. Rate of release of amino acids from α-corticotropin by the action of carboxypeptidase A, estimated as their Dnp derivatives. The carboxy-terminal sequence of the hormone was ...-Pro-Leu-Glu-Phe. Reproduced with permission from Harris, J.I. and Li, C.H. (1955) J. Biol. Chem. *213*, 499–507.

6.7.2. *Leucine aminopeptidase* (Light, 1972)

The peptide (100 nmol) is dissolved in 0.1 ml of 0.1 M Tris–chloride buffer, pH 8.6, containing 2.5 mM $MgCl_2$. Leucine aminopeptidase (5–50 μg; the optimum amount being determined in preliminary experiments) is added, and the solution is incubated at 37 °C. Samples are removed after times from 10 min to 4 h, acidified to pH 2, and stored frozen until applied to the analyzer.

6.7.3. *Aminopeptidase M* (Light, 1972)

The peptide (100 nmol) is dissolved in 0.1 ml of 0.1 M sodium phosphate buffer, pH 7.0. Aminopeptidase M (0.05–1 ml of a 0.1% solution in the phosphate buffer) is added, and the mixture is incubated at 37 °C. Samples are removed at a series of times, acidified, and applied to the analyzer.

The results are interpreted as with the carboxypeptidase A method.

6.8. *Mass spectrometry*

In theory, mass spectrometry offers great promise for the determination of peptide sequences, because of its high speed and sensitivity. To date the method has only been applied in a few laboratories. The reasons for this are the expense of the instrumentation, the experience needed for rapid interpretation of the large number of spectra produced, and, perhaps most importantly, the limitation of the method to small peptides and the present lack of attainment of high sensitivity.

A detailed description of the use of the mass spectrometer in protein sequence analysis has been given by Morris and Dell (1975). The major problem is the conversion of peptides to volatile derivatives; of several methods proposed, acetylation and permethylation (Morris et al., 1971), reduction to polyaminoalcohols (Nau, 1976; Frank and Desiderio, 1978), and trimethylsilylation have been the most useful. Some residues, such as arginine, histidine and cysteine, are not converted to non-polar derivatives by all methods, and in general the yields of volatile derivatives prepared from nanomolar amounts of peptides are low. Contamination of the peptides with materials such as silicone grease or plasticisers must be scrupulously avoided. The sequence information is derived mainly from the 'sequence ions' produced by cleavage at the peptide bonds. Leucine and isoleucine residues are usually indistinguishable.

An advantage of mass spectrometry is that simple mixtures of peptides may be analyzed, through the use of different source temperatures. Another method for the determination of the structures of small peptides in mixtures is the combined use of gas chromatography and mass spectrometry (Nau and Biemann, 1976; Nau, 1976; Sauer and Anderegg, 1978). The method has been applied to the sequence analysis of hydrophobic peptides derived from integral membrane proteins (Gerber et al., 1979), but the information had previously been obtained, more simply, by chemical degradation of purified peptides. The use of gas chromatography/mass spectrometry in conjunction with digestion by dipeptidyl aminopeptidases has been described by Krutzsch and Pisano (1977).

Repetitive sequences in proteins may be difficult to determine by the Edman degradation. Such a structure in a myosin light chain was determined by mass spectrometry (Bridgen and Morris, 1974).

The most useful application of mass spectrometry in protein structure analysis is in the identification of post-translationally modified residues, and some examples of this are given in the next chapter.

6.9. Subfragmentation of peptides

The complete structures of peptides of up to 15 residues can normally be determined by manual degradation from the N-termini, and in favourable cases peptides of 30 residues may be sequenced completely in this way. However, the quality of the data towards the C-terminus of a peptide is often poor. For longer peptides, and peptides with blocked N-termini, subfragmentation with proteases or by chemical methods, followed by the isolation of the peptide fragments and determination of their structures, is necessary. Subfragmentation is also performed for the identification of acid and amide residues, using electrophoretic mobilities (§ 6.1.2.1), and for confirmation of the identification of tryptophan residues, if the dansyl-Edman method is used.

As discussed in Chapter 3, several proteases are suitable for the subfragmentation of peptides. The choice is made on the basis of the information already known about individual peptides, including amino-acid composition and partial sequence data. For example, trypsin will probably be the first choice for cleavage of a peptide of 25 residues with a single lysine or arginine residue at position 15: the decapeptide which will probably be released from the C-terminus will probably give the information required for the completion of the sequence determination. More often, however, the distribution of sites of potential cleavage will not be known, and an element of luck is involved in the choice of methods. Thermolysin (§ 3.3.8) is often useful; the peptides released are usually easily detected with fluorescamine, and no peptides with N-terminal glutamine (which may cyclise, giving problems of detection and sequence analysis) are normally released.

The peptides in the digest of a peptide of about 50 residues may usually be separated with ease by preparative paper or thin-layer peptide mapping and detected with fluorescamine. Digestion of about 20 nmol may be sufficient if a sensitive amino-acid analyzer is available. Yields of 5–15 nmol of the fragments after extraction from two or three thin-layer peptide maps may be expected. For peptides of more than 50 residues, gel filtration followed by peptide mapping or ion-exchange chromatography may be required for the separation of the fragments.

For the deduction of the total sequence of a large peptide, the structures of at least two overlapping sets of fragments are required. Information from other digests of the protein may also be used.

The use of subfragmentation may be illustrated by the following example of the blocked N-terminal 31-residue peptide from the Ca^{2+}-ATPase of rabbit sarcoplasmic reticulum (Allen, 1980).

The peptide, isolated from Fraction T1-C7 from the tryptic digest of the carboxymethylated, succinylated protein (see Fig. 4.7) was detected by its radioactivity; it was ninhydrin-negative. No N-terminal residue could be detected, and the peptide was not degraded by the dansyl-Edman technique. The amino-acid analysis showed that several hydrophobic residues were present (Val_2, Met, Leu_2, Tyr, Phe), indicating that thermolysin digestion would be suitable for subfragmentation.

The peptide (600 nmol; 2.0 mg) was digested in 0.1 M NH_4HCO_3/0.01% (v/v) thiodiglycol (0.3 ml) with thermolysin (50 μg) for 70 min at 45 °C. The digest was chromatographed on a column of Sephadex G-50 (Fig. 4.5). Peptides in each peak were further resolved by thin-layer chromatography and electrophoresis, and the structures of most of them were determined by the dansyl-Edman technique in conjunction with thin-layer electrophoresis at pH 6.5 for the assignment of acid and amide groups.

One small peptide, of composition (Glu, Met), was ninhydrin negative, but was detected with the chlorine/*o*-tolidine reagent. No N-terminal residue was detected. The dipeptide (10 nmol) was digested with CNBr (2 mg) in 50 μl of 70% (v/v) formic acid under N_2 for 20 h at 45 °C. The dried reaction product was dansylated, but without subsequent hydrolysis, and dansyl derivatives were identified by chromatography on polyamide thin layers. Only dansyl glutamic acid was detected. A sample of the dipeptide was subjected to thin-layer electrophoresis at pH 6.5. A ninhydrin-negative spot, detected with the chlorine/*o*-tolidine reagent, had a mobility close to that of glutamic acid, consistent with the structure Ac-Met-Glu for the dipeptide. Another peptide (T1C-7.Th2.2) had a blocked N-terminus. Treatment with CNBr released glutamic acid as an N-terminal residue. Tong (1977) independently identified the N-terminal sequence of the protein to be acetyl-Met-Glu-Ala.

The total sequence of the 31-residue peptide could be deduced by alignment of the sequences of the thermolysin fragments, together with the previously determined sequence (Allen and Green, 1978) of a tryptic peptide from the carboxymethylated protein, peptide T4-D2e, as shown in Fig. 6.13.

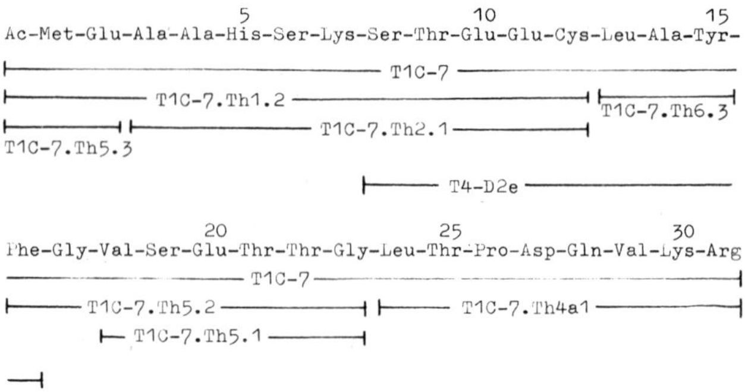

Fig. 6.13. Alignment of the thermolysin fragments of a 31-residue peptide and deduction of the total sequence (see text for details) (Allen, 1980).

6.10. Summary of possible difficulties in peptide sequence determination

Throughout this chapter particular deficiencies of various methods have been noted, and in this section some of the problems which may lead to erroneous conclusions are summarized.

Impurities in peptides, while usually readily detected, may occasionally yield misleading data. Attempts to determine the structures of impure peptides are fraught with dangers, particularly if partial blockage of the peptides occurs.

Blockage of the Edman degradation, which may occur with peptides containing asparagine, glutamine, tryptophan, *S*-carboxymethylcysteine, *O*-acetyl serine and some other modified residues, can lead to difficulties.

An important aspect of manual sequence determination is the agreement of the results of Edman degradation methods with the amino-acid compositions of the peptides. Fortuitous agreement with erroneous results may arise when sequences are determined by the direct method and -Ile-Ile- or -Val-Val- sequences are overlooked, as may occur towards the C-termini of long peptides if overlap is significant. Similarly, a second tryptophan residue may be overlooked in a peptide in which this residue was identified only.

qualitatively by the Ehrlich test. The determination of sequences containing several identical residues together is difficult, particularly if residues giving rise to excessive overlap, such as histidine, proline or glycine, or partial blockage, such as asparagine, are present. Routine amino-acid analysis of peptides is not sufficiently accurate to allow distinction to be made between, for example, Gly_5Lys and Gly_6Lys, although more careful analysis, if necessary after further purification of the peptide, using the average of three or more determinations, may be performed. In addition, repetitive sequences may not provide suitable sites for subfragmentation. A particularly difficult example of this type of structure was found in part of the A1 light chain of rabbit myosin (Frank and Weeds, 1974); the solution to the problem involved the use of automated Edman degradation and mass spectrometry (Bridgen and Morris, 1974).

Serine, threonine, histidine and arginine residues are not easily determined by chromatographic identification of their thiohydantoin derivatives, while tryptophan, asparagine and glutamine are not readily determined by indirect methods: use of both direct and indirect methods is recommended where appropriate. Cysteine residues are not easily determined by either method, and the use of a radiolabelled reagent for the alkylation of these residues in the protein is recommended.

Histidine residues may give problems, resulting from 'pre-cleavage'.

In general, lysine and arginine residues are C-terminal in tryptic peptides (except for the C-terminal peptide in a protein), and these residues are often placed in that position on the basis of the specificity of trypsin. However, trypsin also cleaves other bonds, with a chymotryptic-like activity, and some lysine and arginine peptide bonds are not cleaved. Carboxypeptidase B may be used in cases of doubt.

The presence of modified residues may lead to problems, either because the residues are not detected by amino-acid analysis or because the modifications affect the course of the degradation. The identification of modified residues is discussed in the following chapter.

The reliability of results from the automated spinning-cup

sequencer has been investigated by several workers. Smithies et al. (1971) discuss the importance of quantitation. Air (1976) noted that the manual dansyl-Edman method appeared more reliable than the automated sequencer for determining peptide sequences in ØX-174 proteins. Haslett and Boulter (1976) discuss criteria for the reliable determination of sequences with the automated sequencer.

Structures introduced by post-translational modification

As described in Chapter 1, many unusual amino acids, formed by a variety of post-translational modifications, have been identified in proteins. Modifications which introduce chromophoric groups, such as the attachment of haem or flavin groups, may be determined spectrophotometrically. The amino-acid analysis of the protein may indicate the presence of modified residues, but great care is required in the inspection of the amino-acid analyzer charts if only a single residue of a modified amino acid is present (e.g. Johnson et al., 1967). If the modified residue has the same elution time as one of the normal protein amino acids it will not be detected on the analyzer. In addition, some modified amino acids are destroyed by acid hydrolysis, or fail to react with ninhydrin, fluorescamine or *o*-phthalaldehyde, and will not be detected by the standard methods of amino-acid analysis.

During work on the determination of the amino-acid sequence of a protein, therefore, it is possible that peptides containing modified residues, whose presence is not expected, will be isolated. It is therefore important that observations of peaks at unusual retention times on the analyzer, of electrophoretic mobilities inconsistent with the apparent structures of peptides and of unusual chromatographic spots of derivatives during sequential degradations should be carefully recorded. The study of small peptides is likely to be more reliable than the use of the automated sequencer for the detection of modified residues, since several independent observations may be considered together.

The following examples illustrate some of the approaches which have been used for the determination of modified residues.

7.1. γ-Carboxyglutamic acid

Some peptides from the amino-terminal region of prothrombin were observed to have electrophoretic mobilities inconsistent with their amino-acid compositions (Magnusson, 1973). In dicoumarol-treated animals this discrepancy was not observed. Further study revealed that a vitamin K-dependent modification of glutamic acid residues was responsible, and the new residue, γ-carboxyglutamic acid, was identified by mass spectrometry (Magnusson et al., 1974) and by nuclear magnetic resonance spectrometry (Stenflo et al., 1974). The residue is not detected during sequencing by the dansyl-Edman method, nor by amino-acid analysis, since decarboxylation occurs during the acid hydrolysis:

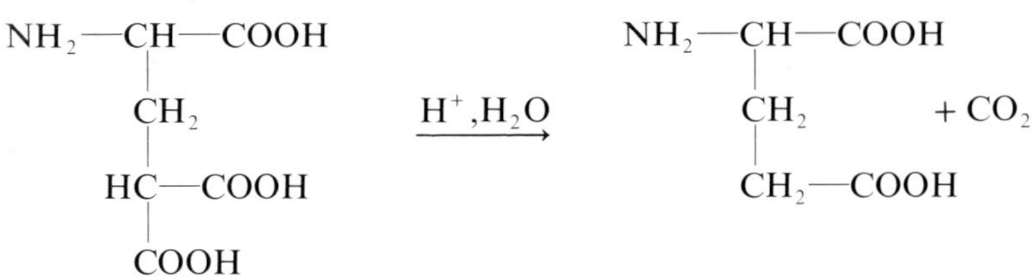

γ-Carboxyglutamic acid may be determined on the amino-acid analyzer after alkaline hydrolysis (Hauschka, 1977) or after reduction by diborane and acid hydrolysis (Zytkovicz and Nelsestuen, 1976). The N-terminal sequence of prothrombin, including several γ-carboxyglutamic acid residues, was determined with the aid of mass spectrometry (Morris et al., 1976).

Vitamin K-dependent proteins have been reviewed by Stenflo and Suttie (1977).

7.2. Phosphorylated amino-acid residues

O^3-Phosphonoserine and O^3-phosphonothreonine residues occur in a wide variety of proteins, including enzymes such as glycogen phosphorylase and phosphorylase kinase, histones and food proteins such as casein and phosvitin. The phosphate ester bond is cleaved during hydrolysis in 6 M HCl at 110 °C for 24 h, and identification of these residues in peptides is not easy. Partial acid hydrolysis may yield limited quantities of *O*-phosphonoserine and *O*-phosphonothreonine, as described by Glazer et al. (1975), or a total enzymic hydrolysis may be used. The electrophoretic mobilities of small peptides may provide a clue to the presence of these residues. Chemical analysis for phosphorus is not sensitive enough to be applied to the minute quantities of peptides typically obtained during sequence analysis. The only satisfactory methods available for the location, on the nanomolar scale, of these residues in peptide sequences utilize radiolabelling with ^{32}P, either in vivo or, as with enzymes such as glycogen synthase, in vitro with γ-[^{32}P]ATP in the presence of a specific protein kinase (Proud et al., 1977).

The purification of [^{32}P]phosphopeptides is readily monitored by liquid-scintillation counting or radioautography, and the position of the phosphorylated residues in peptide sequences may be determined from the release of radioactivity from the peptide, preferably studied by thin-layer electrophoresis of samples of the peptide after each cycle of Edman degradation. Dansyl-serine or -threonine are observed if the dansyl-Edman method is used. The phenylthiohydantoin derivatives of phosphonoserine and phosphonothreonine are unstable, readily undergoing the β-elimination reaction (Proud et al., 1977; Rylatt and Cohen, 1979).

Similar considerations apply to the determination of N^{im}-phosphonohistidine residues in sequences. The identification of this residue in a sequence from phosphoglycerate mutase was described by Rose et al. (1975).

The active-site aspartic acid residues of the $Na^+ + K^+$-ATPase and sarcoplasmic reticulum Ca^{2+}-ATPase are phosphorylated by ATP during each turnover of the enzymes (Degani and Boyer, 1973). The mixed anhydride is very labile, particularly under alkaline conditions, but it was found possible to identify the site of phosphorylation in the sequence of the Ca^{2+}-ATPase protein by the rapid isolation under acidic conditions of tryptic and peptic peptides labelled by γ-[^{32}P]ATP, and comparison with known peptide sequences (Allen and Green, 1976).

7.3. Methylated amino acids

As shown in Table 1.2, several N-methylated amino acids have been identified in proteins. These derivatives are stable under the conditions of acid hydrolysis, and their identification by amino-acid analysis is described by Glazer et al. (1975). Routine analyzer systems give poor resolution of ϵ-N-methyllysines from lysine, and of N^{im}-methylhistidines from histidine, and small amounts of these modified residues may be overlooked in the analysis of the total protein. Close inspection of analyzer charts should therefore be made; abnormal retention times may indicate the presence of methylated residues. Dansyl and phenylthiohydantoin derivatives of methylated basic amino acids also have slightly different chromatographic properties from those of the unmodified amino acids. The lack of tryptic cleavage after a putative lysine residue could indicate that it is methylated.

α-*N*-Methyl amino acids are not readily detected using ninhydrin, fluorescamine or *o*-phthalaldehyde, and detection of peptides with these residues at the amino-termini is also difficult; methods which detect blocked peptides should be used.

The following examples illustrate the methods which have been used for the identification of methylated amino-acid residues in peptide sequences.

7.3.1. Trimethylalanine and ϵ-N-*trimethyllysine*

The identification of *N*-trimethylalanine at the N-terminus of *E. coli* ribosomal protein L11 (Dognin and Wittmann-Liebold, 1977) proved to be a difficult task, solved by mass spectrometry. This amino acid was shown to exist in the 50 S ribosomal proteins by the incorporation of radioactively labelled methyl groups during biosynthesis, followed by amino-acid analysis with determination of radioactive amino acids (Lederer et al., 1977). Since trimethylation prevents reaction with phenylisothiocyanate (and with detection reagents such as ninhydrin), it was necessary to fragment the amino-terminal portion of the protein, giving small peptides which could be studied by mass spectrometry. ε-*N*-Trimethyllysine residues in the same protein were identified as the α-*N*-dansyl derivative, with chromatographic properties similar to those of α-*N*-dansylarginine, and by amino-acid analysis.

7.3.2. N-*Methylalanine*

Two phenylthiohydantoin derivatives, of *N*-methylalanine and of lysine, were released during the first cycle of the Edman degradation of *E. coli* ribosomal protein S11. This led to the proposal of an *iso*peptide bond, involving the ε-NH_2 group of the lysine residue (Chen and Chen-Schmeisser, 1977). However, the use of 4-*N*,*N*-dimethylaminoazobenzene-4′-isothiocyanate revealed that an unusual cyclization and cleavage of the amino-terminal thiocarbamyl-*N*-methylalanine derivative under alkaline conditions was responsible for this observation (Chang, 1978).

7.3.3. N^{τ}-*Methylhistidine*

The single N^{τ}-methylhistidine residue in actin was determined by amino-acid analysis after acid hydrolysis (Johnson et al., 1967), and was shown to be present in a single peptide fraction after tryptic digestion. The residue was placed in the amino-acid sequence of rabbit skeletal muscle actin after the isolation of peptides from several digests of the protein (Elzinga et al., 1973).

Incubation of homogenates of muscle from 5-day-old rabbits with *S*-adenosyl-[*Me*-^{14}C]methionine led to the incorporation of ^{14}C-labelled methyl groups into histidine and lysine residues in both actin and myosin (Hardy and Perry, 1969). This radiochemical method is of general use for the detection of methylated residues, as mentioned above (§ 7.3.1).

7.4. Identification of points of attachment of carbohydrate in glycoproteins

Glycoproteins are widely distributed, particularly as components of cellular or viral membranes and among extra-cellular proteins; the subject is reviewed by Spiro (1970). Two main types of linkage are present: *N*-glycosidic, through asparagine residue (identified in ovalbumin by Marshall and Neuberger, 1964), and *O*-glycosidic, usually through serine or threonine residues, but also through hydroxylysine and hydroxyproline in collagen. The former linkage has the general structure

```
glycosyl-Man\
              >Man-GlcNAc-GlcNAc-┐
glycosyl-Man/                    |          Ser
                           ...-Asn-(X)-  or  -...,
                                            Thr
```

where *X* may be a variety of amino-acid residues. The glycosyl chains often contain galactose, additional mannose and *N*-acetylglucosamine, and terminal sialic acid residues, and sometimes fucose, which may be attached to the core *N*-acetylglucosamine residues. Asparagine-linked glycosylation is inhibited by tunicamycin. *O*-Glycosidic linked carbohydrate structures are often simpler, but there is a less clearly defined peptide sequence at the site of attachment, and several sites may be clustered together. Both types of linkage may be present in the same glycoprotein molecule. Heterogeneity in glycoproteins due to different extents of glycosylation is common, and this may lead to problems during the isolation of peptides.

Methods for the determination of carbohydrate residues in glycoproteins have been described by Glazer et al. (1975). The determination of the structures of carbohydrate chains is outside the scope of this monograph (see Spiro, 1972, for a thorough description); several physical and chemical techniques are applied, including methylation analysis, enzymic degradation, sequential chemical degradation with periodate, nuclear magnetic resonance spectrometry (^{13}C and ^{1}H) and mass spectrometry. Observations of binding to particular lectins and antigenic analysis are also used. The methods are, on the whole, less sensitive than those used in peptide sequence determination.

For the identification of sites of attachment, the glycoprotein is cleaved into small fragments, followed by isolation of the glycopeptides using detection by the phenol/sulphuric acid reagent (§ 5.3.5) and by amino-acid analysis for amino-sugars. Pronase digestion is often employed, generally giving di- and tri-peptides; gel filtration of the digest may be sufficient to separate the relatively large asparagine-linked glycopeptides from the remainder of the small peptides. The glycopeptides may be further purified by ion-exchange chromatography or paper electrophoresis if necessary. Glycopeptides generally have low R_F values on paper or thin-layer chromatography. Moczar (1978) described more polar solvents (e.g. ethanol/nitromethane/acetic acid/water, 5:3:3:3 by vol.) for chromatography of glycopeptides on silica gel thin layers.

A pronase glycopeptide may often have only a single asparagine residue (identified as aspartic acid after acid hydrolysis). Larger peptides, derived for example by tryptic digestion, are required for better location of the attachment sites in the peptide chain.

The determination of the amino-acid sequences of asparagine-linked glycopeptides by the dansyl-Edman method proceeds normally, but the glycosylated asparagine residue is identified as dansyl-aspartic acid. If more than one aspartic acid residue is present in the peptide, subfragmentation, or electrophoretic analysis during the degradation, is required for the identification of the site of attachment. Only a low yield of Pth-aspartic acid is observed when the direct Edman degradation is used: the glycosylated anilino-thiazolinone is not extracted into the organic solvent.

The Edman degradation also proceeds through *O*-glycosyl-serine and -threonine residues, but the glycosylated residues are not usually identified as phenylthiohydantoin derivatives. Products of β-elimination may be observed in low yields. The residues are identified by the dansyl-Edman method as dansyl-serine and -threonine, respectively. Treatment of peptides containing *O*-glycosyl-threonine residues with 0.5 M NaOH at 20 °C for 6 h converted these residues to 2-amino-2-butenoic acid residues. Removal of carbohydrate from *O*-glycosylated peptides with alkaline sodium borohydride (0.8 M $NaBH_4$, 0.1 M NaOH at 37 °C for 68 h) may lead to the identification of *O*-glycosyl-serine residues as Pth-alanine during the Edman degradation (Tomita and Marchesi, 1975; Kessler et al., 1979b). The points of attachment of *O*-glycosidically linked carbohydrate may also be identified by reaction of glycopeptides with alkaline sulphite (0.1 M NaOH, 0.5 M Na_2SO_3, 37 °C for 24–96 h) (Spiro and Bhoyroo, 1974). *O*-Glycosyl-serine and -threonine residues were converted to cysteic acid and α-amino-β-sulphonylbutyric acid residues, respectively.

Anhydrous HF removes carbohydrate from glycoproteins (except for *N*-glycosidically linked *N*-acetylglucosamine) (Mort and Lamport, 1977), and has been used for sequence studies on the extensively glycosylated carcinoembryonic antigen (Glassman et al., 1978).

The following examples illustrate the various methods used to determine the structures of glycopeptides.

7.4.1. Erythrocyte glycophorin (Tomita and Marchesi, 1975)

Fifteen oligosaccharide chains were found to be linked to serine and threonine residues, and one chain to an asparagine residue, in the amino-terminal region of erythrocyte glycophorin. The extensive glycosylation led to incomplete cleavage by proteases, and there were few suitable points of subfragmentation. *O*-Glycosyl residues were identified by the dansyl-Edman method as serine and threonine, and as reduced Pth derivatives after alkaline borohydride treatment.

7.4.2. Antifreeze glycoproteins (Morris et al., 1978)

The structures of proline-rich antifreeze glycopeptides from an antarctic fish were determined by a combination of automated and manual Edman degradation and mass spectrometry. Permethylation led to the β-elimination of carbohydrate. Deuterium labelling was important in the interpretation of the mass spectra.

7.4.3. *Haemoglobin A_{1c}*

A small proportion of human haemoglobin is modified by the condensation of the N-terminal valine residues of the β-chains with glucose, followed by an Amadori rearrangement to give a ketoamine derivative (Bunn et al., 1978). The structure was determined by a variety of chemical and physical techniques.

7.4.4. *Immunoglobulin G* (Smyth and Utsumi, 1967)

One of the carbohydrate chains of rabbit immunoglobulin G was located in the proline-rich 'hinge' region. Galactosamine, present in an *O*-glycosidic linkage with a threonine residue in 35% of the H-chains, was located by amino-acid analysis of papain digestion products and products of partial acid hydrolysis.

7.4.5. *Human chorionic gonadotropin* (Kessler et al., 1979a,b)

Structures of N- and O-linked carbohydrate chains were determined. Four serine-linked glycosyl groups on a proline-rich peptide were located after cleavage with alkaline borohydride. The serine residues were converted to alanine residues, and the positions of the linkages were identified by Edman degradation and subfragmentation of the peptide.

7.4.6. *Other examples*

Some other recent examples of the determination of the structures of glycopeptides are those of fetuin (Nilsson et al., 1979), angiotensin converting enzyme (Hartley and Soffer, 1978) and rabbit transferrin (Strickland and Hudson, 1978).

7.5. *Amino-terminal acyl groups*

The acetyl and pyrrolid-2-one 5-carboxylyl (pyroglutamyl) groups are often found at the N-termini of proteins, blocking the Edman degradation. The presence of these groups significantly increases the difficulty of sequence determination, particularly if automated sequencing methods are used. The pyroglutamyl residue may, however, be removed enzymically, with pyroglutamate aminopeptidase (Doolittle, 1970; Podell and Abraham, 1978).

The nature of the blocking group is most clearly determined by the study of purified amino-terminal peptides from the protein. If a complete amino-acid sequence is being determined, blocked N-terminal peptides will be isolated from various digests of the protein and may be detected on paper or thin layers with the chlorine/*o*-tolidine reagent (§ 5.2.4). Blocked peptides are not detected with ninhydrin (§ 5.2.1) (apart from a weak development of colour with the side-chains of lysine residues), unless first hydrolyzed (§ 5.2.3). If information on the amino-terminal sequence only is required, a method for the selective purification of small peptides lacking a free amino group may be employed, as described in Chapter 4 (§ 4.16).

The structures of short blocked peptides may be determined by a variety of methods, including digestion with carboxypeptidases (§ 6.7.1), mass spectrometry (§ 6.8) and study of the products of further peptide bond cleavage (§ 6.9).

7.5.1. Acetylated N-terminal residues

Some examples of proteins in which the amino-terminal residue is acetylated are given in Table 7.1. An extensive list is given by Narita et al. (1975).

Acetyl groups in proteins and peptides may be determined by hydrazinolysis followed by reaction with dansyl chloride and thin-layer chromatography of extracted *N*-acetyl-*N'*-dansyl hydrazide. The results are usually clearer when the method is applied to small peptides. Great care is required to remove acetate or acetyl compounds from the protein before the analysis. The following method is that of Schmer and Kreil (1969).

The protein (1–2 mg) or peptide (20–50 nmol) is dissolved in 0.5 ml 0.1 M HCl, then dried in vacuo over NaOH to remove traces of acetate. The protein is heated at 100 °C in an evacuated sealed tube for 17 h with anhydrous hydrazine (0.2 ml). The excess hydrazine is removed in vacuo over P_2O_5. The residue is dissolved in 0.2 M sodium citrate buffer, pH 3.0 (0.3 ml), and a 5–10-fold excess, by weight, of dansyl chloride in 0.3 ml ethanol is added. The tube is sealed and incubated at 37 °C for 24 h. The sample is dried, then dissolved in 0.3 ml H_2O. The solution is extracted with three portions of 0.3 ml peroxide-free ether (or chloroform). *N*-Acyl-*N'*-dansyl hydrazides are extracted into the organic phase, while the bulk of the dansic acid remains in the aqueous layer. The organic phase is evaporated to dryness, and a sample (about 1/20) is chromatographed in two dimensions on a silica gel G plate, activated at 120 °C for 30 min, in (I) Benzene/pyridine/acetic acid (16 : 4 : 1, by vol.) and (II) chloroform/*n*-butanol/acetic acid (6 : 3 : 1, by vol.). Reference samples are prepared from acetyl hydrazine, formyl hydrazine, or other appropriate compound. The formyl derivative is obtained in low yield from formyl-proteins.

N-Acetyl-*N'*-dansyl hydrazide is detected with higher sensitivity by chromatography on micropolyamide layers (Takagi and Doolittle, 1974).

A second method for the identification of N-terminal acetyl groups is mass spectrometry of small peptides derived from the amino-terminus of the protein. For example, a blocked dipeptide from a chymotryptic digest of myosin light chain A2 was treated with deuteroacetic anhydride and permethylated. Mass spectrometric analysis showed that the peptide was acetyl-Ser-Phe. The acetyl group was not deuterated: it was therefore present in the protein (Morris and Dell, 1975; Frank and Weeds, 1974).

A third method for the identification of acetyl groups in proteins is hydrolysis followed by gas chromatography (Margoliash and Smith, 1961; Gaetjens and Bárány, 1966); small peptides from the amino-terminus must be studied to ensure that the acetyl groups are not bound to side-chains in the protein.

The acetyl-amino acid bond is relatively resistant to acid hydrolysis, and partial acid hydrolysis (e.g. pH 2, 105 °C, for 90 min) of small peptides may yield an acetyl amino acid, which can be identified chromatographically by comparison with known compounds (Kreil and Tuppy, 1961).

Another method for the identification of blocking groups is nuclear magnetic resonance spectrometry (see Table 7.1).

TABLE 7.1

Some examples of the determination of blocked N-termini

Auffret et al. (1978). Mass spectrometry of the deuteroacetylated, permethylated blocked dipeptide (50 nmol) from the amino-terminus of alcohol dehydrogenase from *Drosophila melanogaster* N-11 showed the structure to be Ac-Ser-Phe.

Van Eerd et al. (1978). An acetyl group was determined in the blocked N-terminal peptide (500 nmol) of troponin C from frog skeletal muscle by nuclear magnetic resonance spectrometry. The peptide was dissolved in D_2O, and a 360 MHz spectrometer was used.

Grey et al. (1979). Pyroglutamic acid was identified at the N-terminus of a proline-rich phosphoprotein from human saliva by nuclear magnetic resonance spectrometry.

Chiu et al. (1979). Pyroglutamate aminopeptidase was used to identify and remove pyroglutamic acid from the N-terminus of the heavy chains of human myeloma cryoimmunoglobulin IgG Hil.

Kluh (1979). Small blocked peptides were isolated specifically on a column of sulphoethyl Sephadex for the determination of pyroglutamyl tyrosine at the N-terminus of hog amylase.

Kreil and Kreil-Kiss (1967). 10% of melittin from bee venom was shown to be formylated at the N-terminus, by hydrazinolysis.

Margoliash and Smith (1961). Gas chromatography was used to identify an acetyl group at the N-terminus of horse heart cytochrome *c*.

Kreil and Tuppy (1961) identified the acetyl group at the N-terminus of horse heart cytochrome *c* by comparison of an acetyl amino acid isolated after partial acid hydrolysis with known compounds.

Nau et al. (1977). The sequence of the acetylated N-terminal tryptic peptide of *Neurospora* tyrosinase was determined by mass spectrometry of the esterified and permethylated derivative.

Narita (1958). The acetylated amino-terminus in tobacco mosaic virus protein was determined using methods including hydrazinolysis and digestion of the blocked peptide with carboxypeptidase A.

Gaetjens and Bárány (1966). Gas chromatography was used in the identification of the N-terminal acetyl-Asp sequence in G-actin.

Hantke and Braun (1973). Fatty acylated *S*-glyceryl cysteine was identified at the amino-terminus of the murein lipoprotein of *E. coli* outer membrane.

7.5.2. Pyrrolidone carboxylic acid residues

Identification of this blocking group may be made by mass spectrometry after the isolation of small blocked peptides from the protein. The conversion of glutamine residues at the N-termini of peptides released by proteases from the interior of the protein to pyrrolidone carboxylyl residues may give misleading results. The isolation of peptides under acidic conditions, which promote the cyclization, should be avoided. The use of thermolysin for the digestion is recommended, since cleavage at the peptide bond involving the amino group of a glutamine residue is unlikely to occur.

The pyrrolidone carboxylic acid residue may be removed with calf liver pyroglutamate aminopeptidase (EC 3.4.11.8) (Podell and Abraham, 1978); the enzyme may be used for the identification of this residue as well as to provide access for the Edman degradation:

The enzyme is stored lyophilized at −20 °C. The reduced and alkylated protein (10 mg in 10 ml) is dialyzed against a pH 8.0 buffer solution, prepared from 0.1 M NaH_2PO_4 adjusted to pH 8.0 with 0.1 M Na_2HPO_4, and containing 5 mM dithiothreitol, 10 mM EDTA and 5% (v/v) glycerol, at 4 °C. Pyroglutamate aminopeptidase (0.5 mg of crude enzyme, containing about 0.025 mg of active enzyme) is added, and the solution is stirred under N_2 at 4 °C for 9 h. A second 0.5 mg of the enzyme is added, and the solution is incubated at 20 °C for 14 h. The solution is dialyzed against 50 mM acetic acid, then freeze-dried and used directly for sequence analysis.

Nearly quantitative deblocking was observed. The enzyme is sensitive to Hg^{2+}, oxidative conditions, and elevated temperatures.

The observation of successful Edman degradation after treatment of a protein with pyroglutamate aminopeptidase, but not without such treatment, is good evidence for the presence of this N-terminal blocking group. The identification should be confirmed by the study of small peptides. The pyroglutamate residue is identified as glutamic acid after acid hydrolysis.

7.5.3. Formylated N-terminal residues

The formyl group is a relatively rare amino-terminal blocking group, although it is present as *N*-formylmethionine in precursor polypeptides of prokaryotic proteins (Adams and Capecchi, 1966; Webster et al., 1966; Osborn et al., 1970). The formyl group may be removed without significant peptide bond cleavage by incubation with 0.5 HCl in methanol for 48 h at room temperature (Sheehan and Yang, 1958). Otherwise the characterization of formylated peptides is made in the same way as that of acetyl peptides.

7.6. Examples of other modified residues

Conversion of the terminal carboxyl groups of proteins to amide groups appears rare, but the peptide hormone oxytocin has a C-terminal amide group (Du Vigneaud et al., 1953).

Hydroxyproline and hydroxylysine in collagen and related proteins are identified by amino-acid analysis and chromatographic identification of their Pth and dansyl derivatives, using appropriate standards (see Gallop et al., 1972 for a review on the structure of collagen).

ϵ-*N*-Formyllysine was identified in a peptide from bee venom by mass spectrometric methods (Doonan et al., 1978).

ϵ-*N*-Acetyllysine residues in trout testis histones were identified by liquid-scintillation counting of anilinothiazolinone derivatives during automated sequence analysis, after the enzymic introduction of ^{14}C-labelled acetyl groups (Candido and Dixon, 1972).

The positions of pyridoxal phosphate binding sites in the amino acid sequence of glycogen phosphorylase were determined after reduction with sodium borohydride, giving stable ϵ-*N*-phosphopyridoxyl-lysine residues (Forrey et al., 1971; Titani et al., 1977). Peptides containing these residues were detected by their fluorescence emission at 395 nm upon excitation at 330 nm.

ADP-Ribosyl groups were found to be attached to rat liver lysine-rich histone, probably through glutamic acid residues 2 and 116 (Riquelme et al., 1979).

7.7. Cross-linking structures

Several post-translational modifications of proteins lead to the formation of cross-links between polypeptide chains or between different points in the same chain. Oxidation of cysteine residues to cystine residues is the most widespread cross-link. Oxidative deamination and condensation of lysine residues gives rise to a series of cross-linking structures in collagen and elastin. *Iso*-peptide bonds, involving the γ-carboxyl groups of glutamic acid residues and the ϵ-amino groups of lysine residues have been identified in many proteins (Folk and Finlayson, 1977). The protein resilin contains cross-links derived from tyrosine residues (Andersen, 1964).

The structures of some of the cross-links which have been found are given in Fig. 7.1.

7.7.1. Cystine residues

For the determination of amino-acid sequences in proteins cystine residues are generally cleaved by reduction and alkylation or by oxidation with performic acid before the protein is digested, although retention of the disulphide bonds during the initial steps of peptide purifications may be found useful. The identification of the cystine cross-links is then made in a separate experiment.

Immediately upon isolation, or upon denaturation prior to digestion, any cysteine residues in the protein should be alkylated to prevent disulphide interchange. Under alkaline conditions, disulphide bonds are slowly hydrolyzed and disulphide interchange then follows. Disulphide interchange is inhibited by the presence of alkylating reagents, such as *N*-ethyl maleimide (Ryle et al., 1955), but acidic conditions are best used for the isolation of cystine-containing peptides, and peptic or cyanogen bromide digestion, at acid pH, are generally preferred.

For relatively small proteins, of less than about 300 residues, the diagonal electrophoresis technique (Brown and Hartley, 1963, 1966) is used. Peptides are separated by high-voltage paper electrophoresis at pH 6.5. The dried paper is exposed to the vapour of

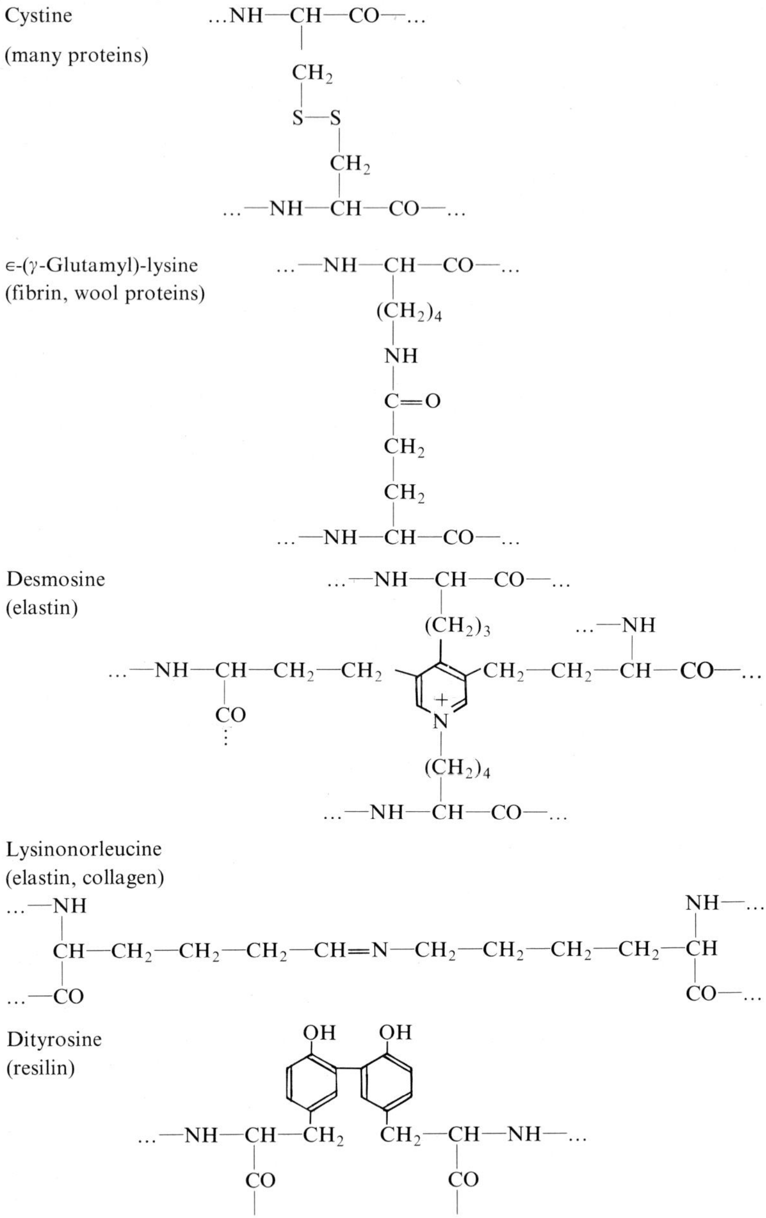

Fig. 7.1. Examples of cross-linking structures in proteins.

performic acid, which cleaves the disulphide bonds, resulting, in the simplest case, in a pair of cysteic acid-containing peptides from each cystine-containing peptide. The performic acid is removed in a stream of air, and the electrophoresis is repeated in the perpendicular direction. All peptides unaffected by performic acid migrate along a line at 45° to each direction of electrophoresis, while the newly-formed cysteic acid peptides migrate off the diagonal, usually on the anodic side in the second dimension. The peptides are detected and identified by the methods described in Chapters 5 and 6. If larger quantities are required, the zones of peptides containing cystine residues (detected by performic acid oxidation and re-electrophoresis of a side strip) are excised and subjected to oxidation and re-electrophoresis, as shown in Fig. 7.2.

Methionine-containing peptides are converted by performic acid to methionine sulphone derivatives, and *S*-alkylcysteine residues are similarly oxidized; peptides containing these residues may have slightly altered mobilities after oxidation. Fortuitous migration of cysteic acid-containing peptides along the diagonal may occur: in such cases a different pH for the electrophoresis may be used.

Complications arise when two or more cystine residues are present in the same peptide, or if a cystine residue links two points in a single peptide. Peptides with N-terminal tryptophan residues also give products lying off the diagonal. Details of the method, including complexities, are discussed by Brown and Hartley (1966). An example of the method is shown in Fig. 7.3.

With larger proteins, which yield too many peptides for adequate separation by paper electrophoresis, prior separation of peptides by gel filtration under acidic conditions is preferable. Cystine-containing peptides may be detected in the column fractions (§ 5.3.4). The pooled fractions containing these peptides are studied separately, either by diagonal electrophoresis (for small peptides) or by reduction and ^{14}C-alkylation of the peptides followed by repeated gel filtration: unless the cross-link joins a large and a very small peptide, the cleaved peptides will both be smaller than the bulk of the peptides in the mixture, and will be eluted later. If the initial gel

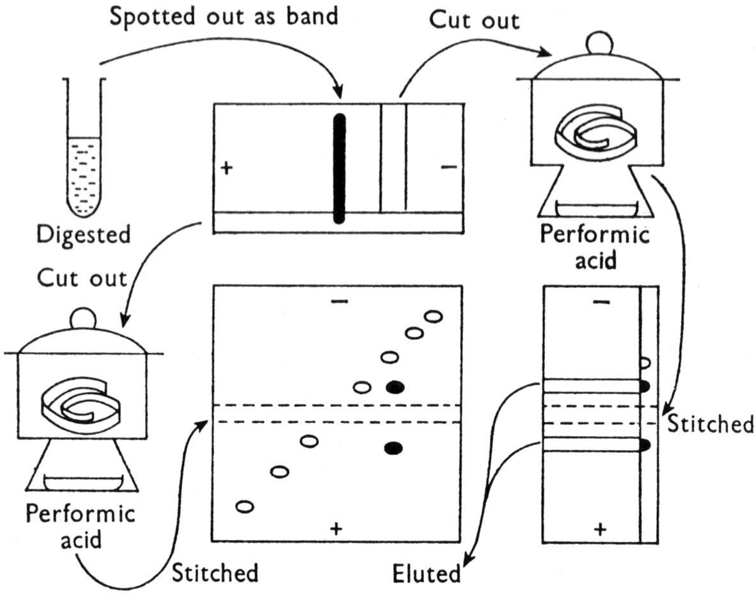

Fig. 7.2. Outline of the preparative diagonal electrophoresis technique for disulphide bridge determination. The digest (see text for suitable enzymes and conditions) is applied as a band across the width of a sheet of Whatman 3MM paper, and subjected to electrophoresis at pH 6.5. A side strip is cut off and treated with performic acid vapour in a vacuum desiccator. The paper strip is aerated, and stitched to the middle of a fresh sheet of paper. Electrophoresis at pH 6.5 is repeated, in a direction at right angles to the first direction on the strip. The paper is dried, and peptides are detected with ninhydrin (§ 5.2.1). Most peptides lie on a diagonal across the paper, but peptides derived by oxidative cleavage of a disulphide-bridged pair of peptides migrate off the diagonal (shown as filled areas). The position on the strip from which these peptides migrated in the second dimension is used to locate the band of disulphide-bridged peptides on the first sheet of paper. The band is cut out, treated with performic acid vapour, aerated and stitched to a third sheet of paper. Electrophoresis at pH 6.5 is repeated. Peptide bands are detected by ninhydrin-staining of a side strip cut from this sheet of paper. Peptides which migrate through the same distances as the off-diagonal spots on the diagonal 'map' are eluted. Analysis of these peptides determines the position of the disulphide bridge in the protein in this simple example. Reproduced with permission from Hartley, B.S. (1970) Biochem. J. *119*, 805–822.

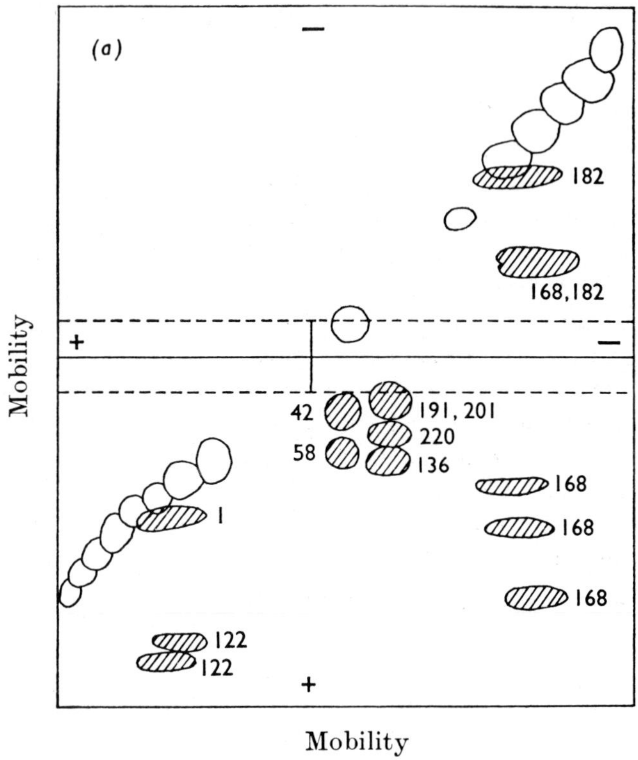

Fig. 7.3. Diagonal electrophoretic map of a peptic digest of α-chymotrypsin. Major ninhydrin spots lying off the diagonal are hatched. Numbers indicate the positions of cystine residues in the sequence of chymotrypsinogen. Electrophoresis at pH 6.5 was carried out horizontally before oxidation and vertically after oxidation. Reproduced with permission from Hartley, B.S. (1970) Biochem. J. *119*, 805–822.

filtration step does not resolve all the cystine-containing peptides from each other, further purification of these peptides must be performed before the cleavage of the disulphide bonds. Conditions leading to hydrolysis of disulphide bonds and disulphide interchange must be avoided. Spackman et al. (1960) describe suitable conditions.

Recent examples of the identification of cystine cross-links in proteins are given by Fox et al. (1979) (myotoxin *a* from Prairie rattlesnake venom), Henschen (1978) (fibrinogen), and Wunderer (1978) (toxin II from *Anemonia sulcata*).

7.7.2. Lysine-derived cross-links in collagen and elastin

As mentioned above, several cross-linking structures have been identified in the extracellular structural proteins collagen and elastin (Fig. 7.1). The subject has been reviewed by Gallop et al. (1972). The cross-linking structures may be identified by amino-acid analysis after acid hydrolysis (see Glazer et al., 1975), but the determination of the amino-acid sequences around the cross-links presents a formidable task, especially for desmosine derivatives, where four separate peptide chains may be involved. Mass spectrometry and nuclear magnetic resonance have been applied to the problem. Fujimoto et al. (1978) used these methods in determining the structure of pyridinoline, suggested as a cross-link in collagen. However, Elsden et al. (1980) have concluded after further investigation that pyridinoline is not the stable cross-link in mature collagen.

CHAPTER 8

Deduction of the total primary structure of a protein from the sequences of constituent peptides

In principle, the alignment of two sets of peptides obtained by specific digestion of the protein at different points in the chain is a simple task, resembling the completion of a jigsaw puzzle. However, some care is needed, and errors have arisen from over-optimistic interpretation of data.

Firstly, with long peptide sequences the probability of random repetition of tri- and tetra-peptide sequences is significant. If the overlapping sequences in the sets of peptides are not long enough, unambiguous alignment of the peptides may not be possible. This difficulty is greatly increased when internal homologies are present, if, for example, the polypeptide has evolved by gene duplication. Similarly, a single small peptide may be derived from two or more positions in the peptide chain. While an unusually high yield of a particular peptide may indicate that it is repeated in the total structure, this is rarely a reliable guide, since yields of peptides vary widely, and are often much less than 50%. The possibility of the presence of a peptide in two positions should therefore be considered.

Secondly, it is important that the overlapping sequences are thoroughly characterized. The amino acid composition of overlapping segments is rarely sufficient to ensure correct alignment, particularly if there is any possibility of impurities affecting the analytical results. Sequence data is much more valuable than amino-acid compositions for this purpose, since the information content is enormously greater. For example, a peptide containing one each of ten different residues may have 10! or 3,628,800 possible sequences.

It is also much easier to recognise overlapping sequences than to fit amino-acid composition data.

Thirdly, even if great care is taken in the isolation of peptides, some peptides from each digest may be lost during the separation procedures, possibly through irreversible adsorption to chromatography columns or streaking out on paper chromatography or electrophoresis. This loss is more likely to occur if chemical changes in the peptide take place during the separation procedures, such as cyclization of N-terminal glutamine residues, oxidation of methionine, *S*-carboxymethylcysteine or tryptophan residues, or reversible formation of lactones in peptides with C-terminal homoserine residues derived by cleavage with CNBr or in peptides derived by oxidative cleavage at tryptophan residues. Incomplete digestion, such as may occur with trypsin when a lysine residue is flanked by acidic residues, or when two or more lysine or arginine residues are contiguous, may reduce the yields of individual peptides to below the level at which they may be characterized. This problem is more serious when proteases of lower specificity are used, when a part of the sequence may be distributed among several peptides, each in low yield.

On the other hand, low yields of peptides apparently inconsistent with the apparent structure of the protein may be obtained. These peptides may arise from impurities in the substrate, or possibly result from microheterogeneity in the protein. Microheterogeneities are not uncommon, particularly if the protein has been isolated from different individuals and not from a pure strain. A recently described example is cytochrome c from hippopotamus, in which one individual was heterozygous (Thompson et al., 1978). However, microheterogeneity should only be assumed after other possibilities have been excluded.

Because of these adverse factors, the structures of at least three, and often more, sets of peptides from different digests of the protein are required. The large amount of redundant information thereby acquired is extremely valuable in highlighting any errors made in the structural analysis, particularly if different sequencing methods have been used for the peptides in different sets.

The difficulty of alignment of small peptides increases greatly with the number of residues in the polypeptide chain, but several factors may aid this process.

Firstly, some amino-acid residues are infrequent in most proteins, including tryptophan, cysteine and histidine, and such residues may serve as markers for the recognition of peptide sequences. Modified residues serve the same purpose.

Secondly, large fragments may be isolated by a variety of methods (Chapter 3), and even if yields are insufficient to allow the use of these as the basis for the sequence determination, peptide mapping studies on large fragments allow known peptides to be assigned to positions in the sequence represented in these fragments. The alignment of a few large fragments may often be deduced from amino-acid analyses and amino- and carboxy-terminal residues.

In the rare cases of large polypeptides for which mutants have been characterized and genetically mapped, information from genetic studies may be used. For example, the alignment of peptide fragments of β-galactosidase from *E. coli* was aided by the study of a chain-termination mutant protein, which consisted of the N-terminal half of the wild-type protein. In addition, intracistronic complementation between a cyanogen bromide cleavage product and an extract of a *lac* Z^- operator-proximal deletion mutant (in which the β-galactosidase gene product lacked part of the N-terminal quarter of the protein) could be used to locate a peptide in the N-terminal region of the protein (Fowler and Zabin, 1978).

Homologous proteins isolated from different organisms often have homologous sequences. For example, the alignment of the constituent peptides of several cytochromes *c* (e.g. that from hippopotamus; Thompson et al., 1978) and of triose phosphate isomerase from the coelacanth (Kolb et al., 1974; § 1.3), was aided by consideration of homology. Caution is required, however, to avoid the superimposition of apparent, but erroneous, homology.

A method of rapidly increasing importance is the use of DNA sequence information. The complete amino-acid sequence of the capsid protein of *E. coli* phage ΦX 174 was determined by a combination of DNA sequencing and partial amino acid sequencing (Air et al., 1978). Peptides from a penicillinase coded for by a plasmid in *E. coli* were aligned by DNA sequencing (Ambler and Scott, 1978). Further examples of the use of DNA sequences for the determination of amino-acid sequences are given in Chapter 10. The combination of information from both DNA and protein sequencing techniques has the advantage that any errors made by either method will be detected and may be corrected. Although the methods for amino-acid sequence analysis should leave no room for error, such errors do occasionally occur, and DNA sequence analysis has already revealed previously undetected mistakes (see for example Beyreuther, 1978). Conversely, DNA sequencing methods are not completely reliable (Konkel et al., 1978).

When satisfactory alignment of the peptides has apparently been achieved, the amino-acid analysis of the protein must be carefully checked against the composition calculated from the proposed sequence. Any discrepancies of more than 5% should give rise to concern. If the discrepancy includes residues which may be detected readily, such as radioactive derivatives of cysteine, selective isolation of peptides containing these residues from additional digests of the protein may be attempted. The almost complete sequence information will enable choice of a suitable cleavage method to be made.

Further investigation of the protein may be made for confirmation of the proposed structure. For example, if a small number of tryptophan residues is present in the protein, and cleavage at these residues has not been used for the generation of primary fragments, cleavage with BNPS-skatole (§ 3.4.3) or iodosobenzoate (§ 3.4.4) may be performed. The products may be rapidly analyzed by SDS gel electrophoresis. Observation of fragments of the predicted sizes lends support to the proposed structure, while if the results are different from those expected further checking of the data will be required. The proposed structure should also be checked for the numbers of acidic and basic groups for comparison with the isoelectric point of the protein. Although it is not possible to calculate exactly the isoelectric point, a neutral protein should have approximately equal numbers of acidic and basic groups.

After the various factors mentioned above and in §6.10 have been taken into account, the task of determining the sequence is complete; the only problem remaining is the preparation of the mass of data for publication. Clear and complete documentation throughout the course of the work is extremely important. The very large amount of supporting evidence has necessitated special arrangements for the reporting of the determinations of primary structures of proteins. Reduced format appendices or information repositories are used by a number of journals, including The Biochemical Journal, Biochemistry and the Journal of Biological Chemistry. Accepted abbreviations, given in *Biochemical Nomenclature and Related Documents* published by the International Union of Biochemistry and available from the Biochemical Society, should be used. Recommendations for the publication of protein sequence data are given in Biochemistry (1976) *15*, 461–462.

A large number of protein sequences are reported each year. A few recent examples of complete sequence determinations, chosen to demonstrate the variety of approaches used, are listed in Table 8.1.

TABLE 8.1

Some examples of recent amino-acid sequence determinations

β-Galactosidase (Fowler and Zabin, 1977, 1978). The longest sequence yet determined (1979), of 1021 residues (mol. wt. 116,248). Peptides from several digests were isolated and sequenced, and information from several sources was used to align the peptides.

Phosphorylase from rabbit muscle (Titani et al., 1977, 1978). A long sequence (841 residues), determined by a combination of methods, including limited proteolysis for the generation of large fragments.

Human fibrinogen, β-chain (Watt et al., 1979). Sequential degradation by thioacetylation was used in the determination of the sequence of this polypeptide, of 461 residues.

A mouse immunoglobulin μ-chain sequence was determined by methods including the use of polybrene in the spinning-cup sequencer (Kehry et al., 1979).

Physarum actin (Vandekerkhove and Weber, 1978b). Sensitive manual methods, with the use of peptide mapping for the isolation of peptides, were used, and

TABLE 8.1 *(continued)*

the alignment of peptides was made with the help of homology with other actin molecules.

λ-Repressor (Sauer and Anderegg, 1978). Methods including gas chromatography/mass spectrometry, Edman degradation and DNA sequencing were used.

Protein L31 from *E. coli* ribosomes (Brosius, 1978). Sensitive manual methods, including thin-layer peptide mapping and the dansyl-Edman degradation, were used in the determination of the sequence, of 62 residues, from 0.7 mg of protein.

Rabbit β_2-microglobulin (Gates et al., 1979). The sequence of this 99-residue protein was determined from about 30 mg, using the spinning-cup sequencer with polybrene for the determination of the structures of large fragments.

Eel calcitonin (Noda and Narita, 1976). The total structure of this 32-residue polypeptide was determined directly by the manual Edman degradation.

ATP-phosphoribosyl transferase of Salmonella (Piskiewicz et al., 1979). Peptide sequences were complemented by DNA sequencing of the hisG gene.

Triose phosphate isomerase of the coelacanth (Kolb et al., 1974). The sequence of this 247-residue protein was determined in less than four months from about 100 mg protein, using both automated and manual methods (§ 1.3).

Protein S8 of *E. coli* ribosomes (Allen and Wittmann-Liebold, 1978). Sensitive manual methods were used to determine the total sequence of this 129-residue protein from less than 8 mg protein in less than three months.

CHAPTER 9

Integral membrane proteins

The methods described in the earlier chapters should be sufficient for the determination of complete amino-acid sequences of all soluble proteins, or proteins which may be dissolved in 8 M urea. However, one class of proteins has so far proved refractory to study by these methods: those in which at least part of the polypeptide chain is inserted into the lipid core of bilayer membranes. Special methods are required for the cleavage of the proteins into suitable fragments, separation of these peptide fragments, and sequential degradation of these fragments. The extremely hydrophobic nature of many of these proteins, and the insolubility of peptides in aqueous solutions lead to considerable difficulties, as discussed by Keefer and Bradshaw (1977) and Walker et al. (1979).

Integral membrane proteins may be classified into Type I, in which a relatively small, but very hydrophobic, region of the polypeptide acts to anchor the protein in the membrane, and the bulk of the protein lies outside the lipid bilayer, and Type II, in which a substantial portion of the protein is inserted into the bilayer, perhaps forming a pore structure for the transport of ions or hydrophilic molecules. Examples of Type I are cytochrome b_5 (Visser et al., 1975; Fleming et al., 1978; Ozols and Gerard, 1977b), human erythrocyte glycophorin (Marchesi et al., 1972), influenza haemagglutinin (Waterfield et al., 1979), major histocompatibility antigens (Shimada and Nathenson, 1969) and small intestinal sucrase/ isomaltase (Frank et al., 1978). Some examples of Type II are bacterial rhodopsin (Ovchinnikov et al., 1979), sarcoplasmic reticulum Ca^{2+}-ATPase (MacLennan, 1970) and the anion-transport protein (band 3) of human erythrocytes (Boxer et al., 1974; Fukuda et al., 1978).

Type I proteins may often be cleaved in situ to give a soluble globular portion, comprising the bulk of the polypeptide chain, and a very hydrophobic peptide which remains bound to the lipid. Some examples of this type of cleavage are given in Table 9.1. The structure of the globular portion may be determined by the methods described in earlier chapters, but the determination of the sequences of the hydrophobic portion requires other methods.

TABLE 9.1

Examples of limited proteolysis of Type I integral membrane proteins

(1) Cytochrome b_5. Tryptic cleavage of the molecule yielded a polar fragment, containing haeme, a hydrophobic fragment, associated with the lipids, and a linking peptide of about 15 residues (Visser et al., 1975).
(2) NAD-cytochrome b_5 reductase. Chymotrypsin converted the 43,000 dalton protein into a core enzyme, mol. wt. 33,000, and aggregated hydrophobic peptides, rich in tryptophan (Spatz and Stritmatter, 1973).
(3) Influenza haemagglutinin. Bromelain cleaved the protein into a soluble portion, retaining the antigenic activity, and a hydrophobic portion, associated with the lipid bilayer in the viral coat (Brand and Skehel, 1972).
(4) Murine histocompatibility antigens, H-2^b, H-2^d. Papain released a soluble part of the molecule, carrying the antigenic activity; the hydrophobic carboxy-terminal part of the heavy chains remained in the membrane (Shimada and Nathenson, 1969).

Type II integral membrane proteins present greater difficulties, and the complete sequence of only one protein of this type has been determined (Ovchinnikov et al., 1979) by protein chemical methods.

9.1. Purification of integral membrane proteins

Advances in the techniques of purification of membrane proteins in the last decade, particularly in the use of non-ionic detergents and affinity chromatography, have made the isolation of many membrane proteins relatively straightforward. Final purification for the purposes of sequence determination may be performed in solutions of denaturing detergents, such as dodecyl sulphate (SDS). Gel filtration in dodecyl sulphate solution may also be used to separate the protein from lipids.

Throughout the purification, reducing conditions should be maintained, by the addition of 2-mercaptoethanol (5 mM) or other thiol. If this precaution is not observed, autoxidation of unsaturated lipids, with the formation of peroxides and other reactive compounds, may occur. These compounds may react with the protein, for example by oxidising methionine, cysteine and tryptophan residues. Transition metals catalyze autoxidation, and the presence of EDTA may be desirable, unless divalent cations are required to stabilize the protein. If organic solvents are used to extract the lipids from the protein (§ 2.4.1), they must be free from peroxides. Treatment of membrane proteins with organic solvents often yields material which will not readily disperse or dissolve subsequently and gel filtration in dodecyl sulphate solution is recommended for delipidation, at least for larger polypeptides.

Another problem encountered during the purification of membrane proteins for sequence analysis is degradation by lysosomal proteases, which may co-purify with the protein in lipid vesicles and be released upon treatment of the preparation with detergents. Heating to 100 °C for 2 min immediately upon the addition of dodecyl sulphate effectively prevents proteolysis.

The delipidation of the Ca^{2+}-ATPase of rabbit skeletal sarcoplasmic reticulum was performed by gel filtration (Allen and Green, 1978). The lipoprotein was prepared by the method of MacLennan (1970), using salt fractionation in the presence of low concentrations of deoxycholate. The protein was reduced with dithiothreitol and carboxymethylated with iodoacetate in 6 M guanidinium chloride containing sodium taurodeoxycholate and EDTA at pH 8.0; in the absence of the detergent, complete denaturation was not observed. The carboxymethylated protein was dialyzed to remove guanidinium chloride, and sodium dodecyl sulphate was added (10 g/g of protein).

The protein was separated from lipids by chromatography on a column of Sepharose 6B in a solution containing sodium dodecyl sulphate (1% w/v), 50 mM Tris, 1 mM EDTA and 1 mM 2-mercaptoethanol, adjusted to pH 7.0 with acetic acid. The protein (estimated by absorbance at 280 nm) was completely separated from the phospholipids (estimated as inorganic phosphate after heating dried samples with $HClO_4$). The protein was separated from sodium dodecyl sulphate by the method of Weber and Kuter (1971), using a short column of Bio-Rad AG1 X2 anion-exchange resin in 8 M urea.

9.2. Specific cleavage methods

Enzymic digestion of membrane proteins is rendered difficult by their insolubility in water. Digestion of the protein in lipid vesicles is often used; soluble peptides released from the extramembranous portions may be separated from the residual part of the protein within the bilayer by ultracentrifugation. Such digestion is also useful in identifying the parts of the polypeptide chain exposed to the aqueous solution. If the protein is asymmetrically distributed in sealed vesicles, limited proteolysis may be used to study the topography, as, for example, with the band 3 glycoprotein of erythrocytes (Jenkins and Tanner, 1975; Drickamer, 1978; Fukuda et al., 1978), and erythrocyte glycophorin (Marchesi et al., 1972).

As mentioned above, the type I integral membrane proteins may often be cleaved into a soluble portion and a small membrane-bound peptide by limited proteolysis. Limited proteolysis of Type II proteins often results in the production of large fragments which remain bound to the lipid bilayer. Such cleavage of the sarcoplasmic reticulum Ca^{2+}-ATPase has been described above (§ 3.2.1.4). Exposed portions of bacterial rhodopsin are cleaved by papain (Ovchinnikov et al., 1979).

For the generation of smaller peptide fragments suitable for sequential degradation, more extensive proteolysis is required, after removal of lipid. The insolubility or aggregation of lipid-free Type II proteins prevents complete proteolysis. For example, the murein lipoprotein of the outer membrane of *E. coli* was not completely digested by trypsin, chymotrypsin or papain; the sequence of the 57-residue polypeptide was determined after more extensive digestion with pronase, subtilisin and thermolysin (Braun and Bosch, 1972).

Chemical modification of the protein to give a soluble derivative improves the yields of digestion products. Reduction and carboxymethylation of cysteine residues and succinylation of amino groups converts the protein to a polyanionic derivative, likely to be soluble in water at alkaline pH and more susceptible to digestion by proteases such as trypsin (at arginine residues), chymotrypsin or thermolysin. Alternatively, the protein may be soluble in concentrated formic acid (90–100%), and dilution into a solution of pepsin may allow digestion to proceed efficiently. Both these approaches were used with the Ca^{2+}-ATPase of sarcoplasmic reticulum (Allen, 1980; Allen et al., 1980).

Many proteases, especially those of microbial origin, such as subtilisin and thermolysin, retain their activity in sodium dodecyl sulphate solution, and efficient digestion of membrane proteins may be achieved in the presence of this detergent. However, the removal of detergents rom hydrophobic peptides is not readily achieved, and subsequent isolation of peptides may be difficult.

Chemical cleavage methods, which may be used under conditions where most proteases are denatured, may be expected to proceed in higher yields. However, even treatment with cyanogen bromide in concentrated solutions of formic or trifluoroacetic acids (§ 3.4.1) gave rather low yields of cleavage at methionine residues in cytochrome b_5 (Ozols and Gerard, 1977b). The methionine residues may not have been fully accessible to the reagent, or able to adopt a conformation required for the cleavage reaction, or possibly they had been partially oxidized. The use of a very large excess of CNBr led to cleavage at tryptophan residues (Ozols and Gerard, 1977a); possibly traces of bromine were responsible.

Cleavage at tryptophan residues (§§ 3.4.3–3.4.5) may be a valuable tool for the study of membrane protein sequences. Partial acid hydrolysis (§ 3.4.6), not often used for the study of soluble proteins (except for the special case of -Asp-Pro- sequences, as described in section 3.2.2.2) may be useful, since particularly sensitive bonds, involving aspartic acid, serine or threonine residues, may be infrequent in the generally hydrophobic sequences. Partial acid hydrolysis followed by sequence determination by gas chromatography/mass spectrometry has been suggested as particularly appropriate for membrane proteins (Gerber et al., 1979).

9.3. Purification of peptides from integral membrane proteins

Enzymic or chemical digests of a simpler, type I, integral membrane protein may consist of a mixture of many soluble peptides and a single insoluble peptide derived from the intramembranous part of the protein. The soluble peptides may be purified by the usual methods (Chapter 4). The insoluble peptide, perhaps after a simple purification step such as gel filtration in concentrated formic acid or counter-current distribution (Furthmayr et al., 1978) may be pure enough for sequence determination. Erythrocyte glycophorin (Tomita et al., 1978) provides an example. It is instructive that earlier reports on the sequence (Tomita and Marchesi, 1975; Segrest et al., 1972) contained errors arising from work with impure peptides.

Some of the peptides derived from the extramembranous portion of the protein may also be insoluble under the conditions of digestion. These peptides may sometimes be extracted from the intramembranous peptides under different conditions, such as low pH, or may be subfragmented by a protease which does not digest the membrane peptides.

The more complex type II integral membrane proteins yield many peptides from the interior of the lipid bilayer, and the separation of these is a problem which has not been satisfactorily resolved except in a very few cases. These peptides characteristically have a strong tendency to aggregate, and if freeze-dried do not dissolve even in 8 M urea or 6 M guanidinium chloride. A large number of different solvents have been studied for the dissolution and separation of these hydrophobic peptides.

Chloroform/methanol mixtures are capable of dissolving very hydrophobic peptides, such as myelin protein (Gagnon et al., 1971) and the proteolipid component (dicyclohexyl carbodiimide binding protein) of mitochondrial ATPase. However, peptides containing both charged and hydrophobic residues do not generally dissolve in these solvents. Organic acids, such as formic acid, acetic acid, propionic acid and trifluoroacetic acid, in high concentrations in water or in alcoholic solution may be effective in dissolving membrane peptides. In particular, 88% (v/v) aqueous formic acid/ethanol (30 : 70, v/v) was found useful (Gerber et al., 1979). Phenol is extremely effective for dissolving proteins, and the system phenol/acetic acid/water (1 : 1 : 1, w/v/v) (Synge, 1957) may be used for gel filtration on Sephadex G-25 (Bagdasarian et al., 1964) or on BioGel P-series columns (Pusztai and Watt, 1970; Allen, 1978). The solvent is viscous and very slow flow (less than 1 cm/h) is used. Chloral hydrate dissolves membrane proteins and peptides (Ballou et al., 1974), but is a reactive compound, combining, for example, with urea (Ballou and Smithies, 1977), and reaction with peptides might be anticipated. Hexamethylphosphoric triamide was found to be a good solvent for membrane polypeptides (Kohl and Sandermann, 1977). The dipolar aprotic solvents, acetonitrile and dimethylformamide, appear not to be satisfactory for dissolving all membrane peptides, but dimethylsulphoxide may be more effective. Concentrated solutions of pyridine are potentially useful; pyridine/1M NH_3 (7 : 3, v/v) has been used for the separation of small hydrophobic peptides (Casey and Rees, 1979).

Many of these solvents dissolve plastic materials and plasticisers, and phenol is particularly potent. All-glass apparatus is used wherever possible.

The dissociating detergents, particularly sodium dodecyl sulphate, which are so useful for separating membrane proteins, are less useful for the separation of peptides. Although the peptides usually dissolve in the detergent solution, they are complexed in micelles, the properties of which are determined largely by the detergent itself. The size of the micelles, about 20,000 Daltons for sodium dodecyl sulphate, is relatively little affected by the binding of peptides of molecular weight below 10,000, and separation of these peptides by gel filtration in SDS solution is not possible. Electrophoresis in polyacrylamide gels of small pore-size is more effective, however (§ 4.11).

The solvents considered above may be used for gel filtration. Acetylated Sephadex G-200 has been used with chloroform/methanol (Monreal, 1976). A single separation system based on differences in the sizes of peptides is not generally sufficient for resolving the mixture present in a typical digest, although the recently introduced high-performance gel permeation chromatography columns (§ 4.12) may be extremely valuable, if they are stable in the more powerful solvents, such as phenol/acetic acid/water.

Ion-exchange chromatography in organic solvents is not generally satisfactory. The formation of strong ion-pairs is common, and ionic equilibration is usually much slower than in water. However, systems used for the separation of classes of phospholipids, such as chromatography on DEAE-cellulose in chloroform/methanol in a gradient of ammonium acetate buffer (Rouser et al., 1969), have been used in the purification of small membrane proteins, including the C_{55}-isoprenoid alcohol phosphokinase of *Staphylococcus aureus* (Sandermann and Strominger, 1972), and the dicyclohexylcarbodiimide reactive protein of *E. coli* ATPase (Fillingame, 1976), and may be useful for the separation of some peptides from larger membrane proteins.

The powerful solvent properties of phenol, and the high dielectric constant of water were combined for the separation of large hydrophobic peptides (of 50–100 residues) from the tryptic digest of succinylated Ca^{2+}-ATPase protein from sarcoplasmic reticulum on a column of DEAE-cellulose (Allen, 1978). Phenol/formamide/water (2 : 1 : 2, w/v/v) was the solvent, and peptides were eluted with a gradient of triethylamine/acetic acid buffer. The high viscosity and low rate of ionic equilibration necessitated the use of low flow rates. Removal of the solvents was effected by gel filtration on Sephadex G-10 in 99% formic acid, after the bulk had been removed by rotary evaporation.

Small hydrophobic peptides may often be purified by thin-layer chromatography on silica gel; R_F values on cellulose or paper are likely to approach unity.

High-performance liquid chromatography, in the reverse-phase mode, with a gradient of ethanol in aqueous formic acid, has been used for the separation of peptides from bacterial rhodopsin (Gerber et al., 1979).

9.4. Determination of the sequences of membrane peptides

The poor solubility of membrane peptides leads to difficulties in sequential degradation. The solvents for the coupling and cleavage reactions in the Edman degradation are fairly effective for dissolving hydrophobic peptides. Formic acid has been suggested for the removal of aliquots for dansylation in the dansyl-Edman method applied to insoluble peptides (Dhaese et al., 1979). Extractive losses of hydrophobic peptides may be high, however. The hydrophobic segments of glycophorin (Tomita et al., 1978) and of cytochrome b_5 (Ozols and Gerard, 1977b; Fleming et al., 1978) were sequenced automatically in the spinning-cup instrument, but repetitive yields were low. The solid-phase technique offers the advantage that extractive losses are avoided, and solubility factors are less important, except during the attachment of peptides to the support. The sequences of dicyclohexylcarbodiimide-binding proteins from mitochondrial ATPases were determined using solid-phase methods (Tzagoloff et al., 1979).

Because of the difficulties in the Edman degradation, more extensive fragmentation of membrane peptides may be required, as for the murein lipoprotein mentioned above (§ 9.2).

Hydrophobic peptides characteristic of intramembranous portions of proteins are particularly suitable for sequence determination by mass spectrometry. Partial acid hydrolysis, followed by conversion of the fragment mixture to volatile derivatives for gas chromatography and mass spectrometry, was used to complement data from the Edman degradation for the determination of partial sequences in bacterial rhodopsin (Gerber et al., 1979).

The study of trace amounts of intrinsically radiolabelled membrane proteins in the presence of soluble carrier proteins may avoid some of the problems encountered on the larger scale, allowing, for example, the purification of peptides by high-resolution SDS-polyacrylamide gel electrophoresis. Very hydrophobic leader sequences of precursor polypeptides of excretory proteins have been studied in this way (§ 10.5).

Perhaps the best way of determining the amino acid sequences of intramembranous peptides will be to determine the nucleotide sequences of DNA or RNA coding for the protein. This method has yielded the primary structures of lactose permease of *E. coli* (Büchel et al., 1980), the surface antigen of hepatitis B virus (Valenzuela et al., 1979), the haemagglutinin of fowl plague virus (Porter et al., 1979), the proteolipid component of a mutant yeast mitochondrial ATPase (Macino and Tzagoloff, 1979), and several viral proteins, which, although not integral membrane proteins, were difficult to study directly because of the low solubility of derived peptides, including proteins of ΦX174 (Air et al., 1978) and simian virus 40 (Ysebaert et al., 1978; Van Heuverswyn et al., 1978; Kempe et al., 1979). The method is particularly suitable for the study of viral or mitochondrial proteins (with allowance made for the different codons used in mitochondria [Barrell et al., 1979]), because of the relative simplicity of the genomes. Post-translational modifications of the protein are, of course, not detected.

CHAPTER 10

Applications of partial sequence analysis

10.1. Identification of positions within the sequence of selectively modified residues

Chemical modification of proteins is often performed as part of investigations into their structures and functions. Particularly reactive functional groups are often present at the active sites of enzymes, and the structures of these portions of the molecules may be of particular interest. Techniques of chemical modification are discussed by Glazer et al. (1975). Peptides containing modified residues are isolated, preferably with the help of radioactive or chromophoric labels, by methods discussed in Chapter 4, and the positions in the sequence of the modified residues are identified as described in Chapter 7.

10.2. Identification of residues involved in chemical cross-linking

Chemical cross-linking may be used for the identification of functional groups in proximity on the surface of a protein. Methods analogous to those used for the determination of disulphide bond cross-links (§ 7.7.1) may be used for the identification of cross-links which are cleaved under mild conditions. 1,3-Dibromoacetone linked cysteine-25 and histidine-158 in papain (Husain and Lowe, 1968). After the isolation of a bridged peptide (using dibromo-[^{14}C]acetone to provide a label) cleavage with performic acid permitted the identification of the two peptide strands.

Non-cleavable reagents have also been used. The identification of points of cross-linking in bridged peptides derived from proteins of known sequence may usually be made by amino acid analysis and partial sequential degradation. The proximity of lysine residues 31 and 37, and 7 and 37, in ribonuclease was shown by cross-linking with dimethyl-[^{14}C]adipimidate (Hartman and Wold, 1966). Lysine-166 of protein S5 and lysine-93 of protein S8 were shown to be in proximity in 30 S ribosomal subunits of *E. coli,* using dimethyl-[^{14}C]suberimidate (Allen et al., 1979). Many cross-linked pairs of ribosomal proteins have been identified after cross-linking by *bis*-imidate reagents, and much information may result from further application of this technique, which is facilitated by the knowledge of most of the sequences of the ribosomal proteins (Brimacombe et al., 1978).

10.3. Comparison of related proteins by peptide mapping

Many techniques for the comparison of closely related proteins by peptide mapping have been described. If more than 2 nmol of the pure proteins are available, the method of choice will probably be two-dimensional thin-layer electrophoresis and chromatography of tryptic peptides (§ 4.3.4). In addition to the pattern of spots, information about the peptides may be obtained by using a variety of staining methods, and peptides may also be excised and further characterized by N-terminal and amino acid analysis. With larger proteins a preliminary fractionation by gel filtration may be useful to reduce the probability of overlap of peptide spots on the thin layers. Paper methods may be used in the same way (§ 4.9).

These techniques have been used widely, and some examples are the study of variant human haemoglobins (Ingram, 1956), actins from various tissues (Vandekerkhove and Weber, 1978a), and ribulose *bis*-phosphate carboxylase from *Oenothera* (Holder, 1976).

Methods using a single separation technique of high-resolving power for the separation of tryptic peptides possess the advantages of easy automation and quantitation. Ion-exchange chromatography on polystyrene resins, with detection by ninhydrin or fluorescamine after alkaline hydrolysis (§ 4.6.3) is one such method. Reverse-phase high-performance liquid chromatography (§ 4.12) is also suitable.

Higher sensitivity is achieved through the use of radioactive labels. Biosynthetic labelling of the proteins with [^{3}H]lysine and [^{3}H]arginine is particularly useful for the detection of tryptic peptides, since all except the C-terminal peptide should be detected with equal sensitivity (apart from partial cleavage products). Different actins were compared by this method (Gruenstein and Rich, 1975).

A useful approach which avoids any ambiguity arising from incomplete reproducibility in peptide mapping is the labelling of two proteins with different isotopes, mixing of the proteins, digesting with trypsin and separating the peptides on one peptide map with determination of the two isotopes by liquid-scintillation spectrometry. Different murine alloantigens were studied in this way (Brown et al., 1974).

If it is not feasible to incorporate a radiolabel biosynthetically, the protein may be chemically modified by radioactive reagents. Two methods which lead to little change in the properties of peptides are reaction with methyl-[^{14}C]acetimidate (§ 10.3.1) and reductive methylation (Nelles and Bamburg,. 1979) with [^{14}C]formaldehyde. Peptides containing lysine residues, and the N-terminal peptide, are labelled. Smaller sets of peptides, such as those containing cysteine residues, labelled with iodo[^{14}C]acetate, or tyrosine residues, labelled with ^{125}I, may suffice for some comparative purposes. Radioiodination has been used frequently for high-sensitivity peptide mapping, although the method has several disadvantages, including the formation of two derivatives of tyrosine residues and the insolubility of many peptides containing iodinated tyrosine residues. However, very high specific activities may be used, and the method is applicable to proteins separated in small amounts by polyacrylamide gel electrophoresis. One method is described below (§ 10.3.2).

Fluorescent labelling of peptides may also be used for high-sensitivity peptide mapping. The introduction of relatively large hydrophobic groups changes the properties of peptides so that different separation methods are required. For example, two-dimensional chromatography of dansylated peptides on polyamide layers has been described (Tichy, 1975). Results are, in general, less satisfactory than those obtained by radiochemical techniques. Reaction of peptides with fluorogenic reagents after separation is more sensitive than staining with ninhydrin. For example, for the detection of arginine residues on cellulose thin layers with the phenanthrene quinone reagent (§ 5.4.3), only about 0.2 nmol of a protein digest is required.

A convenient way of comparing the primary structures of proteins separated by polyacrylamide gel electrophoresis in sodium dodecyl sulphate solution is partial enzymic digestion in the gel slice containing the protein, and repeated gel electrophoresis (Cleveland et al., 1977). The conditions for the digestion of two proteins to be compared must be identical, since the pattern of bands produced is dependent upon the extent of digestion. Cleavage of the protein with CNBr may be used similarly (Luduena and Woodward, 1973); precautions are required to prevent oxidation or other modification of methionine residues which may occur during polyacrylamide gel electrophoresis.

Peptide mapping is most valuable when the proteins to be compared are highly homologous. If the homology in amino acid sequences is less than about 80%, little similarity may be seen in the tryptic peptide maps, apart from the presence of small peptides which may occur by chance in unrelated proteins. Peptide maps alone are of little use for comparing the structures of an enzyme from widely divergent species.

10.3.1. Use of methyl-[^{14}C]*acetimidate* (Bates et al., 1975)

Rabbit muscle aldolase was reduced and *S*-carboxymethylated with iodoacetic acid (§ 2.6.5). A stock solution (0.5 M) of methyl-[^{14}C]-acetimidate · HCl (123 dpm/nmol) was prepared in a mixture of equal volumes of 1 M *N*-ethylmorpholine/acetic acid buffer, pH 8.2, and 1 M NaOH. The pH of the resulting solution is checked immediately with narrow range pH paper and, if necessary, adjusted to be in the range pH 8.0–8.5 with either 0.1 M NaOH or 0.1 M HCl. The reagent solution is used within 30 min. The *S*-carboxymethylated protein was amidinated in 0.1 M *N*-ethylmorpholine/acetic acid buffer, pH 8.2 at 20 °C for 16 h in the presence of 8 M urea or 4 M guanidinium chloride. The protein concentration was about 10 mg/ml and the methyl acetimidate concentration was 0.1 M.

After the reaction, the aldolase was dialyzed exhaustively against 0.5% NH_4HCO_3 and freeze-dried. The protein was digested in 0.5% NH_4HCO_3 with trypsin (1%, w/w) for 4 h at 37 °C; chymotrypsin (1%, w/w) was added, and digestion allowed to continue for a further 4 h. The peptides were recovered by freeze-drying. Peptide maps were prepared by electrophoresis at pH 6.5, followed by chromatography in butan-1-ol/acetic acid/water/pyridine (15 : 3 : 12 : 10, by vol.) in the second dimension on silica gel thin layers. Peptides were detected by radioautography.

It should be noted that tryptic cleavage of amidinated proteins is limited to arginine residues, and therefore other proteases must be used for the generation of peptides sufficiently small for thin-layer peptide mapping.

10.3.2. Use of radioiodination

Davison (1976) recommended the following modification of the method of Bray and Brownlee (1973) for peptide mapping of proteins which may be purified by polyacrylamide gel electrophoresis.

The sample is dissolved at 4–10 mg/ml in a solution containing 1% sodium dodecyl sulphate, 0.02 M borate, pH 8, 10 mM EDTA and 10 mM 2-hydroxyethyldisulphide, to which is added mercapto-ethanol to 0.01%. The solution is heated for up to 5 min at 100 °C to effect complete dissolution, and is then evaporated to dryness in a stream of air at 60 °C to remove the mercaptoethanol and disulphide. The sample is redissolved in water with warming.

To 0.1–0.3 ml protein solution is added 10 μl of 0.1 M sodium phosphate, pH 7, 100–300 μCi of $^{125}I^-$ plus any non-radioactive iodide if appropriate, to an iodine/tyrosine ratio not exceeding 0.05, then 10 μl of 1% chloramine-T in 0.1 M phosphate buffer, pH 7.

After 30 min at 2 °C, 10 μl of 1% sodium metabisulphite in the phosphate buffer is added, and the solution is dialyzed against 0.1% sodium dodecyl sulphate, 1 mM phosphate, pH 7 to eliminate excess iodine and other reagents. Iodinations must be carried out in a well ventilated hood to minimize inhalation of iodine vapour by the investigator.

Horseradish peroxidase and dilute hydrogen peroxide may be used as an alternative to chloramine T (Davison, 1976).

After dialysis, the samples are subjected to polyacrylamide gel electrophoresis (§ 4.3.1). The appropriate stained bands are cut from the gel, sliced to less than 1 mm thickness, and extracted with 2 ml of 0.2% sodium dodecyl sulphate in 0.02 M phosphate, pH 7. After agitating for 24 h, each solution is separated from the gel slices, 5 μl of 1% albumin is added as a carrier, and the solution is mixed with 3 ml of methanol and 30 μl of glacial acetic acid and stored at −10 °C overnight to precipitate the protein. The protein is collected by centrifugation at 60,000 × *g* for 10 min. The protein is dissolved in 70% formic acid, diluted 10-fold with water, and freeze-dried. The samples are then treated with 50–100 μl of 0.1% trypsin or chymotrypsin in 0.1 M NH_4HCO_3, 1 mM $CaCl_2$. After digestion the peptides are recovered by freeze-drying. The radioactivity in the sample is estimated, and thin-layer peptide maps are prepared (§ 4.3.4). Samples containing about 40,000 cpm of ^{125}I usually give satisfactory radioautographs after exposure for 3 days.

A recent example of the use of radioiodination and peptide mapping was given by Green et al. (1979), who compared DNA-linked proteins of five different human adenovirus groups.

10.4. Comparison of partial sequences of apparently unrelated proteins

Similarities in amino acid sequences between two apparently unrelated proteins have occasionally been observed. These similarities reveal interesting evolutionary relationships between proteins, and possibly other relationships. For example, methionine enkephalin was recognized as part of the sequence of a longer polypeptide in pituitary glands (Hughes et al., 1975). Partial amino-acid sequences in streptococcal M protein have been found to be homologous with sequences in mammalian tropomyosin (Hosein et al., 1979); this observation may be relevant to the association of the group A streptococcus with rheumatic heart disease.

10.5. Amino-terminal sequences

The readily accessible amino-terminal sequences of proteins have been of interest in several fields. The variable regions of immunoglobulin light and heavy chains are at the amino-terminal ends, and the study of a large number of amino-terminal sequences allowed the classification of the chains into several groups with related sequences, and the identification of hypervariable sequences around the antigen binding sites. The automated spinning-cup machine has been used in most of these studies. One example is given by Capra and Kehoe (1974).

The study of the amino-terminal sequences of nascent polypeptides has demonstrated that many proteins excreted from cells are synthesized as precursors, frequently with hydrophobic extensions at the amino-terminus. A review of the studies of immunoglobulin precursor polypeptides has been given by Schechter et al. (1979). Highly sensitive radio-chemical methods were required for the study of the minute amounts of protein synthesized in mRNA-dependent cell-free systems. A detailed procedure for the amino-acid sequence analysis of 0.1 pmol of a polypeptide labelled with a single amino acid at a time was given by Schechter and Burstein (1976). The positions of the labelled amino acid in the sequence were determined by liquid-scintillation counting of the extracted anilino thiazolinones, using an automated sequencer. For the complete determination of the amino-terminal sequences (up to about 40 residues), twenty different sequencer runs, each using a different labelled amino acid, are required (e.g. Burstein and Schechter, 1977). In addition, duplicate runs should be performed as a check. Thus the determination of amino-acid sequences by this method is both time-consuming and expensive. A similar method, but involving biosynthetic labelling in tissue slices, was used in the determination of the sequence of the amino-terminal part of human proparathyroid hormone (Jacobs et al., 1974).

The determination of amino-acid sequences by the incorporation of all, or several, amino acids labelled to high specific activity in a single experiment is beset by two problems: a low efficiency of synthesis if all amino acids are present at low concentrations, and the difficulty of identification of phenylthiohydantoins (§ 6.4.2.2). However, Jilka and Pestka (1979) studied immunoglobulin precursors in this way.

A third application of amino-terminal partial sequences is correlation with nucleotide sequences determined from cloned cDNA or genomic DNA (§ 10.6).

Another example of the application of amino-terminal sequence analysis is the comparison of the structures of several variant-specific glycoproteins from *Trypanosoma brucei* (Bridgen et al., 1976); no sequence homology was detected.

10.6. Correlation with nucleotide sequences in DNA

Partial amino-acid sequences are required for the identification of the gene products coded for by nucleotide sequences determined in genomic DNA, and for confirmation of the identification of sequences determined in cloned cDNA derived from purified mRNA. The rapid determination of nucleotide sequences in DNA (Sanger et al., 1977; Maxam and Gilbert, 1977) has already led to the deduction of the amino-acid sequences of several proteins. Some examples are given in Table 10.1.

A minimum requirement for the deduction of the amino-acid sequence of the mature protein from the nucleotide sequence of cDNA is the determination of amino- and carboxy-terminal sequences of the protein, since the initial translation product may have extensions at either end. No information about post-translational

TABLE 10.1
Examples of protein sequences deduced wholly or in part from nucleotide sequences

Gene F (capsid) protein of ΦX174 (Air et al., 1978)
Penicillinase encoded by *E. coli* plasmid R6K (Ambler and Scott, 1978)
Simian virus 40 proteins VPI (Van Heuverswyn et al., 1978), VP2 and VP3 (Ysebaert et al., 1978)
Simian virus 40 structural proteins (Kempe et al., 1979)
Major protein of hepatitis B virus surface antigen (Valenzuela et al., 1979)
Mitochondrial ATPase proteolipid (Macino and Tzagoloff, 1979)
ATP phosphoribosyl transferase of *Salmonella* (Piskiewicz et al., 1979)
Human cytochrome oxidase subunit II (Barrell et al., 1979)
Ribosomal proteins L1, L11, L7/L12 and L10 of *E. coli* (Post et al., 1979)
λ Repressor (Sauer and Anderegg, 1978; Farabaugh, 1978)
Haemagglutinin of fowl plague (influenza) virus (Porter et al., 1979)
Yeast cytochrome oxidase subunit 2 (Coruzzi and Tzagoloff, 1979)
Lactose permease of *E. coli* (Büchel et al., 1980)

modification can be obtained from the DNA sequence, and direct determination of the positions of such modified residues in the protein must be made. The carboxyl-terminal sequence of the protein is particularly useful for correlation with sequences in cloned cDNA, since the nucleotide sequence of this region lies close to the polyA tail in the mRNA, and can be investigated relatively easily.

The presence of intervening non-translated regions in eukaryotic genomic DNA (reviewed by Abelson, 1979) makes the deduction of the amino-acid sequence from the genomic DNA sequence difficult. The recent discovery that the genetic code is not universal (Barrell et al., 1979) also suggests that caution is required.

One of the major problems limiting the application of DNA sequencing techniques at present is the detection of specific genes in restriction endonuclease digests and in colonies of bacteria that contain plasmids with inserts of DNA. The most direct method of identifying clones of interest is hybridisation with a pure oligonucleotide probe. Highly purified messenger RNA is usually only obtainable when the mRNA is a very abundant species, such as globin mRNA in reticulocytes. Oligodeoxynucleotides coding for known peptide sequences in proteins of interest may, in theory, be synthesized for use as probes. The redundancy in the genetic code limits the use of this approach. For example, the gene coding for a typical unique amino-acid sequence of six residues for each of which there are two possible codons may have one of 64 18-nucleotide sequences. The synthesis of this many oligodeoxynucleotides for use as unambiguous probes for the detection and isolation of a gene directly from a human 'gene library' would be a formidable task.

However, a high degree of purification of mRNA may be achieved by the use of relatively small (dodecanucleotide) probes if the protein has sequences containing two or more of the residues for which there is a single codon (tryptophan and methionine) in proximity. Gastrin mRNA could be purified in this way (Noyes et al., 1979; Mevarech et al., 1979). Gastrin contains a -Trp-Met-Glu-Glu- sequence that was suitable for directing the synthesis of such probes. Lower degrees of purification of mRNA species may be attained by the use of shorter probes, which hybridize under less selective conditions to sequences which are likely to be repeated in different mRNAs from a higher organism.

Partial amino-acid sequence determination of a protein, particularly around tryptophan and methionine residues, may thus be useful even when the total sequence is most efficiently deduced subsequently from sequences in DNA.

Appendices

APPENDIX I

Manufacturers and suppliers of equipment and reagents for protein sequence analysis

(1) *Equipment*

Supplier	Products
American Instrument Co. Inc. **Travenol Laboratories Inc.** 8030 Georgia Avenue Silver Spring, MD 20910 U.S.A.	Amino acid analyzers
Beckman Instruments Inc. Spinco Division 1117 California Avenue Palo Alto, CA 94304 U.S.A.	Amino acid analyzers, automated protein sequencers, centrifuges
Beckman-RIIC Ltd. Eastfield Industrial Estate Glenrothes Fife, KY7 4NG Scotland	
Biotronik Wissenschaftliche Geraete GmbH D-6 Frankfurt am Main Borsigallee 22 F.R.G.	Amino acid analyzers

Burkard Scientific (Sales) Ltd.
175A High Street
Uxbridge
Middlesex
England

Small bench centrifuge

Cecil Instruments
Milton Industrial Estate
Milton
Cambridge, CB4 4AZ
England

Spectrophotometers, UV monitors

C. Desaga GmbH
Postfach 101969
D-6900 Heidelberg 1
Maassstr. 26–28
F.R.G.

Thin-layer chromatography and electrophoresis equipment

Dionex Corp. (formerly Durrum)
1228 Titan Way
Sunnyvale, CA 94086
U.S.A.

Amino acid analyzers

Japan Electron Optics Laboratory Co. Ltd.
1418 Nakagami
Akishima
Tokyo
Japan

Amino acid analyzers

Kontron Ltd.
Analytic International
Bernerstrasse-Süd 169
CH-8048 Zurich
Switzerland

Amino acid analyzers

LKB-Producter AB
S-161 25 Bromma
Sweden

LKB Instruments Ltd.
12221 Parklawn Drive
Rockville, MD 20852
U.S.A.

LKB 4400 amino acid analyzer, chromatography equipment, including pumps, fraction collectors, columns, gradient formers and detection systems, LKB 4020 solid-phase peptide sequencer

LKB-Biochrom Ltd. The Science Park Cambridge, CB4 4BH England	
The Locarte Co. 8 Wendell Road London, W12 9RT England	Amino acid analyzers
Pharmacia Fine Chemicals AB Box 175 S-751 04 Uppsala 1 Sweden	Chromatography equipment
Pharmacia Fine Chemicals, Inc. 800 Centennial Avenue Piscataway, NJ 10017 U.S.A.	
Pharmacia (GB) Ltd. Paramount House 75 Uxbridge Road London, W5 5SS England	
Pye-Unicam Ltd. York Street Cambridge England	Electronic gradient former and UV absorbance monitor for column chromatography
Rank-Hilger Westwood Margate Kent, CT9 4JL England	Amino acid analyzer
Raven Scientific Ltd. The Mill Station Road P.O. Box 2 Haverhill Suffolk England	Electrophoresis equipment

Savant Instruments Inc. 221 Park Avenue Hicksville, NY 11801 U.S.A.	High-voltage electrophoresis equipment
Shandon Southern Products Ltd. 93–96 Chadwick Road Astmoor, Runcorn Cheshire, WA7 1PR England	Thin-layer electrophoresis equipment
Shandon Southern Instruments Inc. 515 Broad Street Sewickley, PA 15143 U.S.A.	
Technicon Instruments Corp. 511 Benedict Avenue Tarrytown, NY 10591 U.S.A.	Amino acid analyzers

(2) *Chromatographic materials*

Bio-Rad Laboratories 2200 Wright Avenue Richmond, CA 94804 U.S.A.	Ion-exchange resins, Bio-Gel gel filtration media, Bio-Beads S-X1
Bio-Rad Laboratories, Ltd. 27 Homesdale Road Bromley Kent England	
Camag CH-4132 Muttenz Homburgerstr. 24 Switzerland	Thin-layer chromatography materials and equipment
C. Desaga GmbH (address above)	Thin-layer chromatography materials

Eastman Kodak Co. 343 State Street Rochester, NY 14650 U.S.A. (in England: **Kodak Ltd.** Acornfield Road Kirkby Liverpool, L33 7UF)	Thin-layer chromatography plates, X-ray films, photographic supplies
Machery-Nagel & Co. GmbH D-5160 Dueren Werkstrasse 6–8 F.R.G.	Thin-layer chromatography plates
E. Merck D-61 Darmstadt Postfach 4119 Frankfurter Str. 250 F.R.G.	Thin-layer chromatography plates
Pharmacia (addresses above)	Sephadex and Sepharose gel filtration and ion-exchange materials
Schleicher & Schuell D-3354 Dassel Postfach 4 F.R.G.	Polyamide thin-layer sheets
Whatman Ltd. Springfield Mill Maidstone Kent England	Ion-exchange cellulose materials, chromatography paper
Whatman Inc. 9 Bridewell Place Clifton, NJ 07014 U.S.A.	

(3) *Enzymes*

BDH Chemicals
Poole
Dorset, BH12 4NN
England

Chymotrypsin, papain, trypsin

Boehringer Mannheim GmbH Biochemica
Postfach 310120
D-6800 Mannheim
F.R.G.

Boehringer Corp. (London) Ltd.
Bell Lane
Lewes
East Sussex, BN7 1LG
England

Arginine-specific protease from mouse submaxillary gland, aminopeptidase M, bromelain, carboxypeptidases A, B, Y, clostripain, elastase, leucine aminopeptidase, papain, pepsin, pronase, pyroglutamate aminopeptidase, thermolysin, trypsin

Boehringer-Mannheim Biochemicals
7941 Castleway Drive
P.O. Box 50816
Indianapolis
Indiana, 46250
U.S.A.

Calbiochem-Behring Corp.
10933 No. Torrey Pines Road
La Jolla, CA 92037
U.S.A.

Carboxypeptidase Y, elastase, thermolysin, thrombin

(in England: **CP Laboratories Ltd.**
P.O. Box 22
Bishops Stortford
Herts., CM22 7RQ)

Miles Laboratories Inc.
P.O. Box 2000
Elkhart, IN 46515
U.S.A.

Carboxypeptidase A, chymotrypsin, leucine aminopeptidase, *Staphylococcus aureus* (V8) protease, thrombin

Miles Laboratories P.O. Box 37 Stoke Court Stoke Poges Slough, SL2 4LY England	
Pierce Chemical Co. (addresses below)	Carboxypeptidase Y
Serva Feinbiochemica GmbH & Co. Karl-Benz-Str. 7 D-6900 Heidelberg 1 F.R.G.	Bromelain, chymotrypsin, elastase, papain, pepsin, pronase, thermolysin, trypsin
Sigma Chemical Co. P.O. Box 14508 St. Louis, MO 63158 U.S.A.	Bromelain, aminopeptidase M, leucine aminopeptidase, papain, pepsin, pronase, thermolysin, trypsin
Sigma (London) Ltd. Fancy Road Poole Dorset, BH17 7NH England	
Worthington Biochemical Corp. Freehold, NJ 07728 U.S.A.	Carboxypeptidases A, B and Y, chymotrypsin, clostripain, elastase, leucine aminopeptidase, papain, pepsin, trypsin (TPCK-treated)
Worthington Biochemical Corp. 6078 Neu-Isenberg 2 Am Forsthaus Gravenbruch 73 F.R.G.	

(4) *Other reagents* (selected examples)

Aldrich Chemical Co. 940 W. St. Paul Avenue Milwaukee, WI 53233 U.S.A.	Wide range of reagents, including Polybrene

Aldrich Chemical Co. The Old Brickyard New Road Gillingham Dorset, SP8 4JL England	
J.T. Baker Chemical Co. 222 Red School Lane Phillipsburg, NJ 08865 U.S.A.	Solvents
BDH Chemicals (address above)	Wide range of reagents, including Aristar urea, guanidinium chloride, HCl, etc., Dansyl chloride, *o*-phthalaldehyde, iodoacetic acid, molecular sieve
Beckman Instruments (addresses above)	Reagents for the automated sequencer
Burdick & Jackson Labs Inc. 1953 S. Harvey Street Muskegon, MI 49442 U.S.A.	Purified solvents
Calbiochem-Behring (address above)	Dithiothreitol, tosyl-L-lysyl chloromethane (TLCK), tosyl-L-phenylalanyl chloromethane (TPCK)
Electro-Nucleonics Inc. 368 Passaic Avenue Fairfield, NJ 07006 U.S.A.	Controlled-pore glass beads
Fluka AG CH-9470 Buchs Switzerland	Range of reagents, including *p*-dimethylaminoazobenzene isothiocyanate (Dabitc)
Koch-Light Laboratories 2 Willow Road Colnbrook Bucks, SL3 0BZ England	Cyanogen bromide, butyl-PBD

New England Nuclear 549 Albany Street Boston, MA 02118 U.S.A.	Radiochemicals
Pierce Chemical Co. Inc. P.O. Box 117 Rockford, IL 61105 U.S.A. **Pierce & Warriner (UK) Ltd.** 44 Upper Northgate Street Chester Cheshire CH1 4EF England **Pierce Eurochemie B.V.** P.O. Box 1151, Rotterdam The Netherlands	Wide range of reagents specially purified for sequence analysis, including Sequanal grade phenylisothiocyanate, trifluoroacetic acid, dimethylallylamine, etc., Corning controlled-pore glass, fluorescamine, *o*-iodosobenzoate, Schleicher & Schuell polyamide sheets
The Radiochemical Centre White Lion Road Amersham Bucks England	Radiochemicals, e.g., iodo [^{14}C]acetic acid, Na ^{131}I
Serva (address above)	Wide range of reagents, including *o*-tolidine
Sigma Chemical Co. (addresses above)	Wide range of reagents, amino acids and derivatives

APPENDIX II

Purification of solvents and reagents

Reagents specially purified for use in protein sequence analysis are available from a number of suppliers, notably Pierce Chemical Co. Purification from cheaper grades of the small amounts of most reagents consumed for the small-scale manual techniques is rarely justified. Chromatographic solvents are required in larger amounts, however. Pure reagents are stored in small aliquots sealed under nitrogen in dark glass bottles, preferably at low temperatures.

Solvents

Organic solvents must be free from peroxides and aldehydes. To test for the presence of peroxides, 2 ml of the solvent is added to 1 ml of freshly-prepared 1% aqueous KI solution. The organic solvent is completely evaporated in a stream of N_2. 0.5 ml of 1% starch solution is added. A blue colour indicates that peroxides are present.

Acetic acid Analytical grade is adequate for many purposes. For further purification, the acid is refluxed for 3 h over CrO_3 (10 g/l), then distilled. B.p. 118 °C d^{17} 1.053.

n-*Butanol* Chromatographic grade (BDH) may be used. For removal of any aldehyde impurities, stir with $NaBH_4$ (2 g/l) under N_2 for 6 h. Distil in a nitrogen-flushed apparatus. B.p. 117–118 °C.

n-Butyl acetate Test for peroxides; if positive, pass through a column of activated alumina and distil. B.p. 126 °C.

Diethyl ether Test for peroxides; if positive, pass through a column of activated alumina and distil the first 90%. B.p. 34.6 °C.

N,N-*Dimethylformamide* (DMF) A pure grade should be purchased, as purification is difficult. The solvent should be stored dry, over molecular sieve, 3 or 4 Å (artificial zeolites, from BDH). B.p. 153 °C.

Ethyl acetate As for butyl acetate, but B.p. 77 °C.

N-*Ethylmorpholine* Reflux over ninhydrin (20 g/l) for 1 h, then distil. Redistil from KOH pellets in a nitrogen-flushed apparatus. Other tertiary amines (triethylamine, pyridine, etc.) may be purified in a similar way. Store under N_2 at −20 °C in the dark.

Formamide *Purissimum* grade (Fluka AG) may be used directly. Some grades of formamide contain ammonium formate. The salt is removed by passage through a short column (20 g/l) of mixed-bed ion-exchange resin. The first 5% is discarded. The formamide is dried over molecular sieve (3 or 4 Å pore size). Formamide is partially decomposed by distillation at atmospheric pressure (B.p. 210.5 °C), but may be distilled under reduced pressure.

Formic acid Analytical grade is suitable for most purposes. d_4^{20}1.220. B.p. 100.5 °C.

Phenol Distil, using an air condenser. M.p. 40.9 °C. B.p. 182 °C. Phenol remains colourless if stored in the dark under N_2.

Pyridine See *N*-ethylmorpholine. B.p. 115–116 °C. d_4^{20} 0.978.

Trifluoroacetic acid Sequanal grade (Pierce) or spectroscopic grade (BDH) are suitable. For purification of other grades, the acid is refluxed over CrO_3 for 3 h, then distilled. Store in a double-stoppered bottle.

Water All water used should be the purest available, preferably purified by ion-exchange followed by double distillation in glass. Precautions should be taken to avoid contamination of the water. Hare (1977) has described the purification of water used for sub-nanomolar amino acid analysis.

Other reagents

Citraconic anhydride is distilled under reduced pressure.

Cyanogen bromide Fresh material, completely colourless, should be used. Store sealed at −20 °C.

Guanidinium chloride Aristar grade (BDH) is used directly. The purification of the reagent is described by Nozaki (1972).

Anhydrous hydrazine (Schroeder, 1972a) 95% Hydrazine (100 ml) and NaOH pellets (40 g) are placed in a 250 ml round-bottom flask in a distillation apparatus filled with dry N_2. The hydrazine is heated to the boiling point, then cooled to room temperature. The hydrazine is distilled under vacuum at room temperature, using a capillary leak connected to a dry N_2 supply, and collected in a trap at −70 °C (dry ice/acetone). The hydrazine is stored in sealed ampoules at −20 °C.

Hydrochloric acid For most purposes, Aristar HCl (BDH) is suitable, and is diluted with an equal volume of water for use in hydrolyzing peptides. The 6 M HCl is stored in a dark-glass stoppered bottle fitted with a cap to prevent deposition of ammonium chloride round the rim. Blank hydrolyzates should be tested on the analyzer and should contain no amino acids. If necessary, the 6 M HCl is distilled.

Iodoacetic acid The pure reagent is colourless. If slightly yellow it is recrystallized from heptan-1-ol. M.p. 82–83 °C.

Phenylisothiocyanate Sequanal grade (Pierce) is used. Store in small aliquots at −20 °C under N_2. Allow to warm to ambient temperature before opening, to avoid condensation of moisture.

Quadrol Purification is difficult; details are given by Edman and Henschen (1975).

Thiodiglycol is redistilled under reduced pressure.

Urea Aristar urea (BDH) is used directly. Ammonium cyanate accumulates in aqueous solutions on standing, especially under alkaline conditions, so only freshly-prepared solutions are used. Ammonium cyanate may be removed by passage of the urea solution through a small column (10 g for 1 litre of 8 M urea) of mixed-bed ion-exchange resin. The initial two column volumes are discarded.

APPENDIX III

Safety in the protein chemistry laboratory

These notes are only an outline of particular precautions to be taken when working with certain reagents and apparatus; they are not intended to indicate the measures to be adopted under health and safety legislation.

Reagents

In general, all of the reagents used in protein sequence determination should be treated with the respect appropriate in any chemical laboratory. A well-ventilated environment, preferably with filtered air, is required, and laboratory coats and safety glasses should be worn when working with dangerous materials.

Some particularly hazardous materials are strong acids and alkalis: concentrated H_2SO_4, HCl, HNO_3, P_2O_5, formic acid, trifluoroacetic acid (especially dangerous since volatile: always use in a fume hood), heptafluorobutyric acid, NaOH and KOH. Strongly acidic oxidizing reagents: chromic acid and H_2O_2/H_2SO_4, for cleaning glassware, and performic acid are especially dangerous. Concentrated solutions of phenol can cause considerable skin damage without pain, since phenol can act as a local anaesthetic. Spillage of any of these materials on skin or clothing should be followed immediately by copious flushing with water, and medical attention if necessary. Phenol may also be quickly and effectively removed from the skin with ethanol or ether.

Highly toxic materials include cyanogen bromide (volatile: transfer in a fume hood and weigh in sealed vials), chlorine gas, and cadmium compounds. Many solvents are toxic, including methanol, acetonitrile, dimethyl formamide, butanol, 2-methoxyethanol, etc., but unless ingested harm is unlikely to result from these materials. Benzene is significantly more toxic than toluene, and the latter should be used as a substitute whenever possible. *o*-Tolidine and aziridine (ethyleneimine) are carcinogens. The carcinogenicity of some materials is unknown, e.g. *p*-dimethylaminoazobenzene isothiocyanate.

Chemicals which penetrate the skin and react with proteins should be treated with caution. Fluorodinitrobenzene is particularly unpleasant, since it may induce severe immune responses against the modified tissue.

Glassware

Vacuum desiccators and freeze-driers are possible hazards. Although implosions are unlikely to occur unless the apparatus is mishandled, cages should be used, and the smallest possible desiccators are recommended. Rubber sealed desiccators are less accident-prone than greased ones. The sealing and heating of glass tubes for hydrolyses are potentially hazardous. Thin-walled tubes should not be used, and the seal should be robust. Any glassware with chips or cracks should be repaired or discarded.

Electrophoresis apparatus

High-voltage paper electrophoresis (3–5 kV) is potentially lethal, and strict safety precautions (safety cut-out switches on the lid and cooling water supply) must be observed. The low power required for thin-layer or gel electrophoresis is far less dangerous.

Fire hazards

Many solvents are highly flammable: they should not be stored in unsealed containers in refrigerators or cold rooms not fitted with safety thermostats. Smoking in the laboratory must be prohibited, for reasons of contamination as well as for safety.

Appendix IV

Notes on work with very small amounts of proteins and peptides

(1) Cleanliness

The problem of contamination by particles of dust becomes serious when working with microgram quantities of peptides and proteins. Glass-topped benches are useful, or the bench may be covered with clean paper. Glassware should be thoroughly cleaned, most easily by heating to 500 °C for 24 h. A good cleaning agent is 35% H_2O_2/H_2SO_4 (1 : 9, v/v, freshly prepared). This acid mixture should be prepared in small portions only, by adding conc. H_2SO_4 slowly to H_2O_2 with swirling on ice water; heat is evolved. Thorough rinsing with distilled water after cleaning with this acid is essential. Pipettes, thin-layer plates and other materials with which peptide solutions come into contact should not be touched with bare fingers.

(2) Adsorption to surfaces

Most analytical methods may be made more sensitive by reductions in the volumes of solutions and the sizes of glassware and chromatography plates. Reduction in the concentration of peptides usually leads to losses through adsorption to glassware and the problem of contamination becomes more serious. Suitable quantities of peptides have been indicated in the text for each size of column, etc. Dialysis of small quantities (less than 100 μg) of proteins may lead to serious losses, and gel filtration in small columns (Pasteur pipettes) may give much better recoveries; alternatively, volatile solvents and reagents should be used wherever possible to avoid the requirement for desalting.

Storage of small amounts of peptides (1–10 nmol) in small tubes (5 mm × 50 mm) is recommended; if larger vials are used, larger volumes of solvents are required for dissolving and transferring the material. When larger flasks are used, as for example, for freeze-drying, the glassware may be siliconized.

(3) Transfer of samples

Weighing is inaccurate for the measurement of small amounts (less than 1 mg) of peptides and proteins, particularly since they are often hygroscopic. The peptide or protein is dissolved in a small volume of a suitable solvent, and aliquots are transferred using graduated capillary pipettes. For many purposes, an accuracy of 10% is sufficient, and dissolution in 50 μl, with transfer of 5 μl, if one-tenth the sample is required, is suitable. Many small peptides are soluble in 20% pyridine or 30% acetic acid. Larger acidic peptides may dissolve more readily in 0.1 M NH_3. Some peptides and proteins may be insoluble in these solutions, and strong (e.g. 70%) formic acid or even trifluoroacetic acid may be required for aliquoting the material. The solutions are dried down in vacuo as soon as possible, and stored at −20 °C. Care is required when drying the solutions in small tubes: a bubble of air may eject the whole contents from the tube. Initial application of a water pump vacuum for 10 min until all the dissolved air is removed is recommended.

(4) Reactions

Small volumes of solutions are transferred to the bottoms of tubes by centrifugation. Reagents are mixed by rapid vortexing. Tubes should be sealed quickly, since the composition of solutions of low volume may change rapidly through evaporation (e.g. in the dansyl reaction, with 50% acetone, § 6.2.1). Since pH meters cannot be used with reactions in small volumes, the pH must be controlled with suitable buffers.

(5) Chromatography

A high sensitivity of analytical techniques is obtained through the use of small (e.g. 3 cm × 3 cm) thin-layer chromatography plates. In order to use to full advantage the resolution offered by polyamide layers and high-performance silica gel plates, the sample spot size must be kept small. Drawn-out capillaries are used to apply about 0.1 μl at a time of the sample solution to the origin. Similarly, small chromatographic columns may be used; for good resolution the bead size of the gel filtration media or ion-exchange resins must also be small, and the column fittings must also be reduced in size.

(6) Radioactive proteins

Very sensitive radiochemical techniques have been developed. In order to avoid losses, unlabelled carrier proteins, peptides, and derivatives such as phenylthiohydantoins are added when appropriate, and methods applicable to larger quantities of proteins may then be used.

References

ABELSON, J. (1979) Ann. Rev. Biochem. *48*, 1035–1069.

ABRAHAMSON, M., GRONINGSSON, K. and CASTENSSON, S. (1978) J. Chromatog. *154*, 313–317.

ACHER, R., FROMAGEOT, C. and JUTISZ, M. (1950) Biochim. Biophys. Acta, *5*, 81–88.

ADAMS, J.M. and CAPECCHI, M.R. (1966) Proc. Natl. Acad. Sci. U.S.A. *55*, 147–155.

AFRICA, B. and CARPENTER, F.H. (1966) Biochem. Biophys. Res. Comm. *24*, 113–119.

AIR, G.M. (1976) J. Mol. Biol. *107*, 433–443.

AIR, G.M., COULSON, A.R., FIDDES, J.C., FRIEDMANN, T., HUTCHISON, C.A. III, SANGER, F., SLOCOMBE, P.M. and SMITH, A.J.H. (1978) J. Mol. Biol. *125*, 247–254.

ALLEN, G. (1978) Proceedings of 11th FEBS Meeting, Copenhagen. Vol. 45, Membrane Proteins (Nicholls, P., Moller, J.V., Jorgensen, P.L. and Moody, A.J., eds.), pp. 159–168, Pergamon Press, Oxford.

ALLEN, G. (1980) Biochem. J. *187*, 545–563.

ALLEN, G. and GREEN, N.M. (1976) FEBS Lett. *63*, 188–192.

ALLEN, G. and GREEN, N.M. (1978) Biochem. J. *173*, 393–402.

ALLEN, G. and HARRIS, J.I. (1976) Eur. J. Biochem. *62*, 601–612.

ALLEN, G. and WITTMANN-LIEBOLD, B. (1978) Hoppe-Seyler's Z. Physiol. Chem. *359*, 1509–1525.

ALLEN, G., CAPASSO, R. and GUALERZI, C. (1979) J. Biol. Chem. *254*, 9800–9806.

ALLEN, G., BOTTOMLEY, R. and TRINNAMAN, B.J. (1980) Biochem. J. *187*, 577–589.

AMBLER, R.P. (1963) Biochem. J. *89*, 349–378.

AMBLER, R.P. (1965) Biochem. J. *96*, 32P.

AMBLER, R.P. (1972a) Methods Enzymol. *25*, 143–154.

AMBLER, R.P. (1972b) Methods Enzymol. *25*, 262–272.

AMBLER, R.P. and SCOTT, G.K. (1978) Proc. Natl. Acad. Sci. U.S.A. *75*, 3732–3736.

ANDERSEN, S.O. (1964) Biochim. Biophys. Acta, *93*, 213–215.

ANNAN, W.D. (1979) J. Chromatog. *173*, 194–197.

ASQUITH, R.S. and CARTHEW, P. (1972) Biochim. Biophys. Acta, *278*, 8–14.

ATHERTON, E., BRIDGEN, J. and SHEPPARD, R.C. (1976) FEBS Lett. *64*, 173–175.

AUFFRET, A.D., WILLIAMS, D.H. and THATCHER, D.R. (1978) FEBS Lett. *90*, 324–326.

AUNE, K.C. (1978) Methods Enzymol. *48*, 163–185.

AUSTEN, B.M. and SMITH, E.L. (1976) Biochem. Biophys. Res. Comm. *72*, 411–417.

BAGDASARIAN, M., MATHESON, N.A., SYNGE, R.L.M. and YOUNGSON, M.A. (1964) Biochem. J. *91*, 91–105.

BALIAN, G., CLICK, E.M. and BORNSTEIN, P. (1971) Biochemistry, *10*, 4470–4478.

BALLOU, B. and SMITHIES, O. (1977) Anal. Biochem. *80*, 616–623.

BALLOU, B., SUNDHARADAS, G. and BACH, M.L. (1974) Science, *185*, 531–533.

BALLOU, B., McKEAN, D.J., FREEDLENDER, E.F. and SMITHIES, O. (1976) Proc. Natl. Acad. Sci. U.S.A. *73*, 4487–4491.

BARGER, B.O., WHITE, F.C., PACE, J.L., KEMPER, D.L. and RAGLAND, W.L. (1976) Anal. Biochem. *70*, 327–335.

BARRELL, B.G., BANKIER, A.T. and DROUIN, J. (1979) Nature, *282*, 189–194.

BATES, D.L., PERHAM, R.N. and COGGINS, J.R. (1975) Anal. Biochem. *68*, 175–184.

BAUM, B.J., JOHNSON, L.S., FRANZBLAU, C. and TROXLER, R.F. (1975) J. Biol. Chem. *250*, 1464–1471.

BENSON, J.R. (1975) in Instrumentation in Amino Acid Sequence Analysis (Perham, R.N., ed.), pp. 1–39, Academic Press, London, New York and San Francisco.

BENSON, J.R. (1976) Anal. Biochem. *71*, 459–470.

BENSON, J.R. (1977) Methods Enzymol. *47*, 19–31.

BENSON, J.R. and HARE, P.E. (1975) Proc. Natl. Acad. Sci. U.S.A. *72*, 619–622.

BENTLEY, K.W. (1976) Biochem. J. *153*, 137–138.

BERGER, A. and SCHECHTER, I. (1970) Phil. Trans. Roy. Soc. Lond. Ser. B, *257*, 249–264.

BEYREUTHER, K. (1978) Nature, *274*, 767.

BHOWN, A.S., MOLE, J.E., HOLLOWAY, W.L. and BENNETT, J.C. (1978a) J. Chromatog. *156*, 35–41.

BHOWN, A.S., MOLE, J.E., WEISSINGER, A. and BENNETT, J.C. (1978b) J. Chromatog. *148*, 532–535.

BIRR, C., REITEL, C. and WIELAND, T. (1970) Angew. Chem. Int. Ed. Engl. *9*, 731.

BLOCK, R.J., DURRUM, E.L. and ZWEIG, G. (1955) A Manual of Paper Chromatography and Paper Electrophoresis. Academic Press, New York.

BLOMBÄCK, B., BLOMBÄCK, M., HESSEL, B. and IWANAGA, S. (1967) Nature, *215*, 1445–1448.

BLUMENTHAL, K.M., MOON, K. and SMITH, E.L. (1975) J. Biol. Chem. *250*, 3644–3654.

BORNSTEIN, P. (1969) Biochem. Biophys. Res. Comm. *36*, 957–964.

BORNSTEIN, P. (1970) Biochemistry, *9*, 2408–2421.

BORNSTEIN, P. (1974) Ann. Rev. Biochem. *43*, 567–603.

BORNSTEIN, P. and BALIAN, G. (1977) Methods Enzymol. *47*, 132–145.

BOSEMAN, M., LEVY, M. and SCHENKEIN, I. (1976) Arch. Biochem. Biophys. *175*, 463–476.

BOXER, D.H., JENKINS, R.E. and TANNER, M.J.A. (1974) Biochem. J. *137*, 531–534.

BRADBURY, A.F. and SMYTH, D.G. (1973) Biochem. J. *131*, 637–642.

BRAND, C.M. and SKEHEL, J.J. (1972) Nature New Biol. *238*, 145–147.

BRAUER, A.W., MARGOLIES, M.N. and HABER, E. (1975) Biochemistry, *14*, 3029–3035.

BRAUN, V. and BOSCH, V. (1972) Eur. J. Biochem. *28*, 51–69.

Braunitzer, G. and Pfletschinger, J. (1978) Hoppe-Seyler's Z. Physiol. Chem. *359*, 1015–1018.

Braunitzer, G. and Sorger, U. (1969) Hoppe-Seyler's Z. Physiol. Chem. *350*, 405–406.

Braunitzer, G., Schrank, B. and Ruhfus, A. (1970) Hoppe-Seyler's Z. Physiol. Chem. *351*, 1589–1590.

Braunitzer, G., Schrank, B., Ruhfus, A., Petersen, S. and Petersen, U. (1971) Hoppe-Seyler's Z. Physiol. Chem. *352*, 1730–1732.

Bray, G.A. (1960) Anal. Biochem. *1*, 279–285.

Bray, D. and Brownlee, S.M. (1973) Anal. Biochem. *55*, 213–221.

Bridgen, J. (1975) FEBS Lett. *50*, 159–162.

Bridgen, J. (1976) Biochemistry, *15*, 3600–3604.

Bridgen, J. (1977) Methods Enzymol. *47*, 321–335.

Bridgen, J. and Morris, H.R. (1974) Eur. J. Biochem. *44*, 333–334.

Bridgen, J. and Walker, I.D. (1976) Biochemistry, *15*, 792–798.

Bridgen, J., Graffeo, A.P., Karger, B.L. and Waterfield, M.D. (1975) *In:* Instrumentation in Amino Acid Sequence Analysis (Perham, R.N., ed.), pp. 111–145, Academic Press, London, New York and San Francisco.

Bridgen, P.J., Cross, G.A.M. and Bridgen, J. (1976) Nature, *263*, 613–614.

Briley, P.A., Sibilli, L., Chalvignac, M.-A., Cossart, P., LeBras, G., De Wolf, A. and Cohen, G.N. (1978) J. Biol. Chem. *253*, 8867–8871.

Brimacombe, R., Stöffler, G. and Wittmann, H.G. (1978) Ann. Rev. Biochem. *47*, 217–249.

Brosius, J. (1978) Biochemistry, *17*, 501–508.

Brown, J.R. and Hartley, B.S. (1963) Biochem. J. *89*, 59P–60P.

Brown, J.R. and Hartley, B.S. (1966) Biochem. J. *101*, 214–228.

Brown, J.L., Kato, K., Silver, J. and Nathenson, S.G. (1974) Biochemistry, *13*, 3174–3178.

Brownlee, C.G. (1972) Determination of sequences in RNA, Vol. 3 of Laboratory Techniques in Biochemistry and Molecular Biology (Work, T.S. and Work, E., eds.) North-Holland/American Elsevier, Amsterdam, London, New York.

Bruton, C.J. and Hartley, B.S. (1970) J. Mol. Biol. *52*, 165–178.

Büchel, D.E., Gronenborn, B. and Müller-Hill, B. (1980) Nature, *283*, 541–545.

Bucher, D. (1977) Chromatographia, *10*, 723–725.

Buckley, J.J., Libor, S. and Sundaram, T.K. (1979) Arch. Biochem. Biophys. *192*, 396–404.

Bunn, H.F., Gabbay, K.H. and Gallop, P.M. (1978) Science, *200*, 21–27.

Burstein, Y. and Schechter, I. (1977) Proc. Natl. Acad. Sci. U.S.A. *74*, 716–720.

Butler, W.T. (1969) J. Biol. Chem. *244*, 3415–3417.

Butler, P.J.G. and Hartley, B.S. (1972) Methods Enzymol. *25*, 191–199.

CALLAHAN, P.X., MCDONALD, J.K. and ELLIS, S. (1972) Methods Enzymol. *25*, 282–298.

CANDIDO, E.P.M. and DIXON, G.H. (1972) Proc. Natl. Acad. Sci. U.S.A. *69*, 2015–2019.

CANFIELD, R.E. and ANFINSEN, C.B. (1963) J. Biol. Chem. *238*, 2684–2690.

CAPRA, J.D. and KEHOE, J.M. (1974) Proc. Natl. Acad. Sci. U.S.A. *71*, 845–848.

CASEY, R. and REES, M.W. (1979) Anal. Biochem. *95*, 397–400.

CATSIMPOOLAS, N. and WOOD, J.L. (1966) J. Biol. Chem. *241*, 1790–1796.

CAVADORE, J.-C. (1978) Anal. Biochem. *91*, 236–240.

CHANG, J.Y. (1978) FEBS Lett. *91*, 63–68.

CHANG, J.Y. (1979a) Biochim. Biophys. Acta, *578*, 188–195.

CHANG, J.Y. (1979b) Biochim. Biophys. Acta, *578*, 175–187.

CHANG, J.Y., CREASER, E.H. and BENTLEY, K.W. (1976) Biochem. J. *153*, 607–611.

CHANG, J.Y., CREASER, E.H. and HUGHES, G.J. (1977a) FEBS Lett. *78*, 147–150.

CHANG, J.Y., CREASER, E.H. and HUGHES, G.J. (1977b) J. Chromatog. *140*, 125–128.

CHANG, J.Y., BRAUER, D. and WITTMANN-LIEBOLD, B. (1978) FEBS Lett. *93*, 205–214.

CHEN, R. and CHEN-SCHMEISSER, U. (1977) Proc. Natl. Acad. Sci. U.S.A. *74*, 4905–4908.

CHEN, K.C.S. and KRAUSE, R.M. (1975) Anal. Biochem. *69*, 180–186.

CHEN, B.L. and POLJAK, R.J. (1974) Biochemistry, *13*, 1295–1302.

CHEN, R.F., SCOTT, C. and TREPMAN, E. (1979) Biochim. Biophys. Acta, *576*, 440–455.

CHERBULIEZ, E., BAEHLER, B. and RABINOWITZ, J. (1964) Helv. Chim. Acta, *47*, 1350–1353.

CHIU, Y.-Y.H., LÓPEZ DE CASTRO, J.A. and POLJAK, R.J. (1979) Biochemistry, *18*, 553–560.

CLEGG, J.B., NAUGHTON, M.A. and WEATHERALL, D.J. (1966) J. Mol. Biol. *19*, 91–108.

CLEVELAND, D.W., FISCHER, S.G., KIRSCHNER, M.W. and LAEMMLI, U.K. (1977) J. Biol. Chem. *252*, 1102–1106.

COLIGAN, J.E., KINDT, T.J., EWENSTEIN, B.M., UEHARA, H., NISIZAWA, T. and NATHENSON, S.G. (1978) Proc. Natl. Acad. Sci. U.S.A. *75*, 3390–3394.

COLLIER, H.B. (1973) Anal. Biochem. *56*, 310–311.

CONSDEN, R., GORDON, A.H. and MARTIN, A.J.P. (1944) Biochem. J. *38*, 224–232.

CORRAN, P.H. and WALEY, S.G. (1973) FEBS Lett. *30*, 97–99.

CORUZZI, G. and TZAGOLOFF, A. (1979) J. Biol. Chem. *254*, 9324–9330.

CROMWELL, L.D. and STARK, G.R. (1969) Biochemistry, *8*, 4735–4740.

DARBRE, A. (1977) Methods Enzymol. *47*, 357–369.

DARBRE, A. and RANGARAJAN, M. (1975) *In:* Solid Phase Methods in Protein Sequence Analysis (Laursen, R.A., ed.), pp. 131–137, Pierce Chemical Co., Rockford, Illinois.

DAUTRY-VARSAT, A. and COHEN, G.N. (1977) J. Biol. Chem. *252*, 7685–7689.

DAVISON, P.F. (1976) Anal. Biochem. *75*, 129–141.

DE BERNARDO, S., WEIGELE, M., TOOME, V., MANHART, K., LEIMGRUBER, W.,

Böhlen, P., Stein, S. and Udenfriend, S. (1974) Arch. Biochem. Biophys. *163*, 390–399.
Degani, C. and Boyer, P.D. (1973) J. Biol. Chem. *248*, 8222–8226.
Degani, Y. and Patchornik, A. (1971) J. Org. Chem. *36*, 2727–2728.
Degani, Y. and Patchornik, A. (1974) Biochemistry, *13*, 1–11.
Degen, J. and Kyte, J. (1978) Anal. Biochem. *89*, 529–539.
Devillers-Thiery, A. and Blobel, G. (1978) Anal. Biochem. *87*, 66–70.
Deyl, Z. (1970) J. Chromatog. *48*, 231–237.
Dhaese, P., Vandekerkhove, J.S. and Van Montagu, M.C. (1979) Eur. J. Biochem. *94*, 375–386.
Dixon, H.B.F. and Perham, R.N. (1968) Biochem. J. *109*, 312–314.
Dognin, M.J. and Wittmann-Liebold, B. (1977) FEBS Lett. *84*, 342–346.
Doolittle, R.F. (1970) Methods Enzymol. *19*, 555–569.
Doolittle, R.F., Cassman, K.G., Cottrell, B.A., Friezner, S.J. and Takagi, T. (1977) Biochemistry, *16*, 1710–1715.
Doonan, S., Doonan, H.J., Hanford, R., Vernon, C.A., Walker, J.M., Bossa, F., Barra, D., Carloni, M., Fasella, P., Riva, F. and Walton, P.L. (1974) FEBS Lett. *38*, 229–233.
Doonan, S., Garman, A.J., Hanson, J.M., Loudon, A.G. and Vernon, C.A. (1978) J. Chem. Soc. Perkin Trans. I, 1157–1160.
Doscher, M.S. and Hirs, C.H.W. (1967) Biochemistry, *6*, 304–312.
Doyen, N. and Lapresle, C. (1979) Biochem. J. *177*, 251–254.
Drapeau, G.R. (1977) Methods Enzymol. *47*, 189–191.
Drapeau, G.R., Boily, Y. and Houmard, J. (1972) J. Biol. Chem. *247*, 6720–6726.
Drickamer, L.K. (1978) J. Biol. Chem. *253*, 7242–7248.
Dubois, M., Gilles, K.A., Hamilton, J.K., Rebers, P.A. and Smith, F. (1956) Anal. Chem. *28*, 350–356.
Du Vigneaud, V., Ressler, C. and Trippitt, S. (1953) J. Biol. Chem. *205*, 949–957.

Easley, C.W. (1965) Biochim. Biophys. Acta, *107*, 386–388.
Easley, C.W., Zegers, B.J.M. and de Vijlder, M. (1969) Biochim. Biophys. Acta, *175*, 211.
Edelhoch, H. (1967) Biochemistry, *6*, 1948–1954.
Edman, P. (1950) Acta Chem. Scand. *4*, 283–293.
Edman, P. (1953) Acta Chem. Scand. *7*, 700–701.
Edman, P. (1956) Acta Chem. Scand. *10*, 761–768.
Edman, P. and Begg, G. (1967) Eur. J. Biochem. *1*, 80–91.
Edman, P. and Henschen, A. (1975) *In:* Protein Sequence Determination (Needleman, S.B., ed.), pp. 232–279, Springer-Verlag, Berlin, Heidelberg, New York.
Egorov, T.A., Svenson, A., Rydén, L. and Carlsson, J. (1975) Proc. Natl. Acad. Sci. U.S.A. *72*, 3029–3033.
Elion, J., Downing, M. and Mann, K. (1978) J. Chromatog. *155*, 436–438.

ELLMAN, G.L. (1959) Arch. Biochem. Biophys. *82*, 70–77.

ELSDEN, D.F., LIGHT, N.D. and BAILEY, A.J. (1980) Biochem. J. *185*, 531–534.

ELZINGA, M., COLLINS, J.H., KUEHL, W.M. and ADELSTEIN, R.S. (1973) Proc. Natl. Acad. Sci. U.S.A. *70*, 2687–2691.

FARABAUGH, P.J. (1978) Nature, *274*, 765–769.

FAUPEL, M. and VON ARX, E. (1978) J. Chromatog. *157*, 253–257.

FELIX, A.M. and JIMENEZ, M.H. (1974) J. Chromatog. *89*, 361–364.

FILLINGAME, R.H. (1976) J. Biol. Chem. *251*, 6630–6637.

FISCHER, L. (1980) An introduction to gel chromatography. *In:* Laboratory Techniques in Biochemistry and Molecular Biology (Work, T.S. and Work, E., eds.) 2nd Edition, North Holland/American Elsevier, Amsterdam, London, New York.

FLEISCHMAN, J.B., PAIN, R.H. and PORTER, R.R. (1962) Arch. Biochem. Biophys. Supplement 1, 174–180.

FLEMING, P.J., DAILEY, H.A., CORCORAN, D. and STRITTMATTER, P. (1978) J. Biol. Chem. *253*, 5369–5372.

FOLK, J.E. (1970) Methods Enzymol. *19*, 504–508.

FOLK, J.E. and FINLAYSON, J.S. (1977) Adv. Protein Chem. *31*, 1–133.

FONTANA, A. (1972) Methods Enzymol. *25*, 419–423.

FONTANA, A. (1980) *In:* Solid Phase Methods in Protein Sequence Analysis (Birr, C., ed.) (Proceedings of 3rd International Conference, Heidelberg, Oct. 1979), Elsevier/North-Holland Biomedical Press, Amsterdam.

FORREY, A.W., SEVILLA, C.L., SAARI, J.C. and FISCHER, E.H. (1971) Biochemistry, *10*, 3132–3140.

FOSTER, J.A., BRUENGER, C.L., HU, C.L., ALBERTSON, K. and FRANZBLAU, C. (1973) Biochem. Biophys. Res. Comm. *53*, 70–74.

FOWLER, A.V. (1978) J. Biol. Chem. *253*, 5499–5504.

FOWLER, A.V. and ZABIN, I. (1977) Proc. Natl. Acad. Sci. U.S.A. *74*, 1507–1510.

FOWLER, A.V. and ZABIN, I. (1978) J. Biol. Chem. *253*, 5521–5525.

FOX, B.W. (1976) Techniques for sample preparation for liquid scintillation counting, *In:* Laboratory Techniques in Biochemistry and Molecular Biology (Work, T.S. and Work, E., eds.), North-Holland/American Elsevier, Amsterdam, Oxford, New York.

FOX, J.W., ELZINGA, M. and TU, A.T. (1979) Biochemistry, *18*, 678–684.

FRAENKEL-CONRAT, H., HARRIS, J.I. and LEVY, A.L. (1955) Methods Biochem. Anal. *2*, 359–425.

FRANK, G. (1979) Hoppe-Seyler's Z. Physiol. Chem. *360*, 997–999.

FRANK, H. and DESIDERIO, D.M. (1978) Anal. Biochem. *90*, 413–419.

FRANK, G. and WEEDS, A.G. (1974) Eur. J. Biochem. *44*, 317–332.

FRANK, G., BRUNNER, J., HAUSER, H., WACKER, H., SEMENZA, G. and ZUBER, H. (1978) FEBS Lett. *96*, 183–188.

FREED, J.H., SEARS, D.W., BROWN, J.L. and NATHENSON, S.G. (1979) Molecular Immunol. *16*, 9–21.

FREISHEIM, J.H., BITAR, K.G., REDDY, A.V. and BLANKENSHIP, D.T. (1978) J. Biol. Chem. *253*, 6437–6444.

FRIEDMAN, M., KRULL, L.H. and CAVINS, J.F. (1970) J. Biol. Chem. *245*, 3868–3871.

FRUTON, J.S. (1970) Adv. Enzymol. *33*, 401–443.

FRUTON, J.S. (1974) Accounts Chem. Res. *7*, 241–246.

FUJIMOTO, D., MORIGUCHI, T., ISHIDA, T. and HAYASHI, H. (1978) Biochem. Biophys. Res. Comm. *84*, 52–57.

FUKUDA, M., ESHDAT, Y., TARONE, G. and MARCHESI, V.T. (1978) J. Biol. Chem. *253*, 2419–2428.

FULLMER, C.S. and WASSERMAN, R.H. (1979) J. Biol. Chem. *254*, 7208–7212.

FURTH, A.J., MILMAN, J.D., PRIDDLE, J.D. and OFFORD, R.E. (1974) Biochem. J. *139*, 11–22.

FURTHMAYR, H., GALARDY, R.E., TOMITA, M. and MARCHESI, V.T. (1978) Arch. Biochem. Biophys. *185*, 21–29.

GAETJENS, E. and BÁRÁNY, M. (1966) Biochim. Biophys. Acta, *117*, 176–183.

GAGNON, J., FINCH, P.R., WOOD, D.D. and MOSCARELLO, M.A. (1971) Biochemistry, *10*, 4756–4763.

GALLOP, P.M., BLUMENFELD, O.O. and SEIFTER, S. (1972) Ann. Rev. Biochem. *41*, 618–672.

GALLY, J.A. and EDELMAN, G.M. (1972) Ann. Rev. Genet. *6*, 1–46.

GATES, F.T., III, COLIGAN, J.E. and KINDT, T.J. (1979) Biochemistry, *18*, 2267–2272.

GERBER, G.E., ANDEREGG, R.J., HERLIHY, W.C., GRAY, C.P., BIEMANN, K. and KHORANA, H.G. (1979) Proc. Natl. Acad. Sci. U.S.A. *76*, 227–231.

GIGLIO, J.R. (1977) Anal Biochem. *82*, 262–264.

GLASSMAN, J.N.S., TODD, C.W. and SHIVELY, J.E. (1978) Biochem. Biophys. Res. Comm. *85*, 209–216.

GLAZER, A.N., DELANGE, R.J. and SIGMAN, D.S. (1975) Chemical modification of proteins, *In*: Laboratory Techniques in Biochemistry and Molecular Biology (Work, T.S. and Work, E., eds.), North-Holland/American Elsevier, Amsterdam, Oxford, New York.

GOOSSENS, W., PRÉAUX, G. and LONTIE, R. (1973) Biochimie, *55*, 1199–1207.

GORDON, A.H. (1975) Electrophoresis of proteins in polyacrylamide and starch gels, revised, enlarged edition. *In:* Laboratory Techniques in Biochemistry and Molecular Biology (Work, T.S. and Work, E., eds.) North-Holland/American Elsevier, Amsterdam, Oxford, New York.

GRAY, W.R. (1972) Methods Enzymol. *25*, 121–138.

GRAY, W.R. and HARTLEY, B.S. (1963) Biochem. J. *89*, 379–380.

GREEN, M., BRACKMANN, K., WOLD, W.S.M., CARTAS, M., THORNTON, H. and ELDER, J.H. (1979) Proc. Natl. Acad. Sci. U.S.A. *76*, 4380–4384.

GREY, A.A., WONG, R. and BENNICK, A. (1979) J. Biol. Chem. *254*, 4809.

GROSS, E. (1967) Methods Enzymol. *11*, 238–255.

GROSS, E. and MORELL, J.L. (1974) Biochem. Biophys. Res. Comm. *59*, 1145–1150.

GROSS, E. and WITKOP, B. (1961) J. Amer. Chem. Soc. *83*, 1510–1511.

GROSS, E. and WITKOP, B. (1962) J. Biol. Chem. *237*, 1856–1860.

GROSS, E. and WITKOP, B. (1967) Biochemistry, *6*, 745–748.

GRUENSTEIN, E. and Rich, A. (1975) Biochem. Biophys. Res. Comm. *64*, 472–477.

HANCOCK, W.S., BISHOP, C.A., PRESTIDGE, R.L. and HEARN, M.T.W. (1978) Anal. Biochem. *89*, 203–212.

HANTKE, K. and BRAUN, V. (1973) Eur. J. Biochem. *34*, 284–296.

HARDY, M.F. and PERRY, S.V. (1969) Nature, *223*, 300–302.

HARE, P.E. (1975) in Protein Sequence Determination (Needleman, S.B., ed.), pp. 204–231, Springer-Verlag, Berlin, Heidelberg, New York.

HARE, P.E. (1977) Methods Enzymol. *47*, 3–18.

HARRIS, J.I. (1967) Methods Enzymol. *11*, 390–398.

HARRIS, J.I. and LI, C.H. (1955) J. Biol. Chem. *213*, 499–507.

HARTLEY, B.S. (1970) Biochem. J. *119*, 805–822.

HARTLEY, J.L. and SOFFER, R.L. (1978) Biochem. Biophys. Res. Comm. *83*, 1545–1552.

HARTMAN, F.C. and WOLD, F. (1966) J. Amer. Chem. Soc. *88*, 3890–3891.

HASLETT, B.G. and BOULTER, D. (1976) Biochem. J. *153*, 33–38.

HAUSCHKA, P.V. (1977) Anal. Biochem. *80*, 212–223.

HAYASHI, R. (1977) Methods Enzymol. *47*, 84–93.

HAYASHI, R., MOORE, S. and STEIN, W.H. (1973) J. Biol. Chem. *248*, 2296–2302.

HEILAND, I., BRAUER, D. and WITTMANN-LIEBOLD, B. (1976) Hoppe-Seyler's Z. Physiol. Chem. *357*, 1751–1770.

HEILMANN, J., BAROLLIER, J. and WATZKE, E. (1957) Hoppe-Seyler's Z. Physiol. Chem. *309*, 219–220.

HEINRIKSON, R.L. (1971) J. Biol. Chem. *246*, 4090–4096.

HEINRIKSON, R.L. (1977) Methods Enzymol. *47*, 175–189.

HENSCHEN, A. (1978) Hoppe-Seyler's Z. Physiol. Chem. *359*, 1757–1770.

HENSCHEN, A. and EDMAN, P. (1972) Biochim. Biophys. Acta, *263*, 351–367.

HENSCHEN-EDMAN, A. (1977) *In:* Solid Phase Methods in Protein Sequence Analysis (Previero, A. and Coletti-Previero, M.-A., eds.), pp. 7–18, Elsevier/North-Holland Biomedical Press, Amsterdam.

HENSCHEN-EDMAN, A. (1980) *In:* Solid Phase Methods in Protein Sequence Analysis (Birr, C., ed.) (Proceedings of the 3rd International Conference on Solid Phase Methods, Heidelberg, 1979) Elsevier/North-Holland Biomedical Press, in press.

HERMAN, A.C. and VANAMAN, T.C. (1975) Anal. Biochem. *64*, 550–555.

HERMAN, A.C. and VANAMAN, T.C. (1977) Methods Enzymol. *47*, 220–236.

HERMODSON, M.A., ERICSSON, L.H., TITANI, K., NEURATH, H. and WALSH, K.A. (1972) Biochemistry, *11*, 4493–4502.

HILL, R.L. and DELANEY, R. (1967) Methods Enzymol. *11*, 339–351.

HILL, R.J., KONIGSBERG, W., GUIDOTTI, G. and CRAIG, L.C. (1962) J. Biol. Chem. *237*, 1549–1554.

HIRS, C.H.W. (1967a) Methods Enzymol. *11*, 197–199.

HIRS, C.H.W. (1967b) Methods Enzymol. *11*, 325–329.

HIRS, C.H.W., MOORE, S. and STEIN, W.H. (1960) J. Biol. Chem. *235*, 633–647.

HOLDER, A.A. (1976) Carlsberg Res. Comm. *41*, 321–334.

HORÁKOVÁ, M. and DEYL, Z. (1978) J. Chromatog. *159*, 227–314.

HORN, M.J. and LAURSEN, R.A. (1973) FEBS Lett. *53*, 285–288.

HOSEIN, B. MCCARTY, M. and FISCHETTI, V.A. (1979) Proc. Natl. Acad. Sci. U.S.A. *76*, 3765–3768.

HUGHES, J., SMITH, T.W., KOSTERLITZ, H.W., FOTHERGILL, L.A., MORGAN, B.A. and MORRIS, H.R. (1975) Nature, *258*, 577–579.

HUNKAPILLER, M.W. and HOOD, L.E. (1978) Biochemistry, *17*, 2124–2133.

HUNKAPILLER, M.W. and HOOD, L.E. (1980) Science, *207*, 523–525.

HUSAIN, S.S. and LOWE, G. (1968) Biochem. J. *108*, 855–859.

HUSAIN, S.S. and LOWE, G. (1969) Biochem. J. *114*, 279–288.

ICHIKAWA, H., TANIMURA, T., NAKAJIMA, T. and TAMURA, Z. (1970) Chem. Pharm. Bull. *18*, 1493–1495.

ILSE, D. and EDMAN, P. (1963) Aust. J. Chem. *16*, 411–416.

INAGAMI, T. (1973) Anal. Biochem. *52*, 318–321.

INAGAMI, T. and STURTEVANT, J.M. (1960) J. Biol. Chem. *235*, 1019–1023.

INGLIS, A.S. (1980) *In:* Solid Phase Methods in Protein Sequence Analysis (Birr, C., ed.) (Proceedings of the 3rd International Conference, Heidelberg, 1979), Elsevier/ North-Holland Biomedical Press, Amsterdam.

INGLIS, A.S. and NICHOLLS, P.W. (1973) J. Chromatog. *79*, 344–346.

INGLIS, A.S. and NICHOLLS, P.W. (1974) J. Chromatog. *97*, 289–292.

INGLIS, A.S., NICHOLLS, P.W. and SPARROW, L.G. (1974) J. Chromatog. *90*, 362–364.

INGLIS, A.S., RIVETT, D.E. and MCMAHON, D.T.W. (1979) FEBS Lett. *104*, 115–118.

INGRAM, V.M. (1956) Nature, *178*, 792–794.

ITANO, H.A. and ROBINSON, E.A. (1972) J. Biol. Chem. *247*, 4819–4824.

IVANOV, Ch.P. and MANCHEVA, I.N. (1973) Anal. Biochem. *53*, 420–430.

IVANOV, Ch.P. and MANCHEVA, I.N. (1976) Dokl. Bolg. Akad. Nauk. *29*, 1483–1486.

JACOBS, J.W. and NIALL, H.D. (1975) J. Biol. Chem. *250*, 3629–3636.

JACOBS, J.W., SAUER, R.T., NIALL, H.D., KEUTMANN, H.T., O'RIORDAN, J.L.H., AURBACH, G.D. and POTTS, J.T., Jr. (1973) Fed. Proc. *32*, 648 Abs.

JACOBS, J.W., KEMPER, B., NIALL, H.D., HABENER, J.F. and POTTS, J.T., Jr. (1974) Nature, *249*, 155–157.

JACOBSEN, G.R., SCHAFFER, M.H., STARK, G.R. and VANAMAN, T.C. (1973) J. Biol. Chem. *248*, 6583–6591.

JANÁK, J. (1964) J. Chromatog. *15*, 15–28.

JENKINS, R.E. and TANNER, M.J.A. (1975) Biochem. J. *147*, 393–399.

JILKA, R.L. and PESTKA, S. (1979) J. Biol. Chem. *254*, 9270–9276.

JOHNSON, P., HARRIS, C.I. and PERRY, S.V. (1967) Biochem. J. *105*, 361–370.

JOLLÈS, J., FIAT, A.M., ALAIS, C. and JOLLÈS, P. (1973) FEBS Lett. *30*, 173–176.

JORI, G., GALIAZZO, G., MARZOTTO, A. and SCOFFONE, E. (1968) J. Biol. Chem. *243*, 4272–4278.

JÖRNVALL, H. (1970) Eur. J. Biochem. *16*, 25–40.

JÖRNVALL, H., INMAN, J.K. and APPELLA, E. (1978) Anal. Biochem. *90*, 651–661.

JOYS, T.M. and KIM, H. (1979) Anal. Biochem. *94*, 371–377.

KASPAR, C.B. (1970) *In:* Protein Sequence Determination (Needleman, S.B., ed.), pp. 114–161, Springer-Verlag, Berlin, Heidelberg, New York.

KATO, T., SASAKI, M. and KIMURA, S. (1975) Anal. Biochem. *66*, 515–522.

KATZ, A.M., DREYER, W.J. and ANFINSEN, C.B. (1959) J. Biol. Chem. *234*, 2897–2900.

KEEFER, L.M. and BRADSHAW, R.A. (1977) Fed. Proc. *36*, 1799–1804.

KEHRY, M., SIBLEY, C., FUHRMAN, J., SCHILLING, J. and HOOD, L.E. (1979) Proc. Natl. Acad. Sci. U.S.A. *76*, 2932–2936.

KEIL-DLOUHA, V., ZYLBER, N., IMHOFF, J.-M., TONG, N.-T. and KEIL, B. (1971) FEBS Lett. *16*, 291–295.

KEMPE, T.D., BEATTIE, W.G., WEISSMAN, S. and KONIGSBERG, W. (1979) J. Biol. Chem. *254*, 7561–7569.

KESSLER, M.J., REDDY, M.S., SHAH, R.H. and BAHL, O.P. (1979a) J. Biol. Chem. *254*, 7901–7908.

KESSLER, M.J., MISE, T., GHAI, R. and BAHL, O.P. (1979b) J. Biol. Chem. *254*, 7909–7914.

KEUTMANN, H.T. and POTTS, J.T., Jr. (1969) Anal. Biochem. *29*, 175–185.

KLAPPER, D.G., WILDE, C.E., III and CAPRA, J.D. (1978) Anal. Biochem. *85*, 126–131.

KLUH, I. (1979) Coll. Czech. Chem. Comm. (Engl. Ed.), *44*, 145–147.

KOHL, B. and SANDERMANN, H., Jr. (1977) FEBS Lett. *80*, 408–412.

KOIDA, M. and WALTER, R. (1976) J. Biol. Chem. *251*, 7593–7599.

KOIDE, A., TITANI, K., ERICSSON, L.H., KUMAR, S., NEURATH, H. and WALSH, K.A. (1978) Biochemistry, *17*, 5657–5672.

KOLB, E., HARRIS, J.I. and BRIDGEN, J. (1974) Biochem. J. *137*, 185–197.

KOMATSU, S.K., DEVRIES, A.L. and FEENEY, R.E. (1970) J. Biol. Chem. *245*, 2909–2908.

KONIGSBERG, W. (1972) Methods Enzymol. *25*, 326–332.

KONIGSBERG, W.H. and STEINMAN, H.M. (1977) *In:* The Proteins, 3rd Edn., Vol. *3*, pp. 1–178 (Neurath, H. and Hill, R.L., eds.) Academic Press, New York.

KONKEL, D.A., TILGHMAN, S.M. and LEDER, P. (1978) Cell, *15*, 1125–1132.

KOSTKA, V. and CARPENTER, F.H. (1964) J. Biol. Chem. *239*, 1799–1803.
KOZIARZ, J.J., KOHLER, H. and STECK, T.L. (1978) Anal. Biochem. *86*, 78–79.
KREIL, G. and KREIL-KISS, G. (1967) Biochem. Biophys. Res. Comm. *27*, 275–280.
KREIL, G. and TUPPY, H. (1961) Nature, *192*, 1123–1125.
KRUTZSCH, H.C. and PISANO, J.J. (1977) Methods Enzymol. *47*, 391–404.
KULBE, K.D. (1971) Anal. Biochem. *44*, 548–558.
KULBE, K.D. (1974) Anal. Biochem. *59*, 564–573.
KUPKE, D.W. and DORRIER, T.E. (1978) Methods Enzymol. *48*, 155–162.

LAEMMLI, U.K. (1970) Nature, *227*, 680–685.
LAI, C.Y. (1977) Methods Enzymol. *47*, 236–243.
LANDON, M. (1977) Methods Enzymol. *47*, 145–149.
LARGMAN, C. and MROSS, G. (1975) Anal. Biochem. *69*, 70–75.
LAURSEN, R.A. (1966) J. Amer. Chem. Soc. *88*, 5344–5346.
LAURSEN, R.A. (1971) Eur. J. Biochem. *20*, 89–102.
LAURSEN, R.A. (1972) Methods Enzymol. *25*, 344–359.
LAURSEN, R.A. (ed.) (1975) Solid Phase Methods in Protein Sequence Analysis, Pierce Chemical Co., P.O. Box 117, Rockford, Illinois 61105.
LAURSEN, R.A. (1977) Methods Enzymol. *47*, 277–288.
LAURSEN, R.A., HORN, M.J. and BONNER, A.G. (1972) FEBS Lett. *21*, 67–70.
LAURSEN, R.A., BONNER, A.G. and HORN, M.J. (1975) *In:* Instrumentation in Amino Acid Sequence Determination (Perham, R.N., ed.), pp. 73–110, Academic Press, London, New York, San Francisco.
LEAVIS, P.C., ROSENFELD, S. and LU, R.C. (1978) Biochim. Biophys. Acta, *535*, 281–286.
LEDERER, F., ALIX, J.-H. and HAYES, D. (1977) Biochem. Biophys. Res. Comm. *77*, 470–480.
LEE, K.S. and DRESCHER, D.G. (1978) Int. J. Biochem. *9*, 457–467.
LEE, H.-M. and RIORDAN, J.F. (1978) Biochem. Biophys. Res. Comm. *85*, 1135–1142.
LEQUIN, R.M. and NIALL, H.D. (1972) Biochim. Biophys. Acta, *257*, 76–82.
LIGHT, A. (1972) Methods Enzymol. *25*, 253–262.
LIN, L.J. and FOSTER, J.F. (1975) Anal. Biochem. *63*, 485–490.
LINDLEY, H. (1956) Nature, *178*, 647–648.
LOWE, C.R. (1979) An Introduction to Affinity Chromatography, *In:* Laboratory Techniques in Biochemistry and Molecular Biology (Work, T.S. and Burdon, R.H., eds.) North-Holland, Amsterdam, New York, Oxford.
LOWEY, S. and COHEN, C. (1962) J. Mol. Biol. *4*, 293–308.
LOWEY, S., SLAYTER, H.S., WEEDS, A.G. and BAKER, H. (1969) J. Mol. Biol. *42*, 1–29.
LUDUENA, R.P. and WOODWARD, D.O. (1973) Proc. Natl. Acad. Sci. U.S.A. *70*, 3594–3598.
LUNDBLAD, R.L. (1971) Biochemistry, *10*, 2501–2506.

LUNDBLAD, R.L., KINGDON, H.S. and MANN, K.G. (1976) Methods Enzymol. *40*, 156–176.

LUNDELL, D.J. and HOWARD, J.B. (1978) J. Biol. Chem. *253*, 3422–3426.

McCLARE, C.W.F. (1971) Anal. Biochem. *39*, 527–530.

McKEAN, D.J. and MAURER, R.A. (1978) Biochemistry, *17*, 5215–5219.

McKEAN, D.J., PETERS, E.H., WALDBY, J.I. and SMITHIES, O. (1974) Biochemistry, *13*, 3048–3051.

MACLENNAN, D.H. (1970) J. Biol. Chem. *245*, 4508–4518.

MACHLEIDT, W. and WACHTER, E. (1977) Methods Enzymol. *47*, 263–277.

MACHLEIDT, W., WACHTER, E., SCHEULEN, M. and OTTO, J. (1973) FEBS Lett. *37*, 217–220.

MACHLEIDT, W., HOFNER, H. and WACHTER, E. (1975) *In:* Solid Phase Methods in Protein Sequence Analysis (Laursen, R.A., ed.), pp. 17–30, Pierce Chemical Co., Rockford, Illinois.

MACINO, G. and TZAGOLOFF, A. (1979) J. Biol. Chem. *254*, 4617–4623.

MAGNUSSON, S. (1973) Thromb. Diath. Haemorrh. Suppl. *54*, 31–35.

MAGNUSSON, S., SOTTRUP-JENSEN, L., PETERSEN, T.E., MORRIS, H.R. and DELL, A. (1974) FEBS Lett. *44*, 189–193.

MAHONEY, W.C. and HERMODSON, M.A. (1979) Biochemistry, *18*, 3810–3814.

MANCHEVA, I.N. and VLADOVSKA-YUKHNOVSKA, Y.P. (1978) J. Chromatog. *151*, 207–214.

MANN, K.G. and FISH, W.W. (1972) Methods Enzymol. *26*, 28–42.

MARCHESI, V.T., TILLACK, T.W., JACKSON, R.L., SEGREST, J.P. and SCOTT, R.E. (1972) Proc. Natl. Acad. Sci. U.S.A. *69*, 1445–1449.

MARDIAN, J.K.W. and ISENBERG, I. (1978) Anal. Biochem. *91*, 1–12.

MARGOLIASH, E. and SCHEJTER, A. (1967) Adv. Protein Chem. *21*, 113–286.

MARGOLIASH, E. and SMITH, E.L. (1961) Nature, *192*, 1121–1123.

MARGOLIASH, E. and SMITH, E.L. (1962) J. Biol. Chem. *237*, 2151–2160.

MARGOLIES, M.N. and BRAUER, A. (1978) J. Chromatog. *148*, 429–439.

MARSHALL, R.D. and NEUBERGER, A. (1964) Biochemistry, *3*, 1596–1600.

MARTIN, B., SVENDSEN, I. and OTTESEN, M. (1977) Carlsberg. Res. Comm. *42*, 99–102.

MATSUBARA, H. (1970) Methods Enzymol. *19*, 642–651.

MATSUBARA, H. and FEDER, J. (1971) *In:* The Enzymes 3rd Edn. Vol. 3, pp. 721–795 (Boyer, P.D., ed.), Academic Press, New York.

MAXAM, A.M. and GILBERT, W. (1977) Proc. Natl. Acad. Sci. U.S.A. *74*, 560–564.

MAZUR, R.H., ELLIS, B.W. and CAMMARATA, P.S. (1962) J. Biol. Chem. *237*, 1619–1621.

MENDEZ, E. and GAVILANES, J.G. (1976) Anal. Biochem. *72*, 473–479.

MENDEZ, E. and LAI, C.Y. (1975) Anal. Biochem. *68*, 47–53.

MERRIFIELD, R.B. (1964) Biochemistry, *3*, 1385–1390.

METRIONE, R.M. (1978) J. Chromatog. *154*, 247–250.

MEVARECH, M., NOYES, B.E. and AGARWAL, K.L. (1979) J. Biol. Chem. *254*, 7472–7475.

MICHL, H. (1951) Monatsh. Chem. *82*, 489–493.

MICHL, H. (1952) Monatsh. Chem. *83*, 737–747.

MITCHELL, W.M. (1977) Methods Enzymol. *47*, 165–170.

MOCZAR, E. (1978) J. Chromatog. *146*, 337–343.

MONREAL, J. (1976) Biochim. Biophys. Acta, *427*, 15–19.

MOORE, S. and STEIN, W.H. (1948) J. Biol. Chem. *176*, 367–388.

MORRIS, H.R. and DELL, A. (1975) *In:* Instrumentation in Amino Acid Sequence Determination (Perham, R.N., ed.), pp. 147–191, Academic Press, London, New York, San Francisco.

MORRIS, H.R., WILLIAMS, D.H. and AMBLER, R.P. (1971) Biochem. J. *125*, 189–201.

MORRIS, H.R., BATLEY, K.E., HARDING, N.G.L., BJUR, R.A., DANN, J.G. and KING, R.W. (1974) Biochem. J. *137*, 409–411.

MORRIS, H.R., DELL, A., PETERSEN, T.E., SOTTRUP-JENSEN, L. and MAGNUSSON, S. (1976) Biochem. J. *153*, 663–679.

MORRIS, H.R., THOMPSON, M.R., OSUGA, D.T., AHMED, A.I., CHAN, S.M., VANDENHEEDE, J.R. and FEENEY, R.E. (1978) J. Biol. Chem. *253*, 5155–5162.

MORT, A.J. and LAMPORT, D.T.A. (1977) Anal. Biochem. *82*, 289–309.

MOSER, P.W. and RICKLI, E.E. (1979) J. Chromatog. *176*, 451–455.

MURAMOTO, K., KAWAUCHI, H. and TUZIMURA, K. (1978) Agric. Biol. Chem. *42*, 1559–1563.

NAKAMURA, H., PISANO, J.J. and TAMURA, Z. (1979) J. Chromatog. *175*, 153–161.

NARITA, K. (1958) Biochim. Biophys. Acta, *28*, 184–191.

NARITA, K., MATSUO, H. and NAKAJIMA, T. (1975) *In:* Protein Sequence Determination (Needleman, S.B., ed.), pp. 30–103, Springer-Verlag, Berlin, Heidelberg, New York.

NAU, H. (1976) Angew. Chem. (Internat. Edn. Engl.) *15*, 75–86.

NAU, H. and BIEMANN, K. (1976) Anal. Biochem. *73*, 139–186.

NAU, H., LERCH, K. and WITTLE, L. (1977) FEBS Lett. *79*, 203–206.

NAUGHTON, M. and HAGOPIAN, H. (1962) Anal. Biochem. *3*, 276–284.

NEEDLEMAN, S.B. (ed.) (1975) Protein Sequence Determination 2nd Edn., Springer-Verlag, Berlin, Heidelberg, New York.

NELLES, L.P. and BAMBURG, J.R. (1979) Anal. Biochem. *94*, 150–159.

NEUMANN, N.P. (1972) Methods Enzymol. *25*, 393–400.

NIALL, H. (1973) Methods Enzymol. *27*, 942–1010.

NIALL, H. (1977) *In:* The Proteins, 3rd Edn. Vol. 3, pp. 179–238 (Neurath, H. and Hill, R.L., eds.) Academic Press, New York.

NIALL, H.D., PENHASI, H., GILBERT, P., MYERS, R.C., WILLIAMS, F.G. and POTTS, J.T., Jr. (1969) Fed. Proc. *28*, 661.

NIALL, H.D., JACOBS, J.W., VAN RIETSCHOTEN, J. and TREGEAR, G.W. (1974) FEBS Lett. *41*, 62–64.

NIELSEN, T.B. and REYNOLDS, J.A. (1978) Methods Enzymol. *48*, 3–10.

NIKETIC, V., THOMSEN, J. and KRISTIANSEN, K. (1974) Eur. J. Biochem. *46*, 547–551.

NILSSON, B., NORDÉN, N.E. and SVENSSON, S. (1979) J. Biol. Chem. *254*, 4545–4553.

NISONOFF, A., HOPPER, J.E. and SPRING, S.B. (1975) The Antibody Molecule, Academic Press, New York.

NÓBREGA, F.G. and OZOLS, J. (1971) J. Biol. Chem. *246*, 1706–1717.

NODA, T. and NARITA, K. (1976) J. Biochem. *79*, 353–359.

NOWAK, K., KUCZEK, M., SLOMINSKA, A. and WILUSZ, T. (1977) Acta Biochim. Pol. *24*, 275–280.

NOYES, B.E., MEVARECH, M., STEIN, R. and AGARWAL, K.L. (1979) Proc. Natl. Acad. Sci. U.S.A. *76*, 1770–1774.

NOZAKI, Y. (1972) Methods Enzymol. *26*, 43–50.

OFFORD, R.E. (1966) Nature, *211*, 591–593.

OFFORD, R.E. (1977) Methods Enzymol. *47*, 51–69.

O'HARE, M.J. and NICE, E.C. (1979) J. Chromatog. *171*, 209–226.

OMENN, G.S., FONTANA, A. and ANFINSEN, C.B. (1970) J. Biol. Chem. *245*, 1895–1902.

ORR, H.T., LOPEZ DE CASTRO, J.A., PARHAM, P., PLOEGH, H.L. and STROMINGER, J.L. (1979) Proc. Natl. Acad. Sci. U.S.A. *76*, 4395–4399.

OSBORN, M., WEBER, K. and LODISH, H. (1970) Biochem. Biophys. Res. Comm. *41*, 748–756.

OTIENO, S. (1978) Biochemistry, *17*, 5468–5474.

OVCHINNIKOV, Yu.A., ABDULAEV, N.G., FEIGINA, M.Yu., KISELEV, A.V. and LOBANOV, N.A. (1979) FEBS Lett. *100*, 219–224.

OZOLS, J. and GERARD, C. (1977a) J. Biol. Chem. *252*, 5986–5989.

OZOLS, J. and GERARD, C. (1977b) Proc. Natl. Acad. Sci. U.S.A. *74*, 3725–3729.

PAIK, W.K. and KIM, S. (1971) Science, *174*, 114–119.

PARR, D.M., CONNELL, G.E., KELLS, D.I.C. and HOFMANN, T. (1976) Biochem. J. *155*, 31–36.

PATTHY, L. and SMITH, E.L. (1975) J. Biol. Chem. *250*, 557–564.

PECHÈRE, J.-F. and BERTRAND, R. (1977) Methods Enzymol. *47*, 149–155.

PENKE, B., FERENCZI, R. and KOVÁCS, K. (1974) Anal. Biochem. *60*, 45–50.

PETERSON, E.A. (1970) Cellulosic Ion Exchangers, *In:* Laboratory Techniques in Biochemistry and Molecular Biology (Work, T.S. and Work, E., eds.) Elsevier/North-Holland, Amsterdam, London, New York.

PETERSON, J.D., NEHRLICH, S., OYER, P.E. and STEINER, D.F. (1972) J. Biol. Chem. *247*, 4866–4871.

PETERSON, D.L., GLEISNER, J.M. and BLAKLEY, R.L. (1975) J. Biol. Chem. *250*, 4945–4954.

PIRIE, A. (1971) Biochem. J. *125*, 203–208.

PISANO, J.J. (1972) Methods Enzymol. *25*, 27–44.

PISANO, J.J. and BRONZERT, T.J. (1969) J. Biol. Chem. *244*, 5597–5607.

PISZKIEWICZ, D., LANDON, M. and SMITH, E.L. (1970) Biochem. Biophys. Res. Comm. *40*, 1173–1178.

PISZKIEWICZ, D., TILLEY, B.E., RAND-MEIR, T. and PARSONS, S.M. (1979) Proc. Natl. Acad. Sci. U.S.A. *76*, 1589–1592.

PODELL, D.N. and ABRAHAM, G.N. (1978) Biochem. Biophys. Res. Comm. *81*, 176–185.

PORTER, R.R. (1959) Biochem. J. *73*, 119–126.

PORTER, A.G., BARBER, C., CAREY, N.H., HALLEWELL, R.A., THRELFALL, G. and EMTAGE, J.S. (1979) Nature, *282*, 471–477.

POST, L.E., STRYCHARZ, G.D., NOMURA, M., LEWIS, H. and DENNIS, P.P. (1979) Proc. Natl. Acad. Sci. U.S.A. *76*, 1697–1701.

PREVIERO, A. (1977) Methods Enzymol. *47*, 289–299.

PREVIERO, A. and COLETTI-PREVIERO, M.-A. (Eds.) (1977) Solid Phase Methods in Protein Sequence Analysis, North-Holland, Amsterdam, New York, Oxford.

PREVIERO, A., DERANCOURT, J., COLETTI-PREVIERO, M.-A. and LAURSEN, R.A. (1973) FEBS Lett. *33*, 135–138.

PROUD, C.G., RYLATT, D.B., YEAMAN, S.J. and COHEN, P. (1977) FEBS Lett. *80*, 435–442.

PUSZTAI, A. and WATT, W.B. (1970) Biochim. Biophys. Acta, *214*, 463–467.

RAFTERY, M.A. and COLE, R.D. (1966) J. Biol. Chem. *241*, 3457–3461.

RAMACHANDRAN, L.K. and WITKOP, B. (1967) Methods Enzymol. *11*, 283–299.

RANDERATH, K. (1970) Anal. Biochem. *34*, 188–205.

RASK, L., ANUNDI, H. and PETERSON, P.A. (1979) FEBS Lett. *104*, 55–58.

REED, P.G. and MAKAY, C.M. (1978) Anal. Biochem. *84*, 324–327.

REINDEL, F. and HOPPE, W. (1954) Chem. Ber. *87*, 1103–1107.

RICE, R.H. (1974) Virology, *61*, 249–255.

RICHARDS, F.F. and LOVINS, R.E. (1972) Methods Enzymol. *25*, 314–326.

RICHARDS, F.M. and VITHAYATHIL, P.J. (1959) J. Biol. Chem. *234*, 1459–1465.

RIGHETTI, P.G. and CHILLEMI, F. (1978) J. Chromatog. *157*, 243–251.

RIGHETTI, P.G. and DRYSDALE, J.W. (1976) Isoelectric Focusing, *In:* Laboratory Techniques in Biochemistry and Molecular Biology (Work, T.S. and Work, E., eds.) Elsevier/North-Holland, Amsterdam, Oxford, New York.

RILEY, M. and PERHAM, R.N. (1970) Biochem. J. *118*, 733–739.

RIQUELME, P.T., BURZIO, L.O. and KOIDE, S.S. (1979) J. Biol. Chem. *254*, 3018–3028.

ROBINSON, P.J., DUNNHILL, P. and LILLY, M.D. (1971) Biochim. Biophys. Acta, *242*, 659–661.

ROHVER, H., ZILLIG, W. and MAILHAMMER, R. (1975) Eur. J. Biochem. *60*, 227–238.

ROKUSHIKA, S., OHKAWA, T. and HATANO, H. (1979) J. Chromatog. *176*, 456–461.

ROSE, Z.B., HAMASAKI, N. and DUBE, S. (1975) J. Biol. Chem. *250*, 7939–7942.

ROSEAU, G. and PANTEL, P. (1969) J. Chromatog. *44*, 392–395.

ROSMUS, J. and DEYL, Z. (1972) J. Chromatog. *70*, 221–339.

ROSSI-FANELLI, A., ANTONINI, E. and CAPUTO, A. (1958) Biochim. Biophys. Acta, *30*, 608–615.

ROTH, M. (1971) Anal. Chem. *43*, 880–882.

ROTH, M. and HAMPAI, A. (1973) J. Chromatog. *83*, 353–356.

ROUSER, G., KRITCHEVSKY, G., YAMAMOTO, A., SIMON, G., GALLI, C. and BAUMAN, A.J. (1969) Methods Enzymol. *14*, 272–317.

RUBINSTEIN, M., CHEN-KIANG, S., STEIN, S. and UDENFRIEND, S. (1979) Anal. Biochem. *95*, 117–121.

RUEGG, V.T. and RUDINGER, J. (1977) Methods Enzymol. *47*, 116–122.

RYAN, T.E., WOODS, G.M., KIRKPATRICK, F.H. and SHAMOO, A.E. (1976) Anal. Biochem. *72*, 359–365.

RYDON, H.N. and SMITH, P.W.G. (1952) Nature, *169*, 922–923.

RYLATT, D.B. and COHEN, P. (1979) FEBS Lett. *98*, 71–75.

RYLE, A.P. (1970) Methods Enzymol. *19*, 316–336.

RYLE, A.P., SANGER, F., SMITH, L.F. and KITAI, R. (1955) Biochem. J. *60*, 541–556.

SAMPATH-KUMAR, P.S. and FRUTON, J.S. (1974) Proc. Natl. Acad. Sci. U.S.A. *71*, 1070–1072.

SANDERMANN, H., Jr. and STROMINGER, J.L. (1972) J. Biol. Chem. *247*, 5123–5131.

SANGER, F. (1945) Biochem. J. *39*, 507–515.

SANGER, F. (1959) Science, *129*, 1340–1344.

SANGER, F. and THOMPSON, E.O.P. (1953) Biochem. J. *53*, 366–374.

SANGER, F. and TUPPY, H. (1951) Biochem. J. *49*, 463–481.

SANGER, F., NICKLEN, S. and COULSON, A.R. (1977) Proc. Natl. Acad. Sci. U.S.A. *74*, 5463–5467.

SAUER, R.T. and ANDEREGG, R. (1978) Biochemistry, *17*, 1092–1100.

SAUER, R.T., NIALL, H.D., HOGAN, M.L., KEUTMANN, H.T., O'RIORDAN, J.L.H. and POTTS, J.T., Jr. (1974) Biochemistry, *13*, 1994–1999.

SAVIGE, W.E. and FONTANA, A. (1977) Methods Enzymol. *47*, 459–469.

SCANU, A.M. and EDELSTEIN, C. (1971) Anal. Biochem. *44*, 576–588.

SCHAFFER, M.H. and STARK, G.R. (1976) Biochem. Biophys. Res. Comm. *71*, 1040–1047.

SCHECHTER, I. and BURSTEIN, Y. (1976) Biochem. J. *153*, 543–550.

SCHECHTER, I., WOLF, O., ZEMELL, R. and BURSTEIN, Y. (1979) Fed. Proc. *38*, 1839–1845.

SCHENKEIN, I., LEVY, M., FRANKLIN, E.C. and FRANGIONE, B. (1977) Arch. Biochem. Biophys. *182*, 64–70.

SCHMER, G. and KREIL, G. (1969) Anal. Biochem. *29*, 186–192.

SCHROEDER, W.A. (1972a) Methods Enzymol. *25*, 138–143.

SCHROEDER, W.A. (1972b) Methods Enzymol. *25*, 203–213.

SCHROEDER, W.A., SHELTON, J.B. and SHELTON, J.R. (1969) Arch. Biochem. Biophys. *130*, 551–556.

SCHROEDER, W.A., SHELTON, J.B., SHELTON, J.R. and POWARS, D. (1979) J. Chromatog. *174*, 385–392.

SCHWARTZ, W., SMITH, P.K. and ROYER, G. (1979) Quoted in Pierce Previews, *2* (79) 2, Pierce Chemical Co., Rockford, Illinois.

SEGREST, J.P., JACKSON, R.L., MARCHESI, V.T., GUYER, R.B. and TERRY, W. (1972) Biochem. Biophys. Res. Comm. *49*, 964–969.

SHECHTER, Y., RUBINSTEIN, M. and PATCHORNIK, A. (1977) Biochemistry, *16*, 1424–1430.

SHEEHAN, J.C. and YANG, D.-D.H. (1958) J. Amer. Chem. Soc. *80*, 1154–1158.

SHIMADA, A. and NATHENSON, S.G. (1969) Biochemistry, *8*, 4048–4062.

SHOTTON, D.M. (1970) Methods Enzymol. *19*, 113–140.

SIMONS, S.S., Jr. and JOHNSON, D.F. (1976) J. Amer. Chem. Soc. *98*, 7098–7099.

SMITH, E.L. (1977) Methods Enzymol. *47*, 156–161.

SMITH, I. and SEAKINS, J.W.T. (1976) Chromatographic and Electrophoretic Techniques, 4th Edn. William Heinemann Medical Books, Ltd., London.

SMITHIES, O., GIBSON, D., FANNING, E.M., GOODFLIESH, R.M., GILMAN, J.G. and BALLANTYNE, D.L. (1971) Biochemistry, *10*, 4912–4921.

SMYTH, D.S. and UTSUMI, S. (1967) Nature, *216*, 332–335.

SMYTH, D.G., NAGAMATSU, A. and FRUTON, J.S. (1960) J. Amer. Chem. Soc. *82*, 4600–4604.

SMYTH, D.G., STEIN, W.H. and MOORE, S. (1962) J. Biol. Chem. *237*, 1845–1850.

SMYTH, D.G., STEIN, W.H. and MOORE, S. (1963) J. Biol. Chem. *238*, 227–234.

SPACKMAN, D.H., STEIN, W.H. and MOORE, S. (1958) Anal. Chem. *30*, 1190–1206.

SPACKMAN, D.H., STEIN, W.H. and MOORE, S. (1960) J. Biol. Chem. *235*, 648–659.

SPATZ, L. and STRITTMATTER, P. (1973) J. Biol. Chem. *248*, 793–799.

SPIRO, R.G. (1970) Ann. Rev. Biochem. *39*, 599–638.

SPIRO, R.G. (1972) Methods Enzymol. *28*, 3–43.

SPIRO, R.G. and BHOYROO, V.D. (1974) J. Biol. Chem. *249*, 5704–5717.

STARK, G.R. (1968) Biochemistry, *7*, 1796–1807.

STARK, G.R. (1972a) Methods Enzymol. *25*, 103–120.

STARK, G.R. (1972b) Methods Enzymol. *25*, 369–384.

STARK, G.R. (1977) Methods Enzymol. *47*, 129–132.

STARK, G.R. and SMYTH, D.G. (1963) J. Biol. Chem. *238*, 214–226.

STARK, G.R., STEIN, W.H. and MOORE, S. (1960) J. Biol. Chem. *235*, 3177–3181.

STEINHARDT, J. and BEYCHOK, S. (1964) *In:* The Proteins (Neurath, H., ed.), 2nd Edn., Vol. 2, pp. 139–304, Academic Press, New York.

STENFLO, J. and SUTTIE, J.W. (1977) Ann. Revs. Biochem. *46*, 157–172.

STENFLO, J., FERNLUND, P., EGAN, W. and ROEPSTORFF, P. (1974) Proc. Natl. Acad. Sci. U.S.A. *71*, 2730–2733.

STEVENSON, K.J. (1971) Anal. Biochem. *40*, 29–34.

Strickland, D.K. and Hudson, B.G. (1978) Biochemistry, *17*, 3411–3418.

Summers, M.R., Smythers, G.W. and Oroszlan, S. (1973) Anal. Biochem. *53*, 624–628.

Sutton, M.R. and Bradshaw, R.A. (1978) Anal. Biochem. *88*, 344–346.

Svasti, J. and Milstein, C. (1972) Biochem. J. *128*, 427–444.

Svenson, A., Carlsson, J. and Eaker, D. (1977) FEBS Lett. *73*, 171–174.

Swank, R.T. and Munkres, K.D. (1971) Anal. Biochem. *39*, 462–477.

Synge, R.L.M. (1957) Biochem. J. *65*, 266–273.

Synge, R.L.M. (1957) Biochem. J. *65*, 266–273.

Taborsky, G. (1974) Adv. Protein Chem. *28*, 1–210.

Takagi, T. and Doolittle, R.F. (1974) Biochemistry, *13*, 750–756.

Takagi, T. and Doolittle, R.F. (1975) Biochemistry, *14*, 940–946.

Tang, J. and Hartley, B.S. (1967) Biochem. J. *102*, 593–599.

Tarr, G.E. (1975) Anal. Biochem. *63*, 361–370.

Tarr, G.E. (1977) Methods Enzymol. *47*, 335–357.

Tarr, G.E., Beecher, J.F., Bell, M. and McKean, D. (1978) Anal. Biochem. *84*, 622–627.

Teller, D.C. (1973) Methods Enzymol. *27*, 346–441.

Terhorst, C., Möller, W., Laursen, R. and Wittmann-Liebold, B. (1973) Eur. J. Biochem. *34*, 138–152.

Thompson, R.B., Borden, D., Tarr, G.E. and Margoliash, E. (1978) J. Biol. Chem. *253*, 8957–8961.

Thomsen, J., Kristiansen, K., Brunfeldt, K. and Sundby, F. (1972) FEBS Lett. *21*, 315–319.

Thorley-Lawson, D.A. and Green, N.M. (1975) Eur. J. Biochem. *59*, 193–200.

Tichy, H. (1975) Anal. Biochem. *69*, 552–557.

Titani, K., Koide, A., Hermann, J., Ericsson, L.H., Kumar, S., Wade, R.D., Walsh, K.A., Neurath, H. and Fischer, E.H. (1977) Proc. Natl. Acad. Sci. U.S.A. *74*, 4762–4766.

Titani, K., Koide, A., Ericsson, L.H., Kumar, S., Hermann, J., Wade, R.D., Walsh, K.A., Neurath, H. and Fischer, E.H. (1978) Biochemistry, *17*, 5680–5693.

Tomita, M. and Marchesi, V.T. (1975) Proc. Natl. Acad. Sci. U.S.A. *72*, 2964–2968.

Tomita, M., Furthmayr, H. and Marchesi, V.T. (1978) Biochemistry, *17*, 4756–4770.

Tong, S.W. (1977) Biochem. Biophys. Res. Comm. *74*, 1242–1248.

Töpfer-Petersen, E., Lottspeich, F. and Henschen, A. (1976) Hoppe-Seyler's Z. Physiol. Chem. *357*, 1509–1513.

Tschesche, H. (1977) Methods Enzymol. *47*, 73–84.

Tschesche, H. and Kupfer, S. (1972) Eur. J. Biochem. *26*, 33–36.

Tsung, C.M. and Fraenkel-Conrat, H. (1965) Biochemistry, *4*, 793–801.

Turková, J. (1978) Affinity Chromatography Elsevier, Amsterdam, New York.

TURKOVÁ, J., VAVREINOVÁ, S., KŘIVÁKOVÁ, M. and ČOUPEK, J. (1975) Biochim. Biophys. Acta. *386*, 503–508.

TZAGOLOFF, A., MACINO, G. and SEBALD, W. (1979) Ann. Rev. Biochem. *48*, 419–441.

UDENFRIEND, S., STEIN, S., BÖHLEN, P., DAIRMAN, W., LEIMGRUBER, W. and WEIGELE, M. (1972) Science, *178*, 871–872.

UPHAUS, R.A., GROSSWEINER, L.I. and KATZ, J.J. (1959) Science, *129*, 641–643.

UY, R. and WOLD, F. (1977) Science, *198*, 890–896.

VALENZUELA, P., GRAY, P., QUIROGA, M., ZALDIVAR, J., GOODMAN, H.M. and RUTTER, W.J. (1979) Nature, *280*, 815–819.

VAN BEEUMEN, J., VAN DAMME, J. and DELEY, J. (1978) FEBS Lett. *93*, 373–376.

VANCE, D.E. and FEINGOLD, D.S. (1970) Anal. Biochem. *36*, 30–42.

VANDEKERKHOVE, J. and VAN MONTAGU, M. (1974) Eur. J. Biochem. *44*, 279–288.

VANDEKERKHOVE, J. and WEBER, K. (1978a) Eur. J. Biochem. *90*, 451–462.

VANDEKERKHOVE, J. and WEBER, K. (1978b) Nature, *276*, 720–721.

VAN EERD, J.-P., CAPONY, J.-P., FERRAZ, C. and PECHÈRE, J.-F. (1978) Eur. J. Biochem. *91*, 231–242.

VAN HEUVERSWYN, H., VAN DE VOORDE, A. and FIERS, W. (1978) Eur. J. Biochem. *91*, 415–430.

VAN HOLDE, K.E. (1975) *In:* The Proteins (Neurath, H. and Hill, R.L., eds.) 3rd Edn., Vol. 1, pp. 225–291, Academic Press, New York, San Francisco, London.

VAN ORDEN, H.O. and CARPENTER, F.H. (1964) Biochem. Biophys. Res. Comm. *14*, 399–403.

VISSER, L., ROBINSON, N.C. and TANFORD, C. (1975) Biochemistry, *14*, 1194–1199.

WACHTER, E. (1980) *In:* Solid Phase Methods in Protein Sequence Analysis (Birr, C., ed.), Proceedings of the 3rd International Conference, Heidelberg, 1979, Elsevier/North-Holland Biomedical Press, Amsterdam.

WACHTER, E., MACHLEIDT, W., HOFNER, H. and OTTO, J. (1973) FEBS Lett. *35*, 97–102.

WACHTER, E., HOFNER, H. and MACHLEIDT, W. (1975) *In:* Solid Phase Methods in Protein Sequence Analysis (Laursen, R.A., ed.), pp. 31–46, Pierce Chemical Co., Rockford, Illinois.

WALEY, S.G. and WATSON, J. (1953) Biochem. J. *55*, 328–337.

WALKER, J.E. (1980) *In:* Solid Phase Methods in Protein Sequence Analysis (Birr, C., ed.), Proceedings of the 3rd International Conference, Heidelberg, 1979, Elsevier/North-Holland Biomedical Press, Amsterdam.

WALKER, J.E., CARNE, A.F. and SCHMITT, H.W. (1979) Nature, *278*, 653–654.

WALSH, K.A., McDONALD, R.M. and BRADSHAW, R.A. (1970) Anal. Biochem. *35*, 193–202.

WATERFIELD, M.D. and BRIDGEN, J. (1975) *In:* Instrumentation in Amino Acid

Sequence Analysis (Perham, R.N., ed.), pp. 41–71, Academic Press London, New York, San Francisco.

WATERFIELD, M.D., ESPELIE, K., ELDER, K. and SKEHEL, J.J. (1979) British Medical Bulletin, *35*, 57–63.

WATT, K.W.K., TAKAGI, T. and DOOLITTLE, R.F. (1979) Biochemistry, *18*, 68–76.

WEBER, K. and KUTER, D.J. (1971) J. Biol. Chem. *246*, 4504–4509.

WEBER, K. and OSBORN, M. 1975) *In:* The Proteins (Neurath, H. and Hill, R.L., eds.), 3rd Edn. Vol. 1, pp. 179–223, Academic Press, New York, San Francisco, London.

WEBSTER, R.E., ENGELHARDT, D.L. and ZINDER, N.D. (1966) Proc. Natl. Acad. Sci. U.S.A. *55*, 155–161.

WEEDS, A.G. and HARTLEY, B.S. (1968) Biochem. J. *107*, 531–548.

WEEDS, A.G. and LOWEY, S. (1971) J. Mol. Biol. *61*, 701–725.

WESTHEAD, E.W. (1972) Methods Enzymol. *25*, 401–409.

WILCHEK, M. (1974) Adv. Exp. Med. Biol. *42*, 15–31.

WILCOX, P.E. (1970) Methods Enzymol. *19*, 64–108.

WINGARD, M., MATSUEDA, G. and WOLFE, R.S. (1972) J. Bacteriol. *112*, 940–949.

WITTER, A. and TUPPY, H. (1960) Biochim. Biophys. Acta, *45*, 429–442.

WITTMANN, H.G. and BRAUNITZER, G. (1959) Virology, *9*, 726–728.

WITTMANN-LIEBOLD, B. (1973) Hoppe-Seyler's Z. Physiol. Chem. *354*, 1415–1431.

WITTMANN-LIEBOLD, B., GEISSLER, A.W. and MARZINZIG, E. (1975) J. Supramol. Struct. *3*, 426–447.

WITTMANN-LIEBOLD, B., GRAFFUNDER, H. and KOHLS, H. (1976) Anal. Biochem. *75*, 621–633.

WITTMANN-LIEBOLD, B., BRAUER, D. and DOGNIN, J. (1977) *In:* Solid Phase Methods in Protein Sequence Analysis (Previero, A. and Coletti-Previero, M.-A., eds.), pp. 219–232, Elsevier/North-Holland Biomedical Press, Amsterdam.

WOOD, H.G. and BARDEN, R.E. (1977) Ann. Rev. Biochem. *46*, 385–413.

WOODS, K.R. and WANG, K.-T. (1967) Biochim. Biophys. Acta, *133*, 369–370.

WOOTTON, J.C., BARON, A.J. and FINCHAM, J.R.S. (1975) Biochem. J. *149*, 749–754.

WUNDERER, G. (1978) Hoppe-Seyler's Z. Physiol. Chem. *359*, 1193–1201.

YAMADA, S. and ITANO, H.A. (1966) Biochim. Biophys. Acta, *130*, 538–640.

YOUNT, R.G. (1975) Adv. Enzymol. *43*, 1–56.

YSEBAERT, M., VAN DE VOORDE, A. and FIERS, W. (1978) Eur. J. Biochem. *91*, 431–439.

ZAHLER, W.L. and CLELAND, W.W. (1968) J. Biol. Chem. *243*, 716–719.

ZEEUWS, R. and STROSBERG, A.D. (1978) FEBS Lett. *85*, 68–72.

ZERVOS, C. and ADAMS, E. (1977) Int. J. Peptide and Protein Res. *10*, 1–8.

ZIMMERMAN, C.L., APELLA, E. and PISANO, J.J. (1977) Anal. Biochem. *77*, 569–573.

ZOON, K.C., SMITH, M.E., BRIDGEN, P.J., ANFINSEN, C.B., HUNKAPILLER, M.W. and HOOD, L.E. (1980) Science, *207*, 527–528.

ZYTKOVICZ, T.H. and NELSESTUEN, G.L. (1976) Biochim. Biophys. Acta, *444*, 344–348.

Subject index

N-acetyl group 35, 133, 147, 245, 250
O-acetyl group 178
acetyl CoA carboxylase 19
actin 83, 241, 248, 261, 274, 275
active-site residues 30, 273
ADP-ribosyl group 5, 250
affinity chromatography 130, 205
albumin 49
alcohol dehydrogenase 248
aldehydes 177, 186
aldolase 276
amino acid analysis, peptides 143, 165
  proteins 24
  Pth derivatives 191
amino acid residues 2, 237
N-(2-aminoethyl)-3-aminopropyl glass 207
aminoethylcysteine 30, 53
α-amino group, pK_a 6, 55, 165, 175
aminopeptidase M 32, 228, 230
aminopolystyrene 206, 208
aminopropyl glass 207, 212
amino-terminal residue
  blocked 33, 35, 132, 140, 147
  in peptides 165, 168
  in proteins 31
anylase 35, 133, 248
Anemonia sulcata toxin II 256
angiotensin converting enzyme 245
anilinothiazolinone derivatives
  conversion to Pth derivatives 180
  extraction 180, 186, 200, 205, 217, 224, 243
  formation 175, 178
  reaction with methylamine 181
anion-transport protein 263, 266
anti-freeze glycoproteins 186, 244
arginine residues
  chemical modification 57
  conversion to ornithine 212
  detection 157
  Pth derivative 192, 194, 199
  hydrophilicity 201
  specific proteases 52, 57
  tryptic cleavage 55
Armillaria mellea protease 52
asparagine–glycine peptide bond
  cleavage 49
  rearrangement 64, 183
asparagine residues
  deamidation 88, 181, 192, 194
  glycosyl 186, 242
  mobilities of peptides 164
aspartic acid
  electrophoresis 164
  peptide bond cleavage 58, 70
  Pth, methyl ester 181, 215
  reaction with carbodiimides 209
aspartokinase-homoserine
  dehydrogenase 46
aspartyl–proline peptide bond 50
ATPase from sarcoplasmic reticulum

delipidation 265
digestion 266, 267
limited proteolysis 47
N-terminus 35, 232
peptide separation 93, 94, 102
phosphopeptide 104, 132, 239
ATP-phosphoribosyl transferase 262, 281
avidin 130

BioGel P 84, 269
biotinyl peptides 130
blocked peptides
detection 140, 147, 150
formation 184
isolation 132
BNPS-skatole 68
Braunitzer's reagents 201
Bray's fluid 153
bromelain 46, 264
3-bromopropionate 28
N-bromosuccinimide 62, 68
butan-2,3-dione 57
butan-1-ol, purification 292
n-butyl acetase, purification 292
t-butyl hypochlorite 151
2,5-*bis*-(5-*t*-butylbenzoxazol-2-yl)-thiophen (BBOT) 153
butyl PBD 196

cadmium
ninhydrin reagent 148, 219
toxicity 294
calcitonin 10, 262
carbodiimides 208
carbonyldiimidazole 208
carboxyethylcysteine 28
α-carboxyglutamic acid 5, 219, 238
S-carboxymethylation 30, 267
carboxymethylcellulose (CM-cellulose) 20, 95, 101, 103
S-carboxymethylcysteine
analysis 166
autoxidation 89, 163
cyclization 28, 88, 163, 184
identification 188, 217
carboxymethyl Sephadex (CM-Sephadex) 103
carboxypeptidases 39, 228, 229
carboxy-terminal residues 38
carboxy-terminal peptides 131
carcinoembryonic antigen 244
chloral hydrate 269
chloramine T 277
chlorination 150
chlorobutane 200, 201
chloroform/methanol 269, 270
chloroplatinic acid 159
L-1-chloro-3-tosylamido-4-phenyl-butan-2-one 54
chorionic gonadotropin 245
chymotrypsin 59, 255, 278
citraconic anhydride 293
citraconylation 56
cleavage, in Edman degradation 174, 178
clostripain 52
cobalt (III) chelate 32
collagen 251, 256
collidine 147
columns for chromatography
DEAE-cellulose 97, 101
Dowex 50 108
Sephadex 86, 87
complementation 259
concanavalin A 130
conversion, in Edman degradation 174, 180, 224
Coomassie blue 79, 91, 145
counter-current distribution 20, 75
coupling, in Edman degradation 174, 175, 181
covalent chromatography 75, 131
cross-links 251, 273
cyanate 18, 32
S-cyanocysteine 65

cyanogen bromide 62, 96, 294
cyclohexan-1,2-dione 57
cysteic acid
  dansyl 173
  mobility of peptides 164
  Pth derivative 192
cysteine residues
  attachment to glass beads 212, 214
  chemical modification 28
  covalent chromatography 131
  cyanylation 65
  derivatives
    autoxidation 89, 117, 163, 265
    identification 194
    in iodoplatinate test 159
  oxidation by DMSO 69
cystine residues
  absorbance 152
  destruction by alkali 18
  detection 153
  diagonal electrophoresis 131, 251
  identificaton of cross-links 251
cytochrome b_5 24, 48, 263, 264, 267
cytochrome *c* 35, 107, 248, 258, 259
cytochrome oxidase 281

dansyl amino acids 36, 169
dansyl chloride
  amino acid analysis 166
  detection of peptides 144, 145
  peptide mapping 276
  peptide sequencing 216
  N-terminal residues
    of peptides 168
    of proteins 33
dansyl peptides 171, 276
dehydroserine 194
dehydrothreonine 194, 225
denaturation 52
desmosine 252, 256
diagonal electrophoresis 131, 251
dialysis of peptides 103
dibromoacetone 273
dicyclohexylcarbodiimide 208
diethylaminoethyl cellulose
  (DEAE-cellulose) 95, 97, 102, 270
diethylaminoethyl Sephadex
  (DEAE-Sephadex) 104
diethyl-(2-hydroxypropyl)aminoethyl
  Sephadex (QAE-Sephadex) 104
dihydrofolate reductase 52, 102
dimethyl adipimidate 274
dimethyl allylamine 176, 201
4-dimethylaminoazobenzene-4′-
  isothiocyanate (Dabitc) 37, 173, 221
p-dimethylaminobenzaldehyde 157
dimethylbenzylamine 201
dimethyl formamide 212, 269, 292
dimethyl suberimidate 274
dimethyl sulphoxide (DMSO) 70, 269
2,4-dinitrophenyl amino acids 7
ϵ-*N*-2,4-dinitrophenyl lysine 90, 113, 119,
  126
2,4-dinitrophenyl peptides 130
dipeptidyl aminopeptidase 228, 231
diphenyloxazole (PPO) 80, 156
diphenylthiourea 176, 177, 194
dipping of chromatograms 159
1,5-disulphonyl-naphthalene-3-
  isothiocyanate 201
5,5′-dithio*bis*(2-nitrobenzoate (DTNB) 153
dithionite 192
dityrosine 252
DNA sequences 259, 272, 280
Dowex ion-exchange resins 104, 133

Edman degradation 174–221
  automated 199
  solid-phase 205, 271
  subtractive 221
Ehrlich test 157, 218
elastase 52
elastin 251, 256
electroendosmosis 110, 114, 123, 165

electrophoresis, *see* paper, polyacrylamide gel, thin-layer
β-elimination 179, 186, 194, 215, 239, 243
enkephalin 278
ethanolamine 213
ethyleneimine 28, 295
N-ethylmaleimide 29, 251
N-ethylmorpholine 88, 96, 276, 292
exo-cis-3,6-endoxo-Δ^4-hexahydrophthalic anhydride 56
extinction coefficient of proteins 26

fetuin 245
fibrinogen
  chain separation 19
  cleavage by plasmin 46
  disulphide bonds 256
  sequence 261
flow cells 137, 152
fluorescamine
  amino acid analysis 24
  detection of peptides, in solution 106, 129, 141
    on paper 120, 148
    on this layers 123, 127, 149, 232
fluorodinitrobenzene 7, 32, 295
fluorography 80, 156
formaldehyde 275
formamide 270, 292
formyl group 35, 250
freeze-drying 51, 92

β-galactosidase 9, 103, 259, 261
gamma counting 153
gas chromatography
  acetyl groups 247
  mass spectrometry 231, 268, 272
  peptides 75
  Pth amino acids 191
gastrin 282
gel filtration
  high performance 129
  molecular weights 21
  peptide separation 84, 268
glutamate dehydrogenase 35
glutamic acid
  peptide bond cleavage 58
  Pth, methyl ester 181, 215
glutamine residues
  cyclization 64, 88, 163, 184
  deamidation 88, 181, 192, 194
  mobility of peptides 164, 219
glutathione 154
glycine, Pth 181, 204
glycogen phosphorylase
  limited proteolysis 46
  phosphate groups 238
  pyridoxyl peptide 152, 250
  sequence 261
glycogen synthase 239
glycopeptides
  affinity chromatography 130
  detection 154
  Edman degradation 185
  paper chromatography 117, 243
  structure 242
  thin-layer chromatography 243
glycophorin 186, 244, 263, 266, 268
glycylglycine 202
gradients for chromatography 96, 99, 100, 105
guanidinium chloride 293

haem 24
haemagglutinin 263, 264, 272, 281
haemoglobin 20, 24, 35, 245, 274
heptafluorobutyric acid 178, 200, 294
hepatitis B antigen 272, 281
hexamethyl phosphoric triamide 269
high-performance liquid chromatography (hplc)
  peptides 128, 271, 275
  Pth amino acids 188
histidine residues

detection 158
Edman degradation 181, 182
mobility of peptides 164
oxidation 68
Pth derivative 192, 193, 194, 199
histocompatibility antigens 48, 204, 205, 263, 264, 275
histones 238, 250
homoserine residues
attachment to solid support 209, 212
automated sequencing of peptides 202
formation 63
lactone 96, 163
mobility of peptides 165
Pth derivative 194
hydantoin formation 38
hydrazine 38, 212, 246, 293
hydriodic acid 191
hydrochloric acid 25, 70, 191, 293
hydrogen fluoride 244
hydrolysis
amide groups 88, 181
peptides, alkaline 140, 142
complete 25, 143
partial 50, 70, 150, 267
phosphate esters 239
proteins 25
Pth amino acids 191
hydrophobic peptides
Edman degradation 177, 222, 271
mass spectrometry 231
membrane proteins 263
hydroxylamine 49, 56, 58, 184
hydroxylysine 242, 250
hydroxyproline 242, 250

iminothazolidine-4-carboxylic acid 66
immunoglobulin G
amino-terminal sequences 279
chain separation 19
glycopeptide 245
limited proteolysis 47
precursors 279
immunoglobulin μ chain 261
insulin 7
interferon 203
iodine/azide reagent 194
iodoacetic acid 31, 153, 208, 275, 293
iodoacetyl glass 206, 208, 214
N-(iodoethyl)-trifluoroacetamide 28
iodopeptides 153
iodoplatinate test 159
o-iodosobenzoic acid 69
iodotyrosine residues 87, 275
isatin 159
isoelectric focusing 77, 81, 146
isoleucine
dansyl dipeptides 171
peptide bond hydrolysis 166, 221
Pth derivative 192
isothiocyanato-glass 206, 208, 213

kinase 238

lactose permease 272
lectins 130
leucine aminopeptidase 32, 228, 230
lipids 23, 265
liquid-scintillation counting 91, 106, 153, 156, 275
lysine residues
phenylthiocarbamyl 176, 193, 218, 219
specific proteases 58
tryptic cleavage 57
lysinonorleucine 252
lysozyme 101

maleylamino groups 152
maleylated serine and threonine 186
mass spectrometry
blocked peptides 247
γ-carboxyglutamic acid 238
cross-links in collagen 256
glycopeptides 244

peptide sequencing 230, 272
mellitin 35, 248
methionine residues
analysis 166
autoxidation 64, 89, 117, 163, 194, 265
CNBr cleavage 62
covalent chromatography 131
detection by iodoplatinate test 159
diagonal electrophoresis 132
oxidation with DMSO 69
Pth derivative 192
methionine sulphone 88, 226
methionine sulphoxide 64, 164
methyl acetimidate 275, 276
N-methylalanine 241
S-methylcysteine 65
methyl histidines 221, 240, 241
methyl isothiocyanate 213
methyl lysines 221, 240
N-methyl morpholine 214
β_2-microglobulin 11, 262
microheterogeneity 6, 258
murein lipoprotein 35, 248, 267
myelin protein 269
myoglobin 201
myosin
acetyl peptide 247
chain separation 20
limited proteolysis 48
methylated residues 241
repetitive peptide sequence 235
myotoxin *a* 255
Myxobacter protease II 52

NAD-cytochrome b_5 reductase 264
ninhydrin, for peptide detection
in solution 106, 139
on paper or thin layers 123, 146, 150
Pth derivatives 194
nitrogenase 19
2-(2-nitrophenylsulphenyl)-3-methylindole 68
nitrosonaphthol 158
2-nitro-5-thiocyanobenzoic acid 65
nuclear magnetic resonance 238, 242, 248, 256

Offord plot 164, 219
overlap, in Edman degradation 181
oxygen, in Edman degradation 177, 185, 199
oxytocin 38, 250

papain 46, 47, 61, 264, 266, 273
paper electrophoresis 109
paper chromatography 116
Pauly reagent 146, 158
penicillinase 260, 281
pepsin 46, 47, 59, 133, 251, 267
pepstatin 41
peptide analyzer 106
peptide mapping
colour reactions for peptides 157
comparison of proteins 274
hplc 275
ion-exchange chromatography 274
paper 119
polyacrylamide gel electrophoresis 276
preparative 119, 127
thin layers 81, 127
performic acid 28, 253, 294
periodate 185, 242
peroxidase 278
phage ΦX174 259, 272, 281
phenanthrene quinone 157, 195, 276
phenol 269, 270, 293, 294
phenol/sulphuric acid reagent 154
p-phenylene diisothiocyanate 208, 212
phenyl glyocal 57
phenyl isothiocyanate (Pitc)
hydrolysis 175
N-terminal residue 32
peptide sequencing 174
radiolabelled 203

storage 294
phenylthiocarbamyl (Ptc) derivatives
of lysine residues 176, 219
of peptides 175, 183
phenylthiohydantoin (Pth) derivatives
formation 174, 180
identification 187
phenylthiourea 176, 177, 194
phosphate determination 26
phosphocellulose 96, 101
phosphoglycerate mutase 239
ϵ-*N*-phosphopyridoxyl-lysine 152, 250
pH-stat 54
phosphorylated residues 238
photooxidation 18
o-phthalaldehyde
amino acid analysis 24, 142
peptide detection 106, 142
phthalic anhydride 56
plasmin 46
polyacrylamide gel electrophoresis
analytical 78, 163
detection of peptides 91, 145
preparative 17, 127
polybrene 202
polystyrene ion-exchange resins
blocked peptides 133
peptide separation 104, 275
polystyrene resins for sequencing 205
polyornithine 202
porous glass beads 206
post-proline cleaving enzyme 52
proline residues
Edman degradation 181
fluorescamine reaction 149
isatin test 159
ninhydrin reaction 140, 147
peptide bond cleavage 50, 52
pronase 243
propanol 105, 129
pro-parathyroid hormone 280
proteolipid 269, 272, 281, 293
proteolysis
contaminants 74
limited 17, 44, 264, 266
specific 51
prothrombin 238
purity of peptides for sequencing 161
pyridine, purification 293
pyroglutamate aminopeptidase 249
pyroglutamic acid, *see* pyrrolidone carboxylic acid
pyrrolidone carboxylic acid
formation 64, 133, 184
in ninhydrin reaction 147
N-terminus of protein 35, 133, 249

QAE-Sephadex 104
quadrol 183, 200, 202, 294
quaternary structure 1

radioautography 155, 277, 278
radiolabelled peptides
cysteine derivatives 188
detection 153, 155
phosphopeptides 239
radiolabelled proteins 203, 272, 275, 279
radioiodination 275, 277
Raney nickel 67
reduction of disulphide bonds 30
reliability of sequence data 235, 260
rennin 46
λ-repressor 262, 281
residue weights 2
resilin 251
rhodopsin, bacterial 263, 266, 272
ribonuclease 45, 274
ribosomal proteins of *E. coli* 5, 10, 33, 35, 82, 107, 212, 240, 262, 274, 281
ribulose*bis*phosphate carboxylase 19, 274

salts
in electrophoresis 113
in liquid-scintillation counting 153

removal 23, 86, 103
secondary structure 1, 17, 53
Sephadex 84
serine
β-elimination 179, 194, 215
esterification 178, 186
glycosylated derivatives 185, 242
peptide bond cleavage 70
phosphorylated 238
simian virus 40 272, 281
sodium dodecyl sulphate
attachment, solid-phase sequencing 214
carboxypeptidases 39
delipidation 266
gel electrophoresis 21, 78, 276, 278
gel filtration 84, 88, 269
membrane protein isolation 265
proteolysis 267, 276
removal 92, 266
solubility of peptides 28, 83
SP-Sephadex 104, 133
stannous chloride 191
Staphylococcus aureus V8 protease 58, 96
starch/KI reagent 150, 151
streptococcal M protein 279
submaxillary gland protease 52
subtilisin 45, 46, 61
succinylation 56, 186, 267
succinyl-lysine 171, 193
sucrase/isomaltase 263
sulphanilamide 158
sulphite 244
S-sulphoethylcysteine 29
sulphopropyl Sephadex 104, 133

tertiary structure 1
thermolysin 60, 133, 232
thin-layer chromatography
dansyl amino acids 169
detection of peptides 91, 144
glycopeptides 243
peptide mapping 81, 277
peptide purification 126
Pth derivatives 192
purity of peptides 162
thin-layer electrophoresis
amino acids 144, 167, 168
detection of peptides 144
mobility of peptides 164, 219
purity of peptides 162
separation of peptides 81, 122, 277
thioacetyl thioglycolic acid 227
thiocyanate 228
thiodiglycol 89, 166, 294
threonine
β-elimination 179, 194, 215
esterification 178, 186
glycosylated 185, 242
peptide bond cleavage 70
phosphorylated 238
thrombin 46, 52
tobacco mosaic virus 248
o-tolidine 150, 151, 295
α-*N*-tosyl-L-lysyl chloromethyl ketone (TLCK) 55, 59
N-tosyl-L-phenylalanyl chloromethyl ketone (TPCK) 54
transferrin 245
trichloroacetic acid 47, 53
triethylamine 96, 104, 149, 212
triethylenetetramine polystyrene beads 207
trifluoroacetic acid
CNBr reaction 64
Edman degradation 178, 215, 216, 224
homoserine lactone formation 212
peptide solubility 180
purification 293
trimethylalanine 35, 240
triose phosphate isomerase 11, 259, 262
tropomyosin 279
troponin C 46, 107, 248
trypanosome variant glycoproteins 108, 280